Muscles Alive

Fourth Edition

Muscles Alive
Their Functions Revealed
by Electromyography

J. V. Basmajian,
MD, FACA, FRCP(C)

*Director, Rehabilitation Centre, Chedoke Hospitals
and McMaster University, Hamilton, Ontario*

THE WILLIAMS & WILKINS COMPANY
Baltimore / 1978

Library of Congress Cataloging in Publication Data

Basmajian, John V   1921-
   Muscles alive, their functions revealed by electromyography.

   Bibliography: p.
   Includes index.
   1. Muscle. 2. Electromyography. I. Title. [DNLM: 1. Mus-
cles— Physiology. 2. Electromyography. WE500.3 B315m]
QP321.B33 1978   612'.74   78-7360
ISBN 0-683-00413-1

First Edition, 1962
Reprinted, 1966
Second Edition, 1967
Reprinted August, 1968
Reprinted December, 1969
Italian Edition, 1971
Reprinted January, 1972
Third Edition, 1974
Reprinted December, 1974
Spanish Edition, 1976

Composed and printed at the
Waverly Press, Inc.
Mt. Royal and Guilford Aves.
Baltimore, Md. 21202, U.S.A.

Preface to Fourth Edition____

A N EXCERPT from the preface to the third edition deserves repeating here: "Electromyography has now come of age. Technique has caught up to enthusiasm. Investigators no longer are forced to choose the easiest way, but more and more they choose the best electrodes, best apparatus and best analytic methods. Inserted electrodes, multiple-channel apparatus, multifactorial recordings, computer technologies all are becoming routine. Electromyographers are now truly electrophysiologists. All of us connected with *Muscles Alive* can only be delighted by the revolution."

As in 1973–74, I have been forced by the flood of new reports and by the demands of the scientific community to produce a new edition only four years after the thoroughly revised third edition. Again some 400 new references have been woven into the narrative. Sometimes whole pages have been added; but occasionally only a word or two suffices for newcomers to these pages. As before, the test of what enters anew and what is not deleted is its usefulness to the readers of this book.

A special bonus appears in a new Chapter 3. Written by my former graduate student, Dr. Carlo DeLuca, this contribution replaces the former Chapter 3 written by my wonderful old friend Professor Emeritus A.D. Moore. The passage of time has confirmed many of his original ideas and his paper in the American Journal of Physical Medicine (from which Chapter 3 was derived) is a classic. However modern needs demand not just a revision but a completely new chapter; hence Dr. DeLuca's special contribution.

Hamilton, Ontario   *J.V.B.*
1978

*Preface to First Edition*___

A LMOST a hundred years ago, Duchenne concluded the preface of his epoch-making *Physiology of Motion* with the remark that after ten years of continuous research his task was completed with the writing of his book. After a similar period of some ten years of intense research on muscle function by a modern electronic technique that Duchenne himself would have enthusiastically employed, I have brought together into this book my own findings and those of many others from all over the world. Pleased as I am to be completing my present toil, my main feeling is one of renewed and profound admiration for the old master who, single-handed and against considerable opposition, produced a work that has stoutly withstood a century's buffeting. True, many modifications of his teaching have been dictated by clinical observations (particularly by Beevor) and by the electro-myographic findings to be described in this volume. What impresses me most, however, is that this book will in no way replace or even subordinate his. On the contrary, it will complement and illuminate it as did Beevor's Croonian Lectures *On Muscular Movement* to the Royal College of Physicians of London in 1903. I would like to hope, then, that this work will be accepted kindly as the direct and vigorous lineal descendant of the works of Duchenne and Beevor.

In particular, this book is intended for all those who deal with living muscles and movement, and I have consciously tried to make it indispensable for such workers. These include physiologists, zoologists, and anatomists, and their students; orthopedic surgeons, kinesiologists, physical medicine specialists and therapists; neurologists; and physical educationists. Indeed, it is difficult to stop the list there because the chapter on normal pharyngeal and laryngeal muscles and another on eye muscles will be of special interest to specialists in those fields. Such chapters were especially meant to be comprehensible to ordinary scientific readers as well.

Finally, it is my fond hope that the reader will soon discover that this book is not just another treatise on standard kinesiology, a subject that is already quite adequately dealt with by an impressive (and sometimes oppressive) series of books. Nonetheless, the informed reader will soon detect that it includes a great deal of both old and new information that rightfully belongs in standard kinesiology textbooks but which has not been generally available to their authors.

Acknowledgments

T HE author of any book of this type owes a great deal of gratitude to many people. Friends, colleagues and assistants have all contributed, sometimes quite unawares, to this publication. First contributing to its initiation through their inspiration for research and then to the pursuit of research through practical help and direct encouragement, they finally forced me into the actual writing of this book. To them all I am truly grateful—not the least to my good friend Otto Mortensen of the University of Wisconsin who added the final straw by his authoritative insistence that I *must* write it.

Through the years and sometimes at critical periods the following friends and colleagues have made my researches possible: J. C. B. Grant (first my teacher and then, until his retirement, my "chief"), R. G. MacKenzie, W. A. Hawke, W. T. Mustard, A. W. Ham and A. N. Mitchell. My association with Philippe Bauwens at St. Thomas's Hospital in London during 1953 (after a number of years of unguided floundering) was not only a turning point in my electromyographic work, but it was also a most pleasant experience which has become a cherished memory.

Many of my individual research projects, although they have never been large and stultifying "team-efforts," nonetheless have been carried out cooperatively with one or two colleagues and graduate students. Where publications have resulted, these are noted in suitable parts of the text and in the list of references, but here I must thank in particular the following persons: J. F. Murray (my first and perhaps most stimulating companion-in-research), Johanna W. Bentzon, W. B. Spring, Abdul Latif, Robert Boyko, Alex Szatmari, Wilma E. K. Brown, Rita M. Harland, H. J. Lawrence, R. S. Lewis, F. J. Bazant, Anthony Travill, C. R. Dutta, G. M. Lyons, M. D. Low, G. A. Stecko, W. J. Forrest, M. Baeza, C. Fabrigar, T. G. Simard, V. Janda, R. K. Greenlaw, H. E. Scully, D. P. Cunningham, M. Milner, A. O. Quanbury, J. F. Lovejoy, Jr., T. P. Harden, R. R. Munro, E. M. Regenos, A. W. Hrycyshyn, E. W. Donisch, R. Tuttle, C. E. Johnson, W. Dasher, W. R. Griffin, Jr., S. L. Wolf, G. A. Super, C. B. Chyatte, C. L. Isley, E. R. White, W. J. Newton, J. Samson, M. Fujiwara, M. Hoogmartens, M. Iida, M. Vitti, W. D. McLeod, M. P. Smorto, Mary Baker, Paul Fair and Gary DeBacher.

My debt is immeasurable to the following associates and assistants in electronics: Glenn Shine, James Hudson and Dr. Neal Nunnally. The assistance of various undergraduates has generally gone unheralded. Nevertheless it has been deeply appreciated. I regret that I cannot list the names of the thousands who have acted as our "guinea pigs."

Electronic research on a large scale is expensive. Ours has been generously supported at various times by the following organizations: Bickell Foundation, Banting Research Foundation, Muscular Dystrophy Association of Canada, Medical Research Council of Canada, Rehabilitation Foundation for Poliomyelitics and Orthopedically Disabled ("March of Dimes"), Alcoholism Research Foundation, Poulenc, Ltd. (Montreal), Lederle Company, and Stanley Cox, Ltd., of London (which built expensive apparatus at cost), the U.S. Department of Health, Education and Welfare, Emory University, the University of Toronto and Queen's University.

Many authors have been drawn upon freely. In quoting then, I hope that I have not misrepresented their views. Many are personal friends and acquaintances and have helped me through their letters and conversations. I have also used a substantial number of illustrations from the works of others. Permissions for reproduction have been obtained from the publishers or editors and the sources are individually indicated in their proper places. I am especially grateful to Dr. Carlo J. DeLuca, my former student, who wrote a new chapter 3 to replace the now classic (but aging) chapter by my wonderful old friend, Professor A. D. Moore. Happily "A.D." is still very active in his own special field of *"Electrostatics."*

The enduring tolerance and active assistance of my wife, the unstinting attention to detail of my former secretary at Emory University Rehabilitation Research and Training Center, Arlene DeBevoise, and the pleasant relationships throughout with my publishers have all made continued work on this book a pleasure.

Contents

1.	Introduction	**1**
2.	Apparatus and Technique	**23**
3.	Towards Understanding the EMG Signal	**53**
4.	Muscular Tone, Fatigue and Neural Influences	**79**
5.	Conscious Control and Training of Motor Units and Motor Neurons	**115**
6.	More on EMG and Myoelectric Biofeedback	**131**
7.	Nerve Conduction Velocity and Residual Latency	**141**
8.	Muscle Mechanics	**151**
9.	Posture	**175**
10.	The Upper Limb	**189**
11.	Wrist, Hand and Fingers	**213**
12.	Lower Limb	**235**
13.	The Back	**281**
14.	Human Locomotion	**295**
15.	Anterior Abdominal Wall and Perineum	**319**
16.	Muscles of Respiration	**339**
17.	Mouth, Pharynx and Larynx	**359**
18.	Muscles of Mastication, Face and Neck	**379**
19.	Extraocular Muscles and Muscles of Middle Ear	**401**
20.	The Future of Normal Electromyography	**413**
	Appendix: Some Useful Commercial Equipment	**417**
	References	**421**
	Index	**487**

Chapter 1

Introduction

I NHERENT movement is the prime sign of animal life. For this and many other reasons, man has shown a perpetual curiosity about the organs of locomotion in his own body and in those of other creatures. Indeed, some of the earliest scientific experiments known to us concerned muscle and its functions.

With the reawakening of science during the Renaissance, interest in muscles was inevitable. Leonardo da Vinci, for example, devoted much of his thought to the analysis of muscles and their functions. So, too, did the acknowedged "father" of modern anatomy, Andreas Vesalius, whose influence through his monumental work, the "Fabrica," extends down to this day. In one sense, however, the heritage of Vesalius was unfortunate because it stressed the appearance and the geography of dead muscles rather than their dynamics (fig. 1.1). During the subsequent years, the first scientist to give life back to the muscles was Galvani who at the end of the eighteenth century reported his epoch-making experiments with nerve-muscle preparations and animal electricity (fig. 1.2). For more than two centuries, then, biologists have known and acted on Galvani's revelation that skeletal muscles will contract when stimulated electrically and, conversely, that they produce a detectable current or voltage when they contract from any cause.

Of course Galvani's findings formed the beginning of neurophysiology and the study of the dynamics of muscular contraction, but the world had to wait for the Frenchman, Duchenne, in the middle of the past century, to apply the use of electricity for the systematic determination of the dynamics of intact skeletal muscles (fig. 1.3). His immortal work, *Physiologie des mouvements,* is now, fortunately, available again in a new English edition (translation by E. B. Kaplan). It was based on numerous studies of the movements produced by muscles that were stimulated through the skin by electric currents (fig. 1.4). No one before or after has contributed so much to our understanding of muscular function, although Beevor's (1903) contributions cannot be ignored.

In this introductory chapter, I have studiously avoided a general discourse on the history of muscle function that dates back to Aristotle and Galen and runs through Galileo, Borelli, Volta, Du Bois Raymond and others, because

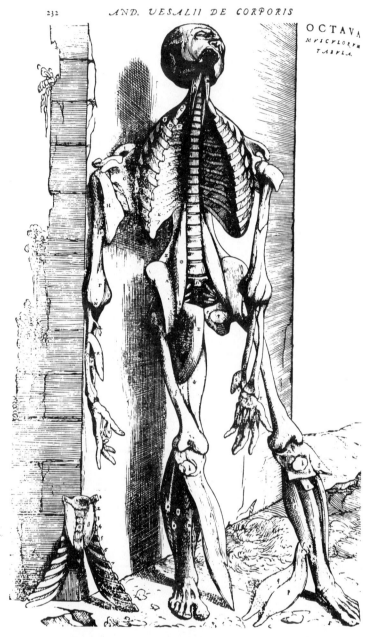

Fig. 1.1. A "muscle-man" from Vesalius' *Fabrica*. (Reproduced by permission from a rare 1555 edition in the Library of Queen's University.)

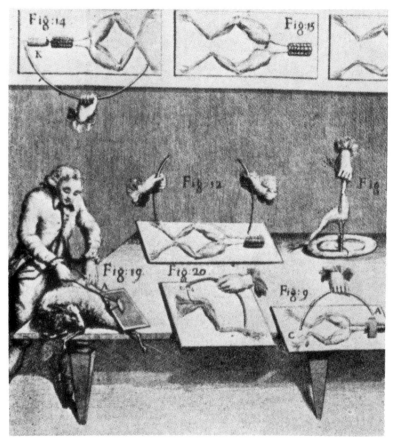

FIG. 1.2. Galvani's demonstrations of the effects of electricity on muscles of frogs and sheep. (From Fulton's reproduction of a plate in Galvani's *De viribus electricitatis in motu musculari commentarius*, 1792.)

the facts are now freely available in other books. Particularly useful and readable is the historical account given by Rasch and Burke in their textbook (1974) which has a wide distribution.

The second aspect of Galvani's discovery, namely, that muscles produce electricity, proved to be largely a scientific curiosity until the twentieth century when improved methods of detecting and recording minute electrical discharges became widely available. The main (though certainly not sole) credit for launching almost 40 years ago the new technique that deals with electrical potentials produced by muscle—or electromyography—must go to the English and American physiologists (Adrian and Bronk; and Denny-

FIG. 1.3. G. B. Duchenne, father of medical electrophysiology.

Brown) and to several Scandinavians. Being neurophysiologists, these men and their colleagues did not concern themselves with the use of the new techniques for unravelling the functions of individual muscles and their neural controls. Moreover, it must be admitted that the earliest techniques were really not appropriate for such detailed studies. For two decades, whenever electromyography was applied to man it was more for diagnostic and clinical reasons than for basic kinesiology.

Toward the end of the Second World War, with a marked improvement of electronic apparatus and the increasing availability of such tools, anatomists, kinesiologists and orthopedic surgeons began to make increasing use of electromyography. The first study that gained wide acceptance was that of Inman, Saunders and Abbott (1944) who reported their work on the movements of the shoulder region.

During the decade of the fifties, electromyography for kinesiological studies became widespread. Frequent reports from American, British, Canadian, Scandinavian, French and German sources became commonplace in the literature. It may be noted, however, that most papers on human electro-

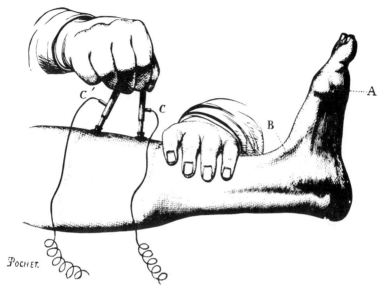

FIG. 1.4. Duchenne's illustration of electrical stimulation of muscles.

myography are purely clinical and those on the functional aspects of muscles are scattered here and there in many hundreds of journals in a dozen different languages. This has made the job of kinesiology textbook writers and editors most difficult and, apparently, sometimes overwhelming. As a result, many of the newer findings are not reaching the textbooks, particularly those findings that are reported in obscure journals. To a large extent, that very fact led to the writing of the first edition of this book.

In his book, Duchenne freely admitted that localized faradization of muscles by his technique was "insufficient to throw light on the physiology of voluntary motion" because "isolated action of the muscle is not in the nature of things." To overcome this inherent defect in his technique, Duchenne supplemented it with many clinical observations. His conclusions form the basis of all our textbook descriptions of muscle action, but they are not dogma. I am certain that Duchenne himself would have enthusiastically embraced the technique of electromyography if he were alive today.

Basis of Electromyography

The Motor Unit

The reader must have a clear knowledge of the structural and functional units in striated muscles to appreciate fully much of the literature in electromyography. The structural unit of contraction is, as everyone knows, the muscle cell or muscle fibre (fig. 1.5). Best described as a very fine thread, this muscle fibre has a length of up to 30 cm but is less than 100 μ (or 0.1 mm) wide. On contracting it will shorten to about 57% of its resting length (Haines, 1932, 1934).

By looking at the intact normal muscle during contraction one would believe, quite erroneously, that all the muscle fibres were in some sort of continuous smooth shortening. In fact, this is not true; instead there is a virtual buzzing of activity in which the fibres are undergoing very rapid changes. The apparently smooth contraction is a summation of all these rapid changes (to be described below).

In normal mammalian skeletal muscle, the fibres probably never contract as individuals. Instead, small groups of them contract at the same moment. On investigation, one finds that all the members of each of these groups of muscle fibres are supplied by the terminal branches of one nerve fibre or axon whose cell body is in the anterior horn of the spinal grey matter. Now, this nerve cell body, plus the long axon running down the motor nerve, plus its terminal branches and all the muscle fibres supplied by these branches, together constitute a motor unit (fig. 1.6). The motor unit is, then, the functional unit of striated muscle, since an impulse descending the nerve

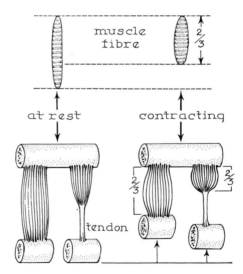

Fig. 1.5. The structural unit of contraction is the muscle fibre. The greatest amount a whole muscle can actively shorten is dependent on the maximum contraction of its contractile units. (From Basmajian, 1970.)

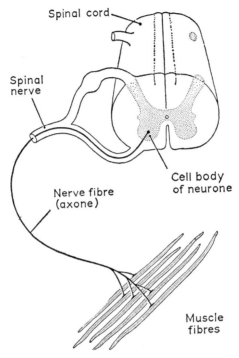

Spinal cord

Spinal nerve

Nerve fibre (axone)

Cell body of neurone

Muscle fibres

Fig. 1.6. Scheme of a motor unit. (After Basmajian, 1955a.)

axon causes all the muscle fibres in one motor unit to contract almost simultaneously.

Motor units normally contract sharply upon the arrival of such nervous impulses at various frequencies, usually below 50 per second. This frequency seems to be the upper physiological limit for the frequency of propagation of axonal impulses and, apparently, such factors as a necessary recovery period and the threshold of fatigue in nerves and muscle must be involved in determining it. As interesting as this matter is, it must be postponed to a later chapter (p. 83).

The number of muscle fibres that are served by one axon, i.e., the number in a motor unit, varies widely, but certain rules have been established in recent years. Generally, it has been agreed that muscles controlling fine movements and adjustments (such as those attached to the ossicles of the ear and to the eyeball and the larynx) have the smallest number of muscle fibres per motor unit. On the other hand, large coarse-acting muscles, e.g., those in the limbs, have larger motor units. The muscles that move the eye have small motor units with less than 10 fibres per unit, as do the human tensor tympani muscle of the middle ear, the laryngeal muscles and the pharyngeal muscles. These are all rather small delicate muscles which apparently control fine or delicate movements.

Krnjević and Miledi (1958) report 7 to 17 fibres per motor unit in the rat diaphragm, which suggests that this muscle, too, has a fine or delicate control. The size of motor units in the rabbit pharyngeal muscles is also quite small—ranging from as few as two to a maximum of only six (Dutta and Basmajian, 1960). The size of the motor units in our study was determined by tracing the individual nerve fibres along their final distribution to the muscle fibres (figs. 1.7 and 1.8). Other observers have calculated the total

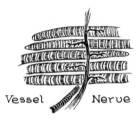

Vessel   Nerve

FIG. 1.7. Drawing of a nerve bundle ending on muscle fibres—teased specimen (low power, phase contrast microscope). (From Dutta and Basmajian, 1960.)

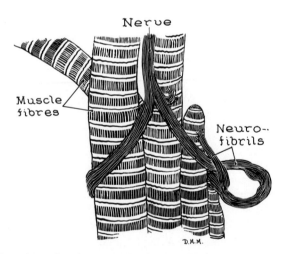

FIG. 1.8. Drawing of a photograph of nerve fibres ending on muscle fibres. (Magnification: about 500×.) (From Dutta and Basmajian, 1960.)

number of muscle fibres in a muscle and the total number of nerve fibres in its motor nerve. Then, by dividing the former by the latter figure, they have calculated the size of the motor units. The latter method is rather questionable because we know that the motor nerve of a muscle contains many sensory and sympathetic fibres as well as motor fibres (fig. 1.9). Nonetheless, it is a method that does produce reasonable approximations.

Tergast (1873) estimated that the motor units of the sheep extraocular muscles have 3 to 10 muscle fibres; Bors (1926) estimated 5 to 6 for human extraocular muscles. More particularly, Feinstein *et al.* (1955) reported 9 muscle fibres per motor unit in the human lateral rectus, 25 in platysma, 108 in the first lumbrical of the hand and 2000 in the medial head of gastrocnemius. Van Harreveld (1947) reported 100 to 125 muscle fibres per motor unit in the sartorius of the rabbit; Berlendis and De Caro (1955), 27 in the stapedius and 30 in the tensor tympani of the rabbit; Wersäll (1958), 10 in

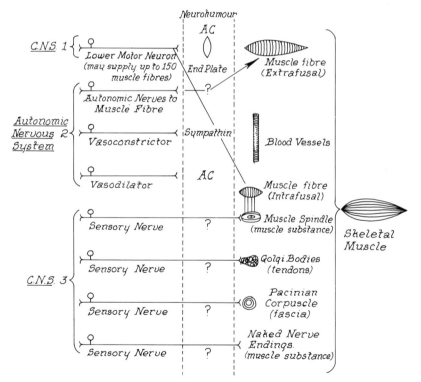

FIG. 1.9. Scheme of multiple innervation of skeletal muscle. (After Solandt, from Dutta and Basmajian, 1960.)

the human tensor tympani; and Rüedi (1959), 2 to 3 muscle fibres per motor unit in the human laryngeal muscles.

Now, it is apparent that even the larger bundles of muscle fibres are quite small and that a strong contraction of a skeletal muscle must require the contraction of many such motor units. A fundamental principle governing such contraction is that there must be a complete asynchrony of the motor unit contractions imposed by asynchronous volleys of impulses coming down the many axons. All the motor units are contracting and relaxing with twitch-like action at differing rates of up to 50 per second. The result of a continuous shower of twitches with different frequencies within a muscle is a smooth pull. In certain disturbances, however, the contractions become synchronized, resulting in a visible tremor.

Van Harreveld (1946, 1947) working with the rabbit's sartorius concluded that the fibres in a motor unit may be scattered and intermingled with fibres of other units. Thus the individual muscle bundles one sees in cross section in routine histological preparations of normal striated muscles rarely if ever correspond to individual motor units as such. Having paid attention to this, I am convinced from my own studies and those of others that this is indeed true in man as well. Norris and Irwin (1961) went farther with their conclusion (supported by excellent evidence) that in rat muscle the fibres of a motor unit are widely scattered.

Buchthal, Guld and Rosenfalck (1957), using an elegant 12-lead multielec-trode technique, finally demonstrated quite conclusively that (in the human biceps brachii) the spike potentials of each motor unit were localized to an approximately circular region, with an average diameter of 5 mm to which the fibres of the unit are confined. (However the potentials could be traced in their spread to over 20 mm distance.) That the area of 5 mm includes many overlapping motor units has been equally convincingly proved by Buchthal *et al.*

Buchthal and Rosenfalck (1973) concluded from histochemical and emg studies that the intermingling of several motor units means that the individual motor unit is not characterized by a single action potential; rather it is a multitude of action potentials which can be picked up separately at different sites within the motor units. Each of these potentials represents all fibres (not subunits, as Buchthal had earlier supposed). They differ in amplitude and shape according to the grouping of the fibres near the recording special multielectrode. The difference in shape of the different potentials is due to the temporal dispersion—as great as 5–7 msec—of the spike components which, in turn, is caused by the spatial distribution.

Motor endplates are located near the middles of the muscle fibres (fig.

1.10). This has been shown by Coërs and Woolf (1959) in human skeletal muscle, by Gurkow and Bast (1958) in the trapezius and sternomastoid of the hamster, by Jarcho *et al.* (1952) in the gracilis of the rat and by Dutta and myself (1960) in the pharyngeal constrictors of the rabbit.

A question that arises occasionally concerns the actual amount of physical work produced by a single motor unit. I have observed on several occasions patients in whose various hand muscles all the motor units but one are paralyzed. In such cases, repetitive firing of the one unit is capable—but only rarely—of producing a slight visible movement of the joint spanned. Others, including Philippe Bauwens, have told me of similar experiences.

FIG. 1.10. Bundle of parallel muscle fibres with endplates (dark dots) stained by cholinesterase technique (From Coërs and Woolf, 1959.)

MOTOR UNIT POTENTIAL. When an impulse reaches the myoneural junction or motor endplate where the axonal branch terminates on a muscle fibre, a wave of contraction spreads over the fibre resulting in a brief twitch followed by rapid and complete relaxation. The duration of this twitch and relaxation varies from a few msec to as much as 0.2 seconds, depending on the type of fibre involved (fast or slow; see p. 17). During the twitch a minute electrical potential, with a duration of only 1 or 2 (or possibly 4) msec, is generated which is dissipated into the surrounding tissues. Since all the muscle fibres of a motor unit do not contract at exactly the same time—some being delayed for several milliseconds—the electrical potential developed by the single twitch of all the fibres in the motor unit is prolonged to about 5 to 12 msec. (One millisecond is a thousandth of a second.) The electrical result of the motor unit twitch then is an electrical discharge with a median duration of 9 msec and a total amplitude measured in microvolts (μv) (or millionths of a volt) with standard needle electrodes. With surface electrodes the durations are prolonged as the potentials are "blurred" and rounded out; with bipolar fine-wire electrodes (see p. 32) the potentials are considerably shorter, the mean duration being 5 msec (Basmajian and Cross, 1971). In comparing four widely-separated muscles of series of normal subjects, we found no marked differences although there is a statistically significant difference between a hand muscle (shorter duration) compared with biceps brachii and two large lower limb muscles. In addition, Simard (1971) has described various minor anomalies and instabilities in the shape of myopotentials.

The majority of these motor unit potentials is around 500 μv or 0.5 millivolts (mv) (fig. 1.11). When displayed on a cathode-ray oscilloscope or other display device the result is a sharp spike that is most often tri- or biphasic, but it may also have a more complex form. Generally, the larger the motor unit potential registered, the larger is the motor unit producing it. However, complicating factors, such as distance of the unit from the electrodes, the types of electrodes and equipment used, etc., enter into the final size of individual motor units recorded by the investigator. For further details, the reader should consult the papers of Håkansson (1956, 1957a,b) and Buchthal (1959).

Even in the same muscle, motor unit potentials in different areas have significant differences in duration. For example, Kaiser and Petersén (1963, 1965) found that the durations (with a median of about 9 msec) are shorter in the long head than in the short head of biceps brachii. They also reported that average durations are about 0.3 msec longer for men than for women.

Although the motor unit potential is relatively brief, the mechanical twitch time is surprisingly prolonged. Even the class of *fast-fibre motor units* have a twitch lasting several times the duration of the potential accompanying it.

Slow-fibre motor units may take as long as a tenth of a second or more to relax with each twitch (see Buchthal and Schmalbruch, 1969). As noted below (under "Fast and Slow Fibres"), electromyography has not shown any clear relationship between twitch durations and potential durations.

Petersén and Kugelberg (1949 *et seq.*) of Stockholm and Bauwens (1948 *et seq.*) of St. Thomas's Hospital in London have given the best descriptions of the characteristics of motor unit potentials. The former authors showed that the electrode type affects the recorded duration and amplitude of the action potentials. They demonstrated characteristic variations, e.g., the smallness of potentials in facial muscles as compared with those in muscles of the extremity. With changes in muscle tension, there is a change in the signal characteristics of averaged emg potentials picked up by surface electrodes (number of spikes, amplitude, rise-time and amplitude-rise time) (Komi and Viitasalo, 1976).

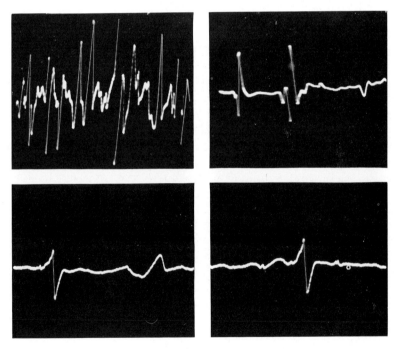

Fig. 1.11. Sample normal electromyograms showing one, two and many superimposed motor unit potentials ("interference pattern"). The single potential in the upper left corner had a measured amplitude of 0.8 mv and duration of 7 msec.

Tanji and Kato (1972) discovered that in rapid isometric contractions individual motor units would initially burst into activity at rates of up to 90 Hz and fall quickly to about 10 to 20 Hz as maximum tension was achieved. During slow contractions of the same muscle (abductor digiti minimi) the rates of the same motor units rose to only 40 Hz at about the 70% level of the tension and fell after that to the same "hold" level of 10 to 20 Hz. Meanwhile, Gillies (1972), in a study of the effects of changing from one level of voluntary isometric contraction to another, demonstrated that with gradual (10% per second) increases of force, newly recruited units first discharge at lower rates than those recruited at low force. When ramp increases of force (e.g., 30% per second) were called for, there was a temporary increase in frequencies before new levels were maintained. Discharge frequencies study varied from 8 to 35 Hz. Thus both the level of force and its rate of change determine the discharge frequency of motor units in a voluntary isometric contraction. Goldberg and Derfler (1977) found a positive correlation between force of recruitment, spike amplitude, and peak twitch tension.

Maton and Bouisset (1975) found in their study of voluntary dynamic movements that a myoelectric burst *precedes the onset of movement.* As soon as a motor unit is set into activity at the start of the myoelectric burst, there is a consistent relation between the work to be accomplished and the interval between the two consecutive discharges of this motor unit. Thus, the peak velocity of the movement is preprogrammed when the amplitude of movement is voluntarily limited.

In a study of the mechanisms for increased force during voluntary contractions, Milner-Brown *et al.* (1972, 1976) found that during an isometric contraction of increasing magnitude, recruitment of additional motor units is the chief mechanism for raising tension at low forces. At higher force levels, the predominant mechanism is to increase the rates of firing of motor units. Units are recruited in order of increasing strength as the total force output is increased. Also units with smaller twitch tensions recruited at low forces show smaller fractional increases in their rates of firing when compared with units which start discharging at higher forces.

In the medial gastrocnemius of the adult cat, Reinking, Stephens and Stuart (1975) of Tucson, Arizona showed that motor units with long contraction times (>45 msec) were non-fatigable and small. In contrast, fast-twitch units ($\leq$45 msec) have a broad range of tetanic tensions and fatigability with a tendency to develop more tetanic tension.

MOTOR UNIT FIRING RATES. Since the 1920's, it generally has been accepted that the normal *upper limit* of activation of motor units in man is about 50 per second. Adrian and Bronk (1928, 1929) found this to be the case and it was confirmed by Smith (1934) and Lindsley (1935). There is some evidence that higher rates occur in other mammals (e.g., our unpublished findings suggest a rate of over 100 as a maximum in the lower limb muscles in rabbits). Marg, Tamler and Jampolsky (1962) believe that in human extraocular muscles, single motor units can fire at rates as high as 270 per second or even more. This needs further investigation for their finding may be due to irritation of motor units by a probing needle.

As we shall see in Chapter 5, man can consciously control the rate of firing of individual units. Moreover, in recruitment (see below), the rate of firing of motor units is increased with stronger contractions. Petajan and Williams (1972) have shown that the frequency characteristics of motor units that initiate contraction are altered by cooling. The recruitment pattern is changed so that in shivering, higher threshold units are recruited earlier than with normal recruitment. Apparently the tonic bursts of higher threshold units serve to pace the shivering tremor, possibly through activation of the myotatic loop.

My former colleagues, De Luca and Forrest (1973), using an elegant needle-electrode technique, recorded myopotentials from human deltoid muscles during the complete time duration of constant-force isometric contractions varying in discrete steps from mimimum to maximum force levels. They found the inter-pulse intervals between adjacent motor unit potentials of a particular train of potentials to be distributed according to the Weibull probability distribution function with time- and force-dependent parameters. They were able to derive an equation that would generate a real continuous random variable whose properties would be identical with those of the inter-pulse intervals.

A motor unit remains active throughout the complete time duration of the constant-force contraction. Near the end of a sustained contraction, the amplitude of the myopotential decreases somewhat and the time duration tends to increase; however, one-third of the potentials actually showed a decreased duration. The generalized firing rate or expected firing rate of a typical motor unit decreases with time as it does for isometric contractions at lower constant-force levels. The probability of a motor unit firing after a previous firing has occurred increases exponentially with respect to elapsed time (De Luca and Forrest, 1973).

MOTOR UNIT RECRUITMENT. It is now common knowledge that under normal conditions, the smaller potentials appear first with a slight contraction and as the force is increased larger and larger potentials are recruited, and all motor units increase their frequency of firing, as mentioned above (Henneman *et al.*, 1965; Olsen *et al.*, 1968; Ashworth *et al.*, 1967; Grimby and Hannerz, 1968, 1970, 1974 a and b; Hannerz, 1973; Grimby *et al.*, 1974). This is called the normal pattern of recruitment. It is absent in cases of partial lower motor neuron paralysis or, to be more specific, the small potentials never appear, apparently because only the larger motor units have survived.

Using microelectrodes, Norris and Gasteiger (1955) showed that action potentials tended to increase in amplitude with excitation and tension during isometric contractions of normal muscles. They also elaborated on the concept of normal recruitment of motor units. Hodes, Gribetz, Moskowitz and Wagman (1965) found that the stimulation thresholds for human motor fibres in the same nerve were lower for those of small diameter than for those of large diameter. The smaller fibres supply the smaller motor units which appear to be the most easily recruited in normal voluntary contraction. In spite of this, we have found that man can be trained to suppress the small, low-threshold units (Basmajian, 1963; Basmajian, Baeza and Fabrigar, 1965). This last matter is discussed further in Chapter 5. Ashworth, Grimby and Kugelberg (1967) confirmed (as have many others since) that subjects with biofeedback assistance could reverse the recruitment order of motor units from the normal pattern. However, this reversal cannot be systematized by the subject except in the form of recruiting larger units in *solo* firing without the normally earlier recruited smaller units' firing first. While Freund *et al.* (1975) and Thomas *et al.* (1976) believe that rather late-recruiting, large motor units cannot be activated without smaller units being recruited first, we have succeeded in teaching this trick to many of the best motor-unit-training subjects. But it is a trick the secret of which is to have an excellent subject and to focus on a large unit from the start of training; the task is almost impossible if obtrusive (early) units are trained initially (see also Chapter 5).

Muscle Fibre Potentials

Generally, it is agreed that a motor unit potential represents the fusion of all accessible individual fibre potentials within a set limit of time (Fleck, 1962). Therefore, studies of single fibre potentials give promise of clarifying the mechanisms of contraction. Buchthal and Engbaek (1963) determined the refractory period and conduction velocity of the transmembrane potentials in single frog muscle fibres at various temperatures. At 25°C the absolute refractory period is 2 msec and the conduction velocity 2.8 metres per second. Applying the voltage-clamp techniques for studying isolated giant axons to single, surface fibres of frog muscles, Jenerick (1964) has been investigating the ionic currents associated with the propagated impulse. He also is studying the relationship between membrane voltage and membrane ionic current as these are reflected in the phase plane trajectory of the response. As yet no conclusive results have come from this promising approach.

Håkansson (1957b) recorded action potentials and the mechanical response of single muscle fibres, finding that the rising phase of the intracellular action potential had traversed the whole length of a fibre before the first sign of twitch tension appeared. Conduction velocities increase up to as much as 50% when the fibre is stretched—perhaps because of an increase of capacitance of the fibre membrane—while the twitch tension falls sharply.

Ekstedt (1964) has recorded single fibre potentials with special techniques *in vivo*. They are smooth biphasic spikes, often followed by terminal phases of low amplitude and long duration. His median value for voltage was 5.6 mv (with maximum of 25.2 mv). The median spike duration was 470 μsec. He also demonstrated intermingling of fibres belonging to different motor units.

Although the subject of single-fibre EMG is outside the scope of this book, mention should be made of two techniques. The first, intracellular EMG, is described by Brooks (1968) and others. The second employs the EMG jitter phenomenon described a decade ago by Ekstedt and in recent years gaining increasing recognition (see Ekstedt *et al.*, 1971).

Jitter Phenomenon

Ekstedt (1964) drew attention to a useful phenomenon with his use of the colorful adjective "jitter." He applied it to the fact that recordings of single-fibre potentials from two fibres in a motor unit always have a variable time interval in their firing patterns. This variability, the *jitter*, is expressed as standard deviation. In normal persons it is about 20 μsec (Ekstedt *et al.* 1974). The chief use of this discovery has been in clinical diagnosis.

Fast and Slow Fibres

Electromyography in mammals offers only limited evidence for the two types of contraction among the muscle fibres, one twitch-like and the other much slower. Indeed the weight of evidence in man is negative, in spite of morphology showing mixture of two or three types of fibres. In invertebrates there is no question that two types of fibres exist and are important. Dorai Raj (1964) showed that in one muscle of the crab (the distal head of the accessory flexor) the muscle fibres of both types are innervated by the same single axon. In those vertebrates in which two types of muscle fibres have been found (e.g., the frog) there are separate axons for fast and slow fibres (Kuffer and Vaughan Williams, 1953). This also seems to be the case in those mammals (e.g., the cat) in which two types of fibres are proved to exist morphologically (Buller, Eccles and Eccles, 1960).

Whether a muscle is of the "slow" or "fast" type may depend to a large degree on its innervation. Slow soleus in rabbits shows continuous electromyographic activity while fast tibialis anterior is active only when brought into reflex action. It is now generally accepted that the cat soleus muscles—a "red" muscle—consists of primarily slow-contracting motor units (McPhedran *et al.,* 1965). Contrasting the findings of cat soleus with tibialis anterior Mosher *et al.* (1972) found that not only were tibialis motor units faster contracting but also that their axons were faster conducting. At low stimulation rates the soleus units developed motor tension. However, Mosher *et al.* warned that the contractile properties of motor unit types may vary from muscle to muscle. If continuous motoneuronal activity in soleus is abolished either by cutting its tendon of Achilles or by tenotomy combined with section of the spinal cord, then soleus becomes a fast muscle (Vrbovà, 1963).

In mammals, fibres are always of the same type morphologically (and apparent physiologically) within any one motor unit. This rule is true even with the highly mixed muscles of man.

Effects of Age and Sex on EMG

Petersén and Kugelberg (1949) first reported a slight prolongation of the motor unit potential with advancing age. Later, Sacco, Buchthal and Rosenfalck (1962) proved, in a systematic study of abductor digiti quinti, biceps brachii and tibialis anterior of normal infants (3 months of age) and adults, that the duration of the action potentials was significantly shorter in the muscles of the infants (fig. 1.12). This they explained in terms of the increase in width of the endplate zone with growth. In persons of from 20 to 70 years of age, the mean duration of the action potentials increased a further 25% in the brachial biceps, but they remained unaltered in the abductor digiti quinti. Whenever the duration of the action potentials increased with age, there was an increase in mean amplitude. The increase in duration at advanced age was attributed to an increased fibre density within the motor units caused by a decrease in the volume of the muscle.

With random insertion of concentric needle electrodes, the incidence of potentials with an initial negative deflection was two to three times greater in the abductor digiti quinti than in the biceps brachii. This difference was most pronounced in infants and indicated a relatively greater extent of the endplate zone in the abductor digiti quinti than in the biceps.

In a study of the influence of age and sex on the emg contraction pattern, Visser and de Rijke (1974) of Amsterdam used a computer for analysis of

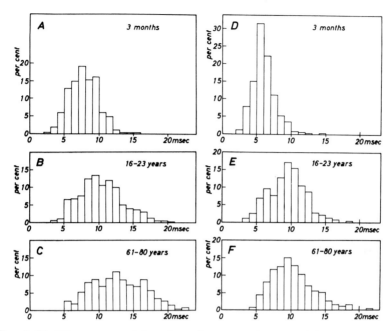

Fig. 1.12. Histograms of duration of motor unit potentials recorded with concentric electrodes from biceps brachii (*A–C*) and abductor digiti quinti (*D–F*) of subjects of different ages. (From Sacco *et al.,* 1962.)

isometric contractions of a thumb muscle. In order to produce a given tension, women as a rule required a higher amplitude and a larger peak number of potentials than did men. In other words, women activate more motor units to produce the same contraction force. There was only a slight (insignificant) decrease in amplitude and peak number with age.

In aged persons, the Russian physiologists Fudel-Osipova and Grishko (1962) noted low voltage potentials (200 to 400 μv), a gradual (rather than an immediate) increase in amplitude on sustained muscular contraction; they also found considerable numbers of polyphasic, prolonged potentials of over 10 msec's duration. Carlson, Alston and Feldman (1964) reported consistent deviations from normal electromyographic findings in aging skeletal muscle. On maximal contraction, they found an obvious decrease in amplitude in the elderly age group when compared with those of the younger normal controls (fig. 1.13). This was interpreted on the basis of a decrease in the number of muscle fibres comprising the motor unit, a decrease in size of the individual muscle fibres within the motor unit, or both. Perhaps, as Carlson

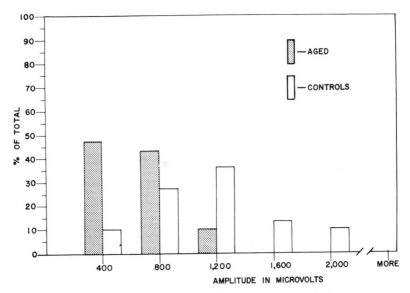

F<small>IG</small>. 1.13. Comparison of EMG amplitudes of aged individuals and controls. (From Carlson *et al.*, 1964.)

and his colleagues suggest, with advanced age there is an inability of at least a portion of the fibres to maintain a sustained contraction.

They also found a considerable number of highly complex and long duration motor unit potentials in more than half of the older age group (fig. 1.14), suggesting that such a deviation from normal motor unit activity is a characteristic of aged skeletal muscle. (No polyphasic or long-duration potentials were noted in the young normal individuals who made up their control group.)

In view of the absence of denervation (fibrillation) potentials in all aged subjects and the finding of normal motor nerve conduction velocities, Carlson *et al.* could not relate the presence of complex potentials to a neurogenic disturbance. A delay in transmission at the endplate of a number of motor unit fibres could account for such asynchronous motor unit behaviour. Similarly, they thought that this could be explained on the basis of physiological alteration of the muscle fibre with an associated delay in fibre response.

Mitolo's (1964) limited investigation dealt with the decrease of muscular "tonus" and the training capacity for physical exercise in old age, providing only general conclusions on the benefits of careful exercises.

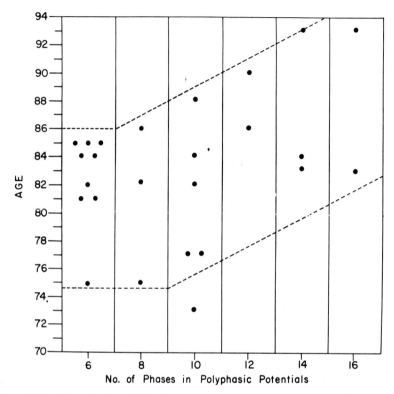

FIG. 1.14. Complexity of EMG as a function of age (increase in polyphasic potentials). (From Carlson *et al.*, 1964.)

That changes in the electromyogram occur with advancing years is now undeniable in man. These also occur in other species as illustrated by work of Holliday, Van Meter, Julian and Asmundson (1965): hyperexcitability to mechanical stimuli (movement of the needle electrode) is the rule in chicks, but it diminishes with increasing age until it disappears completely in adult chickens.

Smooth (Unstriated) Muscle

A word should be addressed here to any reader who seeks information about the involuntary muscle of the vessels and hollow organs. While smooth muscle is clearly not the subject of this volume, it does provide electrical changes arising from the contraction of its cells. These are quite slowly developing and so require DC amplifiers. For further information and references, consult Hakansson and Toremalm (1967), Daniel (1968), Kelly *et al.* (1969), Christensen and Hauser (1971).

The Place of Electromyography in Biology and Medicine

A brief review of the special role that electromyography has assumed in scientific research is a proper ending for this introductory chapter. As with other scientific techniques, this one arose in response to a need. Obviously, the actions and functions of muscles and their nervous control had not been fully understood in spite of continued interest and investigation. Notwithstanding the admirable zeal of many investigators, serious limitations in the classical methods of muscle evaluation account for the gaps and errors in our knowledge.

The classical methods of study on which most of our knowledge of muscle function have been based are: (1) topographical study of dead muscles combined with mechanistic calculations of what they "ought to do," (2) direct electrical stimulation, (3) visual observation and palpation through the skin of the muscles in action and (4) study of paralyzed patients and an evaluation of the deficits.

Except in some obvious applications, the above methods are incomplete, whether they are taken alone or all together. They cannot adequately reveal—as electromyography can—the function of deep, impalpable muscles and the exact time-sequences of activity. It is not enough to estimate by classical methods what a muscle *can do* or *might do*. Electromyography is unique in revealing what a muscle actually *does* at any moment during various movements and postures. Moreover, it reveals objectively the fine interplay or coordination of muscles; this is patently impossible by any other means.

Chapter *2*

Apparatus and technique

E LECTROMYOGRAPHY is so rapidly changing that techniques, electrodes and apparatus keep changing daily. Some standards are required and these have been offered by two groups, the Second International Congress of IFSECN (published by Guld, Rosenfalck and Willison, 1970) and the Second Congress of the International Society of Electromyographic Kinesiology in 1972 (unpublished).

Electrodes

The electrodes used in electromyography could well be—and actually are—of a wide variety of types and construction. Their use depends on the first principle that they must be relatively harmless and must be brought close enough to the muscle under study to pick up its electrical changes.

The two main types of electrodes used for the study of muscle dynamics are surface (or "skin") electrodes and inserted (wire and needle) electrodes. Each has its advantages and its limitations, and they will now be described.

Surface Electrodes

Most often one finds that the simple silver discs used widely in electroencephalography are adapted for electromyography (fig. 2.1). Their advantages revolve around one point: convenience. For example, they are readily obtained from supply houses; they can be applied to the skin after very little training and with reasonable success (within the limitations to be discussed); and they give little discomfort to the subject.

Extremely important in the technique of applying surface electrodes is the ensuring that the electrical insulation between muscle and electrode is

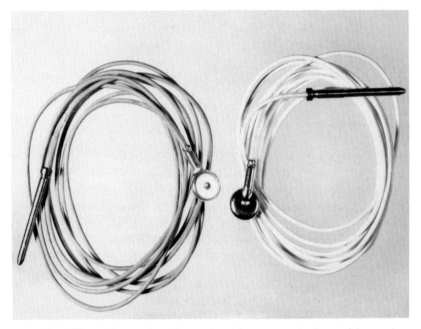

Fig. 2.1. Silver-disc surface electrodes (electroencephalographic type).

reduced to a minimum. Obviously, since a poor contact must be avoided, continued pressure is important. Fortunately, the pressure provided by the adhesive strips used for the securing of the electrodes is usually adequate. Electrical contact is greatly improved by the use of a saline "electrode jelly"; this is retained between electrode and skin by making the silver disc slightly concave on the aspect to be applied to the skin. The dead surface layer of the skin along with its protective oils must be removed to lower the electrical resistance to practical levels (of around 3000 ohms). This is best done by light abrasion of the skin at the site chosen for electrode application. In recent years we have found that it is best produced by "rubbing in" those types of electrode jelly that have powdered abrasive included in their formula. Kramer, Frauendorf and Küchler (1972) showed that, although electrode area was not important, electrode pressure and interelectrode distance were very important factors. The optimal distance varies with the project. The closer they are, the more localizing is their pickup (fig. 2.2).

Modified types of surface electrodes may be made or bought, e.g., small plastic suction cups incorporating two electrodes; the air is evacuated from the cup after it has been applied with jelly to the skin. Another type of

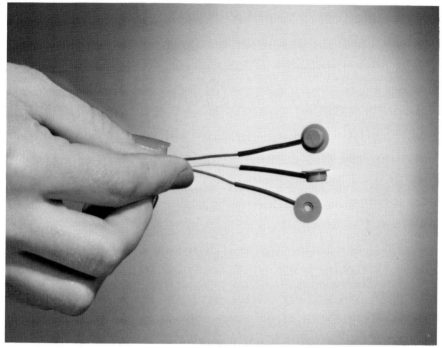

Fig. 2.2. Miniature silver-silver chloride electrodes (Beckman type).

suction cup electrode sometimes available commercially is made of rubber and can be applied with moderate ease. Harris, Rosov, Cooper and Lysaught (1964) recommend a system that provides suction to multiple, metallic-cup electrodes through a manifold. According to these authors, the electrodes successfully grip the surface of skin or mucosa with a minimum of preparation. An excellent, inexpensive suction cup electrode of very simple design has been described by Moore (1966). The elegant silver-silver chloride commercial electrodes (e.g., Beckman type) are widely used and are excellent (fig. 2.2).

Surface electrodes may be used in pairs for localizing the pick-up or they may be used with a more distant common ground or earthing electrode. In either case, the chief disadvantages of surface electrodes are that they can be used only with superficial muscles and that their pick-up is generally too widespread. Thus, many of the results obtained could be deduced with reasonable care from palpation and direct inspection of the muscles in action. Perhaps the chief usefulness of surface electrodes appears where the simultaneous activity or interplay of activity is being studied in a fairly large group of muscles under conditions where palpation is almost impossible (e.g., in the muscles of the lower limb during walking).

Another obvious use of surface electrodes is where a *global* pickup is desirable, e.g., the integrated EMG of all the flexor muscles of a joint as an indirect measure of force. Bouisset and Maton (1972) found that integrated EMG from surface EMGs are linearly related to those of fine-wire intramuscular electrodes from the appropriate muscles. Surface electrodes are clearly the method of choice in psychophysiological studies of general gross relaxation or tenseness, such as in biofeedback research and therapy.

Surface electrodes may be used in pairs for localizing the pick-up or they may be used with a more distant common ground or earthing electrode. In either case, the chief disadvantages of surface electrodes are that they can be used only with superficial muscles and that their pick-up is generally too widespread. Thus, many of the results obtained could be deduced with reasonable care from palpation and direct inspection of the muscles in action. Perhaps the chief usefulness of surface electrodes appears where the simultaneous activity or interplay of activity is being studied in a fairly large group of muscles under conditions where palpation is almost impossible (e.g., in the muscles of the lower limb during walking).

Another obvious use of surface electrodes is where a *global* pickup is desirable, e.g., the integrated EMG of all the flexor muscles of a joint as an indirect measure of force. Bouisset and Maton (1972) found that integrated EMG from surface EMGs are linearly related to those of fine-wire intramuscular electrodes from the appropriate muscles. Surface electrodes are clearly the method of choice in psychophysiological studies of general gross relaxation or tenseness, such as in biofeedback research and therapy. (fig. 2.3).

O'Connell and Gardner (1963) gave an excellent review of the special problems of technique when using surface electrodes for kinesiological studies; and Grossman and Weiner (1966) cautioned psychologists against facile acceptance of integrated surface electromyography, because there are many technical errors that creep in. If their warnings are heeded, substantial progress can be made with skin electrodes in uncomplicated, general investigations. Yet, as indicated above, one should not reduce the effectiveness of research because of imagined limitations. Inserted electrodes are no longer the imposing instruments they once were and are to be preferred for most kinesiology. The fine-wire electrodes described on p. 32 are easy to insert and they are as easy to tolerate as skin electrodes are.

We must condemn the exclusive use of surface electrodes to study fine movements, deep muscles, the presence or absence of activity in various postures and, in short, in any circumstances where precision is desirable (fig. 2.3).

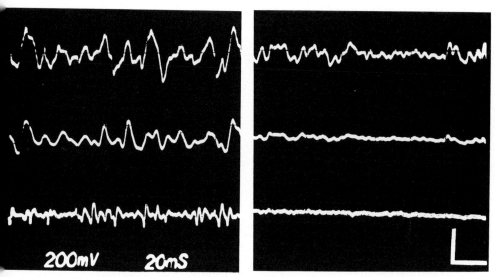

FIG. 2.3. Comparison of surface electrodes and fine-wire electrodes. *Upper channel:* EMG from Beckman miniature skin electrodes 5 cm apart over biceps brachii; *middle channel:* same, 2 cm apart; *lowest channel:* fine-wire electrodes in the underlying part of biceps. *Left tracings*—elbow flexion; *right tracings*—elbow extension (biceps silent).

Inserted Electrodes

NEEDLE ELECTRODES. By far the commonest inserted electrode is the needle electrode, but occasionally other types are also used to advantage. The commonest needle electrode, in turn, is the bipolar concentric-needle electrode first described by Adrian and Bronk (1929) and used widely by clinical electromyographers and to a lesser extent by anatomists (Basmajian, 1958a; Becker and Chamberlin, 1960). The concentric-needle electrode consists of a simple stainless steel hypodermic needle which contains an insulated wire in its barrel (fig. 2.4). The tip of the wire is bared and acts as one electrode while the barrel of the needle acts as the other. The outer needle may be insulated (except for its tip). A second wire can be included and used as the second electrode; in this case the barrel of the needle is used as an insulator or "isolator." This last type of electrode is extremely localizing and so it is seldom used in kinesiology.

Many clinicians prefer the unipolar type of electrode introduced by Jasper and his colleagues in Montreal and consisting basically of little more than a fine insulated sewing-needle (fig. 2.5) (Jasper and Ballem, 1949). Though it

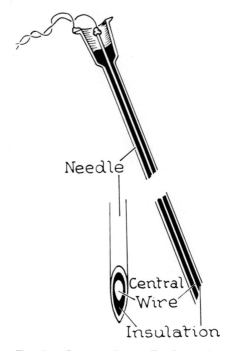

Needle

Central
Wire

Insulation

FIG. 2.4. Concentric needle electrode.

is excellent for clinical investigation (where I once used it frequently), the unipolar electrode usually proves rather clumsy for multiple simultaneous pick-ups because it requires to be paired with a neighboring surface electrode or another unipolar needle.

Sterilization of needle electrodes may offer some problems. No difficulties have been encountered after many years' experience in the use of ethanol as a bath in which the entire electrode is emersed for several hours before use. Equally effective are autoclaving and dry heat.

Lundervold and Choh-Luh Li (1953) of Oslo concluded after comparing different kinds of needle electrodes that the unipolar electrode is the best for studying fibre potentials and motor-unit potentials but the difference is not significant for kinesiological studies. Concentric-needle electrodes were proved to be preferable in some applications. Independently, Landau (1951) arrived at essentially the same conclusions.

Jarcho *et al.* (1952, 1958) demonstrated conclusively that potentials from contractions produced by indirect stimulation *via* the motor nerve vary with the position of the recording electrodes. Admittedly, however, the potentials

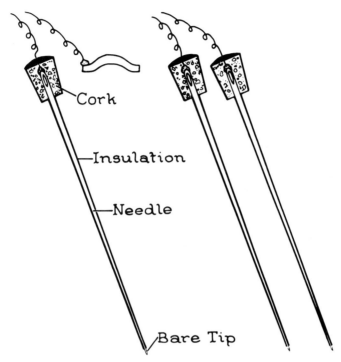

Fig. 2.5. Unipolar needle electrodes (modified Jasper type). *Left,* paired with silver-disc skin electrode; *right,* paired.

resulting from nerve stimulation are quite different from those obtained during a normal voluntary contraction, being a summation of many units contracting simultaneously, which is decidedly not the case with volitional muscle contraction.

On occasion, as a particular problem on hand demanded, we have used various pliable electrodes made of stainless steel wire. For example, to study the very deeply placed human sphincter urethrae, such a wire was implanted in that muscle (see p. 331). Insulated except for its tip, the wire was first loosely threaded through a long "spinal puncture needle" and its very end bent back into a hook. The spinal needle was used to direct the wire to its desired location and then withdrawn leaving the hook in proper place and the other end of the wire dangling and ready to be connected to the electromyograph. A firm tug sufficed to withdraw the electrode at the end of the experiments.

We have made other wire electrodes of a similar type but modified in detail for study of pharyngeal muscles in rabbits and in man (figs. 2.6, 2.7

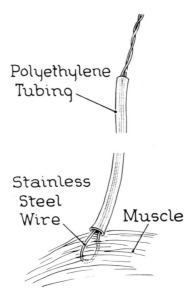

FIG. 2.6. Insulated, stainless steel, wire electrode for pharyngeal muscles of rabbits. (From Basmjian and Dutta, 1961a.)

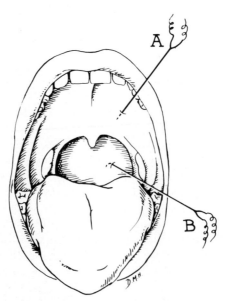

FIG. 2.7. Bipolar, insulated, wire electrodes for transmucosal palatal and pharyngeal electromyography in man. (From Basmajian and Dutta, 1961b)

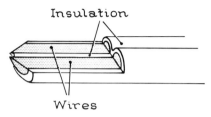

Insulation

Wires

FIG. 2.8. Cut-away diagram of close-up of tip of electrodes in Figure 2.7. (From Basmajian and Dutta, 1961b.)

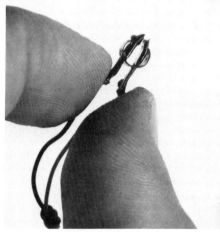

FIG. 2.9. Bipolar clip-on electrode for diaphragm. (Basmajian and Boyd.)

and 2.8) (see p. 367). Similar electrodes have been also used by Long *et al.* (1960) for a study of intrinsic muscles in the hand and by Close *et al.* (1960). For animal experiments, we devised special tiny bipolar clip-on electrodes for the diaphragm (fig. 2.9) (Boyd and Basmajian, 1963). Schoolman and Fink (1963) have also devised electrodes which may be chronically fastened to the diaphragm of experimental animals. Surely there is no limitation to the possible electrodes one might invent for special applications.

In experiments where chronic intramuscular electrodes are required, it is sometimes desirable to avoid wire emerging through the skin. Oganisyan and Ivanova (1964) of Moscow described a novel technique in making recordings from such chronically implanted electrodes in dogs. Subcutaneously they bury a rubber capsule containing an electroconductive gelatinous mass from which a wire electrode runs into the muscle. Whenever recordings are desired, connection to amplifier input wires is completed by thrusting a needle through the skin into the capsule. I have had the pleasure of seeing scientists at the Institute of Higher Nervous Activity and Neurophysiology employing this technique successfully.

FINE-WIRE ELECTRODES. Since 1961, for routine multielectrode studies my colleagues and I have almost abandoned both surface and needle electrodes in favour of our fine-wire, bipolar electrodes (Basmajian and Stecko, 1962). Similar electrodes that differ only in the details of their construction have been independently developed by a number of other research centres, notably Highland View Hospital in Cleveland. Fine-wire electrodes have proved a boon to kinesiological studies because they are: (1) extremely fine (and therefore painless), (2) easily implanted and withdrawn, (3) as broad in their pick-up from a specific muscle as are the best surface electrodes, and yet, (4) they give beautiful, sharp spikes similar to those from needle electrodes. At the present time this type of electrode approaches the ideal for detailed kinesiology. As Sutton (1962) has shown, inserted electrodes with 1 mm of exposed tip record the voltages from a muscle much better than surface electrodes. But in clinical types of studies and in studies where the electrode must be moved about in an exploratory fashion, needle electrodes remain the obvious choice.

In most serious scientific work, fine-wire electrodes are casting a progressively deeper pall on surface electrodes which will of course be useful in crude or generalized studies or studies in which the output of a whole large area or group of muscles is desired. Such applications do exist, but, for accurate indication of activity in single muscles, the fine-wire electrode is preferred because it gives as great an output as surface electrodes along with an incomparable isolation of pick-up to an individual whole muscle. Occasional artifact from high impedance is reduced by the current of an ohmmeter used for testing (Basmajian, 1973). Nordh *et al.* (1974) have shown that varying interelectrode distance has little significant effect on the recorded motor unit potentials. McLeod *et al.* (1976) found the myoelectric power density spectra of 25-μ wires are significantly higher in peak and have a broader bandwith than that of 75-μ wires.

Our bipolar fine-wire electrodes are made now from a nylon-insulated, Evanohm alloy wire, 25 μ in diameter (manufactured by Wilbur B. Driver Co., Newark, New Jersey). An excellent alternative is Stabilohm (manufactured by Johnson Matthey Metals, Ltd., 81 Hatton Garden, London EC1P 1AE, England) insulated with polyurethane enamel (Styck and Hoogmartens, 1975). Teflon-coated fine-wires of platinum, iridium, silver and stainless steel are also available. One distributor, A-M Systems, Inc. (Box 7332, Toledo, Ohio 43615), has four fine silver threads embedded in one "wire" only 0.0045 inches (0.1 mm) in diameter.

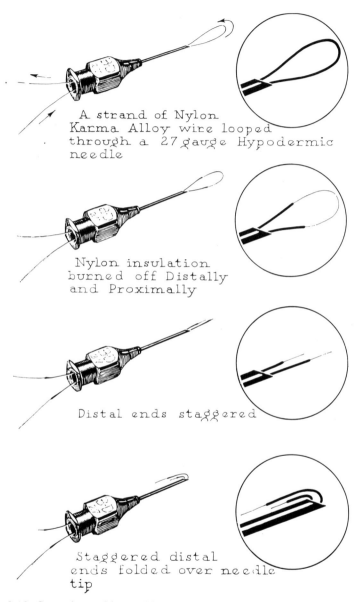

A strand of Nylon
Karma Alloy wire looped
through a 27 gauge Hypodermic
needle

Nylon insulation
burned off Distally
and Proximally

Distal ends staggered

Staggered distal
ends folded over needle
tip

FIG. 2.10. Steps in making a bipolar fine-wire electroode with its carrier
needle used for insertion. (Basmajian and Stecko, 1962.)

The steps for making an electrode (fig. 2.10) are as follows: (1) A double strand of the nylon-coated wire is passed through the shaft of a hypodermic needle. A small loop is left distally, and 5 to 7 cm of wire is left proximally. (2) A small amount of the insulating nylon at the distal tip and 3 to 4 cm of each strand proximally are burnt off in a Bunsen or alcohol-lamp flame—although my former colleague, William McLeod, as well as others, finds chemical removal of the nylon results in a cleaner surface (personal communication). (3) The loop is cut, leaving 1 to 2 mm of bared wire distally on each strand. These bare ends are staggered so that they will not come in contact. They are then bent sharply back to lie against the needle shaft for a short distance. Scott and Thompson (1969) recommend twisting the wires together.

The assemblies are dry-sterilized in a simple paper folder for 60 minutes at 130°C (or autoclaved at 15 lb pressure for 30 minutes). Dry sterilization is preferable because it avoids condensation of moisture in the needle, but temperatures must be controlled to avoid melting off the nylon insulation. Electrode assemblies can be prepared in large numbers and kept in sterilized folders.

Such electrode assemblies may be driven easily into a muscle without anesthesia, and the attendant pain is the usual pain resulting from the needle puncture. If fresh, sharp, 27-gauge needles are used, the pain is minimal and transitory. The needle withdraws easily, and its removal only rarely dislodges the electrodes for they are retained by the hooks at their ends. The electrodes are taped to the skin at the site of emergence to ensure that an accidental tug does not remove them. At the end of an experiment a gentle pull brings the electrodes out painlessly, for each wire is so fine that the barb straightens out on traction and offers little if any palpable resistance. We have had no accidental breakage in many hundreds of uses; nor would we be disturbed if we had, because the fine nylon-clad wire is innocuous. Jonsson and Bagge (1968) have described the various deformations and dislocations of fine-wire electrodes and report that very vigorous exercise may break 25-μ wires. Fortunately the wire is available in 50- and 75-μ sizes also and these have been shown to be much tougher. Jonsson and Reichmann (1968) and Komi and Buskirk (1970) have described the reproducibility of EMG from fine-wire electrodes; and Jonsson and Reichmann (1970) have developed a radiographic technique for controlled insertion of fine wires that otherwise cast no x-ray shadow.

If one wishes to insert individual unipolar fine-wire electrodes, the modification of our technique by Scott (1965) is satisfactory. It depends upon the use of the hypodermic needle not only to insert the electrode but also to act as a cutting instrument once the electrode is deep in place. A single strand of

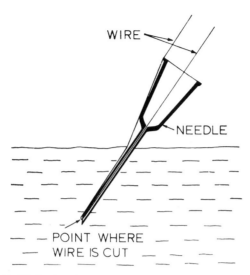

WIRE

NEEDLE

— — POINT WHERE — — —
— WIRE IS CUT — — — —

FIG. 2.11. Unipolar fine-wire electrode. (From Scott, 1965.)

fine wire is passed through the hypodermic needle and a long loop turned back from the needle tip (fig. 2.11). The unit is sterilized and inserted into the muscle in the usual manner. The wire is cut by pulling on its free ends, one of which emerges from the skin alongside the needle and one out of the needle. Both needle and its contained wire are then withdrawn leaving one unbarbed wire deep in place.

After a period of use, we found only one tedious step or complication with fine-wire electrodes, *viz.*, the connection of the almost invisible filament to the larger, braided, lead-in wire of the EMG amplifiers. Others have met and overcome this problem in different ways. For example, Long and his colleagues at Highland View Hospital in Cleveland rely on pre-connection of the fine-wires to standard wires to produce a unit. However, when their needles are withdrawn, they cannot be discarded because they cannot be drawn over the connections. Because we prefer disposable needles and electrodes that are used only once and discarded, we make connections after the needle is entirely removed. Making the connections can be a tedious procedure when it must be done a dozen times for one experimental set-up. Therefore, being lazy, we developed a simple method.

After finding that soldering, microwelding and miniature alligator clamps all have drawbacks, we devised a spring-wire coil connector (Basmajian, Forrest and Shine, 1966) which has proved itself in 15 years of extended usage. It is a brass spring (about 4 mm in diameter by 12 mm in length) soldered permanently to the free ends of each amplifier lead-in wire. The spring is tightly wound from a resilient 22 "spring-brass" wire which gives considerable pinch between adjacent coils. This type of hard brass wire is available through ordinary commercial channels.

To make connection to the fine-wire electrodes after they have been inserted and the needle discarded, the spring is bent slightly between the thumb and index finger. This spreads the coils and allows the bared end of the electrode wire to be slipped between one or two pairs of coils (fig. 2.12). Released, the spring clamps the fine wire and gives good electrical connection instantly. Wrapping a bit of adhesive tape around the connection (for protection and insulation) completes a procedure that takes only moments and saves many tedious minutes required by other methods.

Another ingenious approach to the problem of connecting fine wires to coarser wires for the purpose of kinesiological experiments has been devised by Rinker (personal communication, 1966). He makes the connection with the quick drying paint or ink that he found being used in industry to repair broken connections in etched electronic circuits. A minor disadvantage of his technique is the need for a small plaque against which the connections can be made. However, we have taken advantage of this drawback by incorporating the plaques in the first stage of the input, the mini-amplifiers described on p. 38. Now that fine-wire electrodes are gaining wide acceptance, no doubt other useful innovations are to be expected.

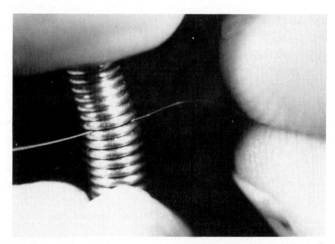

FIG. 2.12. Special spring-connector for fine-wire electrodes (Basmajian, Forrest and Shine, 1966). Here a coarser electrode-wire is used for visibility; it is slipped between two adjacent opened turns of the coil which will clamp down on it when the pressure of the fingers is removed from the ends of the coil.

Multifilament Electrodes

Various complex special-purpose electrodes have been developed. To follow single motor unit activity through strong contractions, De Luca and Forrest (1972), my former associates at Queen's University in Canada, described a special four-filament, shielded, intramuscular electrode that is very efficient. Steiner *et al.* (1972) elaborated a multifilament, regularly spaced electrode based on our bipolar fine-wire technique. Another type has been described by Shiav (1974), while a special variation for recording locally from single motor units was described by Hannerz (1974) of the Karolinska Hospital in Stockholm.

Wire-Electrode Implants

Caldwell and Reswick (1975) described an excellent electrode which survives for many months *in vivo.* The fine wires are wound into a fine continuous coil filled with silicone rubber and these coiled wires are inserted with a hypodermic needle.

Apparatus

Electromyographs range from the most makeshift homemade equipment—some of which is first rate—to fine looking, massively impressive, beautifully engineered, custom-made (and expensive) "machines"—some of which are actually as good as they look. Many kinesiologists are confined (or apparently choose to confine themselves) to using discarded EEG or ECG equipment; sometimes this apparatus is used for EMGs without proper adjustments having been made to record what is a completely different phenomenon. This practice can only tarnish electromyography. On the other hand, with proper adjustments modern EEG equipment can be adapted nicely to most routine non-clinical EMG recording.

Basically, an electromyograph is a high-gain amplifier with a preference or selectivity for frequencies in the range from about 10 to several thousand Hz (cycles per second). Keith J. Hayes (1960) suggests that the sharply peaked spectra of motor unit potentials derived with surface electrodes make the use of amplifiers with limited frequency response practical. He finds several advantages in rejecting frequencies below 20 Hz and above 200 Hz. Then, amplifier "noise," general non-muscular "tissue noise" (which he found to be even more disturbing) and movement artifact would be largely eliminated without significant loss of motor unit potentials. He suggests an upper limit of 200 Hz as satisfactory but admits that a somewhat higher frequency response might be desirable. I would agree that 1000 Hz as the upper limit of the band width is excellent but would prefer the provision of high and low frequency cut-off switches.

In the first two editions of *Muscles Alive* an "ideal research apparatus" was described. Times have changed and the custom-built Stanley Cox six-channel electromyograph which was the back-bone of many of the earlier studies of my group is now in storage, replaced by a variety of more versatile and powerful tools. The biggest improvement is in the available recording systems; e.g., FM tape-recording of signals, tentatively suggested in the second edition, are now widely used (see below).

For research purposes, no longer is there an "ideal" multipurpose electromyograph. Thus, if an investigator is interested in one or two channels of EMG data he can either devise apparatus from standard off-the-shelf amplifiers, oscilloscopes and recorders or purchase a clinical electromyograph (such as those mentioned in the Appendix). However, for multichannel kinesiological studies, he is well-advised to assemble apparatus around an FM tape recorder with the number of channels deemed adequate and financially feasible. Thus, the amplifiers—the essence of electromyographic technique—have become rather secondary in the consideration of equipment. Our group uses several types of commercial and custom built amplifiers. These may or may not be mounted in a rack and they may or may not be wholly or partly transistorized. Cost, convenience and performance specifications determine the amplifiers used (figs. 2.13 and 2.14).

Reducing Noise

While the use of shielded rooms (p. 49) reduces most unwanted signals from the atmosphere and surrounding electrical fields, a major source of irritation always remains for the investigator: *movement artifact.* With flexible cables (see below) much of this has been cured in recent years. Yet, if the work must be done in unshielded rooms (which is almost universal) some additional technological improvements are necessary. We have taken advantage of the high input impedance of our fine-wire electrodes by placing a small cathode-follower-like device right on the subject where the fine wires are connected (either by springs or paint-on). The device is a *dual field-effect transistor source follower* (or "FET") (fig. 2.15). While it gives no amplification at the source, it completely removes the interference of noise coming from the cables leading to the main amplifiers. An improvement is to use miniature amplifiers (say 100 ×) instead (p. 419), greatly enhancing the signal-to-noise ratios (Basmajian and Hudson, 1974).

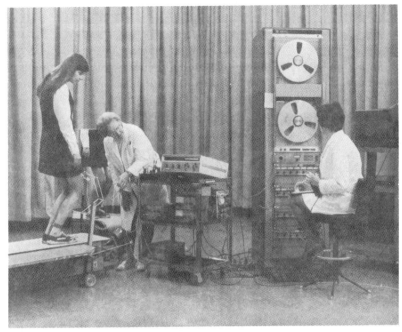

FIG. 2.13. Multichannel amplifying and recording apparatus for electrogoniometry and foot-switches. (From Basmajian and Lovejoy, 1971.)

Transistorized electromyographic amplifiers are coming into wider use, and transistors have replaced most electronic tubes. One now can build an electromyograph so small that it may be combined with an FM transmitter and implanted within the body of experimental animals (figs. 2.16 and 2.17). Without any doubt, engineering and technology can and will reduce the size of such equipment even further while greatly improving general performance.

Multiwire Cables

In kinesiological studies where gross movements are under study, movement artifacts present a constant annoyance. These may arise from the movement of skin electrodes; but, more important, they arise from the movements of the wires and cables leading from the electrodes to the amplifiers. The greater the number of leads, the greater the difficulties. Various attacks have been made on the problem. Kamp, Kok and de Quartel (1965) built and described a multiwire cable depending on the pliant shielding and protection of as many as 16 wires. We have had the good fortune of coming upon a good

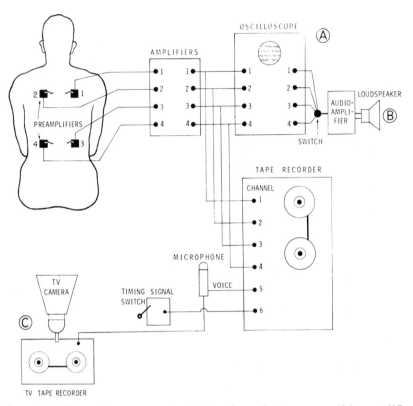

FIG. 2.14. Recording set-up employing four electromyographic amplifier channels and simultaneous videotaping of behavior. (From Donisch and Basmajian, 1972.)

commercial multiwire cable material (Super-flex Conductor manufactured by Cicoil Corp. Van Nuys, Calif.) This cable has proved to be particularly useful and free from trouble. It consists of many parellel wires embedded in a thin ribbon of soft plastic (fig. 2.18). We also frequently use flexible, shielded, multilead cables that are very light, servicable and economic.

Telemetering of EMG

As with any electric signals, EMG potentials can be transmitted for long distances either by telephone lines (Levine *et al.*, 1964) or by FM radio. The latter is particularly useful in field studies of wild animals or in experiments in which the subjects should be completely unfettered by dragging cables (as in running, jumping, etc.). A number of excellent telemetering systems have

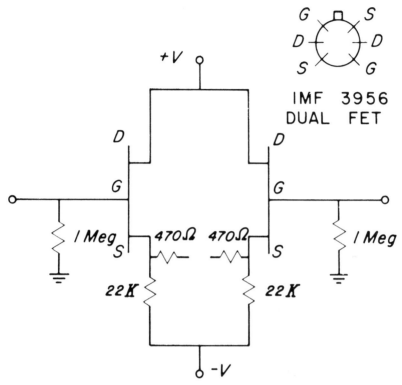

Fig. 2.15. Dual FET source follower as constructed at Emory University Regional Rehabilitation Research and Training Center, Atlanta.

been described and good standard equipment is available from commercial sources. Our laboratory has had limited experience with multichannel telemetering until recently when we started more extensive trials. Single-channel telemetering of EMG is much simpler for obvious reasons and many groups including ours have accumulated convincing evidence of its value. One might expect that within a few years extensive use of multichannel EMG telemetry will be quite common. In particular, good systems are being made available in locomotion laboratories, the best ones employing multiplexing systems for many channels (Perritt and Milner, 1976).

Recorders and Their Evaluation

Recording of motor unit potentials are made by many devices with varying success. Most convenient for the beginner and least wasteful is the pen-

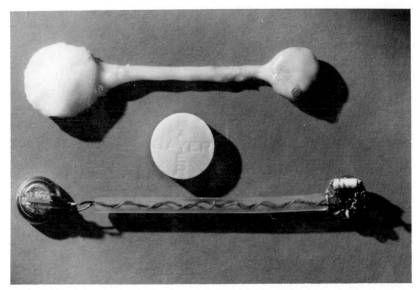

FIG. 2.16. The K-5 implantable miniature EMG plus FM radiotransmitter with and without its protective silastic coating, developed at the Case Institute of Technology (compared with an aspirin tablet). The battery is the larger mass to the left. (Photograph through the courtesy of Dr. W. H. Ko.)

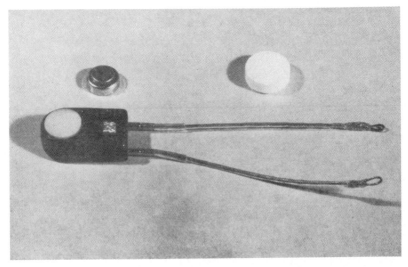

FIG. 2.17. Modern version of Ko implantable transmitter before coating in silastic, with battery in place and another spare battery alongside, plus an aspirin for comparison.

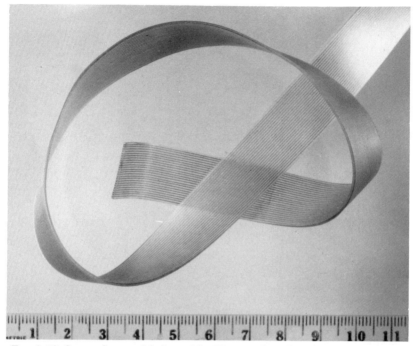

Fig. 2.18. Super-flex multiwire cable. Twenty-four fine wires are embedded in a pliable ribbon of plastic.

writing recorder such as one finds usually on EEG equipment (fig. 2.19). Writing on a moving strip of paper with as many as 24 independent channels simultaneously, this type of recorder is cheap and is easily serviced. However, it has one serious drawback when one is attempting to do accurate work in that the inherent responsiveness of the mechanical system is usually inadequate. Frequencies of over 100 Hz are not faithfully reproduced by almost all routinely used ink writers. In recent years, it is true, modified pen recorders have been produced that have proved to be efficient for many experiments.

The most accurate and elegant technique which should certainly be available in all emg* laboratories requiring first-rate data amenable to manipulation by various analytic methods (including computers) is the multichannel FM tape recorder. There are a growing number of such scientific instruments being marketed with prices ranging from 1000 to 2000 dollars per channel. As noted before, this cost is considerably greater than the cost of the amplifiers. However, the versatility of FM tape recorders is a boon to EMG.

* Hereafter, the adjective electomyographic will be abbreviated to emg and the nouns electromyograph and electromyogram to EMG.

Fig. 2.19. One model of an ink-writing recorder suitable for simple kine-siological (but not clinical) EMG's. (Photo through courtesy of Grass Instrument Co., Quincy, Mass.)

At present, data stored on magnetic tape are processed further into a quantitative or semi-quantitative state in various ways:

A. selected portions are turned into visual tracings by—(1) storing on the face of a storage cathode-ray oscilloscope from which a photograph may be taken and then analyzed by experts; (2) writing-out on paper with pen writers of various sorts (fig. 2.19) for analysis; and (3) writing-out on photographic direct recorders (see below); and—

B. the analogue signal is processed by analogue computers, or after various steps (e.g., integration of spikes over time) the signal is converted into a form suitable for a digital computer. The latter can be programmed to make judgments that are quantitatively as accurate as the best human eye and much less tedious than human rating of data (see below).

Direct recorders that produce immediate photographic paper records are now available and can be adapted quickly to any presently used amplifier system. The light-beam galvanometer recorders described in the Appendix (p. 418) are by no means the only ones available.

Routine motion-picture "frame" photography of the face of the cathode-ray tubes has only a limited usefulness in kinesiological studies but may be combined with simultaneous "movies" of the subject's movements by split-frame photography. Various ingenious motion pictures have been produced in this way, but the technique is not flexible enough for ordinary use. The techniques of Vreeland *et al.* (1961) and Perry *et al.* (1971) are noteworthy.

For locomotion studies, we devised an inexpensive technique of combining colour motion-picture photography of the foot while walking with simultaneous magnetic-tape recording of the signal from single channels of a 6-channel EMG (Gray and Basmajian, 1968). The device is an inexpensive camera (and companion projector) that uses 8 mm film with a magnetic sound track on its edge. Instead of sound, we record the output of the amplifiers. This can then be played back on an oscilloscope in exact synchrony with the 24 frames-per-sec film. Modified, the technique probably could be used for synchronous recording of several channels.

The growing enthusiasm for computer analyses of human gait employing television has taken our group (along with other centers) into combined TV-EMG recording. There is no question that it is feasible, but the real problem may be one of cost. Our studies of computer analysis of EMGs in gait (Milner *et al.*, 1971) have their counterpart in the computer analysis of the gait pattern (Kasvand and Milner, 1971). At the time of writing (in 1978), we find that computer analysis of the emg channels related to signals from other transducers (such as electrogoniometers, contact switches, etc.) an easier and more convenient methodology. We combine the FM tape recording with simultaneous videotaping of the subject's behaviour; the raw emg signals that are being stored on FM tape also are displayed on a monitor oscilloscope which has a videocamera aimed at it. Using a split-screen or mixing device, we place the emg signals alongside or superimposed on the picture of the subject. This technique gives an excellent reference for later analysis of *exactly* what the subject was doing; but it is not a fully automated method of analysis which may or may not be deemed necessary by 1980.

Evaluation of Records

Evaluation of the recordings, whatever their type, is the most abused part of electromyography. Part of the difficulty stems from the use of poor records and part from inexperience. Records can be improved, of course, and experience gained. Nonetheless, one should consider carefully the reputation of the author of an electromyographic paper (or of the laboratory in which he is being trained) before evaluating his results and conclusions. Computers have not altered this warning.

Evaluation of any record is, of necessity, of two types—quantitative or qualitative. Many investigators have confused these two classes while others have assumed that any but quantitative results are not reliable. Most electromyographic records must be considered by both criteria and, although this approach imposes a discipline that quantitative results alone appear not to demand, it is more reliable.

Many attempts have been made to make electromyography purely quantitative. For example, integration of the electrical potentials mechanically or electronically has been used (see below). Bergström (1959) emphatically claimed that the simple counting of spikes is a quantitative reflection of the amount of muscular activity, and we are presently trying to confirm his claim which appears to be true within narrow limits.

Experience has shown that the easiest and, in most cases, most reliable evaluation is by the trained observer's visual evaluation of results colored by his knowledge of the technique involved. Indeed, not only are the number of spikes a factor but the amount of superimposition (summation) of spikes, and their height and type, are important too. We may classify, after some experience and training, the activity reliably into various levels, e.g., nil, negligible, slight, moderate, marked and very marked. Such a classification has proved extremely practical. Furthermore, when arbitrary units are assigned to each class (e.g., 0, ±, +, + +, + + + and + + + +, respectively) tabulation becomes simple. In our hands this technique has eventually proved the most useful and least fraught with self-deceptive pseudo-quantitation. Hirose, Uono and Sobue (1974) showed that "manual" evaluation (visual rating) compares favorably with computer analysis.

Computer Analysis

As noted before, the virtues of computers are that once an adequate program is written for a project, the analysis becomes semi-automatic and relieves the primary investigator of much tedious visual analysis. The fallacy is a belief that no human judgment or tedium are involved; of course they come in when the program is being written. The lack of writing adequate judgment into the program will allow the machine to quantify everything including artifacts and electrocardiograms which may have crept in. Thus, computer analysis without visual monitoring by an experienced person is fraught with error.

What to analyze? If grading of emg channels over time is the concern, then sampling methods with appropriate time periods must be agreed upon. The programmer must know whether all spikes both below and above the base-line are his target. Shall he count them or integrate the total voltage in specific time frames? Have you given him coded signals on secondary channels of the tape recording which he can use to activate runs of computing? Is he to relate integrated EMG at different angles of a joint (recorded from an appropriate transducer)? What will you do with the data next?

Perhaps the most vexing problem is: what is the standard for the quantification? With computers (as with visual inspection) grading must depend on a standard of maximum to zero with any system of grading desired in between. Thus, a reference maximum contraction for each channel must be placed on the record at least at the start and the end of each recording session. The computer (or human observer) relates all signals to that standard. But it cannot be stressed too strongly that the investigator can only relate the activity levels within a channel to the standard for *that* channel. The temptation to look across a series of channels at any moment in time is great but for absolute accuracy it can lead to difficulty. Of course this points up one of the advantages of the computer, which, once it is given a maximum, will normalize the results against it without temptations or fatigue.

In this chapter, no attempt will be made to cover the many problems of computer applications. They would fill a book—in fact they have! Interested readers should consult our monograph in the Butterworth "Computers in Medicine Series" (Basmajian, Clifford, McLeod and Nunnally, 1975).

Power Spectra Analysis

The 1960's introduced a growing need for bioengineers charged with the task of designing improved myoelectrically controlled artificial limbs—i.e., a clearer categorization of the signals coming out of specific muscles. Thus, various electrical engineering methods were applied to the myoelectric output. The most widely used is power spectra analysis to determine those frequency spectra in individual EMGs in which the greatest concentrations of power lie. Power spectra may be measured by the method of Kaiser and Petersén (1965) and their various colleagues. They pass the raw myoelectric signal through four octave bandpass filters centered at frequencies of 50 Hz, 200 Hz, 800 Hz and 1600 Hz. After rectifying the outputs, the resulting DC voltages are compared, using the 200-Hz filter output as the reference. The three ratios provide a measure of the shape of the power spectrum of each in dB. The specialized reader requiring further details should consult Herberts *et al.* (1969), and our book, mentioned in the previous paragraph.

Integration of EMG Potentials

Inevitably at some stage in their investigation, workers in electromyography become dissatisfied with visual analysis of primary records and turn to integration of potentials or other techniques as a solution. Integrators are electronic devices which can produce, when the emg or any other potentials are fed into them, an arbitrary quantitative figure derived from the variables of amplitude, frequency and spike shape. Various integrators are available since they are widely used in other electronic applications. Integrated EMGs have become more widely used and the initial scepticism about their reliability has given way to cautious acceptance. There are many applications where integration of the output may be preferred (Tursky, 1964).

Its greatest dangers lie in: (1) failure to discriminate between artifacts and unit potentials (where naked-eye examination of standard records is superior), and (2) the self-deception of novices who think they can compare the integrated potentials from one channel with those from another. One should always bear in mind that one can only compare the levels of the integrated curve with parts of the same curve.

The chief virtue of integrated outputs is the convenience of an immediate numerical read-out. Further, such output may be fed directly to automatic devices and requires no intermediate human interpretation. This great virtue must not blind us to the shortcomings of integration when it is used unwisely, as often it is (Grossman and Weiner, 1966).

The first useful information produced by integration of the emg potentials resulted from the work of Bigland and Lippold (1954a) who showed that integrated potentials vary directly with the strength of a contraction in a muscle. These workers had earlier shown that mechanical integration with a planimeter (such as used by geographers) was equally effective though much more tedious (Lippold, 1952). A promising technique of "mean voltage recording" (Rosenfalck, 1960) may prove very useful in the future. Another technique based on electronic counting of spikes (Close *et al.*, 1960) also shows much promise in some applications.

Vector Analysis

Pauly (1957) attempted to apply to muscle potentials vector analysis with a cathode-ray oscilloscope in the manner of vector electrocardiography. No special virtues are apparent or were claimed for this technique and it has been largely ignored in electromyography. However, Gydikov and his colleagues in the Bulgarian Academy of Sciences (see, for example, Gydikov and Kosarov, 1974) have applied vector electromyographic techniques to the analysis of individual motor-unit and muscle-fibre potentials in the hope of revealing their microfunctions. Their work is interesting and must be consulted by all specialists in that narrow field of EMG.

Microvibrations

If one applies suitable mechanical transducers to the surface of the body, a summation of tiny mechanical oscillations can be amplified and recorded. These bear a rough resemblance to EMGs; indeed the main source of the vibrations comes from muscular contraction (Williams, 1963). But other sources cause vibrations too, such as vascular pulses, phonics caused by the heart beat, etc. When integrated EMGs and integrated microvibrations are recorded over the same contracting muscle there is often a remarkable similarity between the two (fig. 2.20). The technique deserves wider investigation for in some applications it may prove to be preferable to electromyography.

Shielded Rooms

Much good electromyography is done in ordinary rooms. Often, however, extraneous electromagnetic and electrostatic interference creates chaos in the

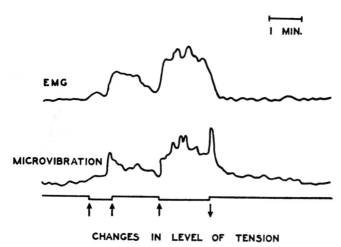

FIG. 2.20. Comparison of amplitude-integrated EMG and microvibrations over same muscle-group. (From Williams, 1963.)

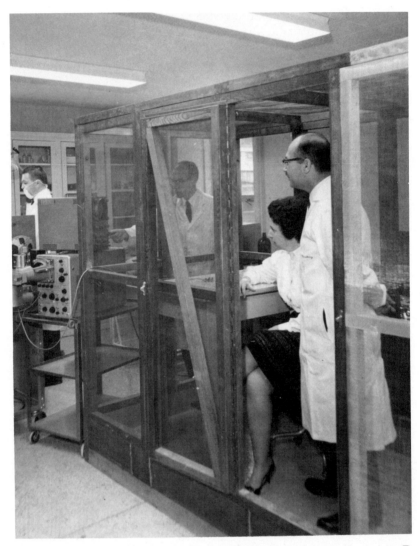

Fig. 2.21. An inexpensive, sectional copper-screened cage (Anatomy Department, Queen's University). The door is left open here, but normally should be shut.

amplifiers and recorders. Subjects, cables and the apparatus all may act as antennae for radio signals and the ubiquitous electromagnetic effects of power lines. Ideally, the laboratory should be isolated from such effects.

A useful provision in any laboratory is an isolation transformer which feeds all the outlets in the laboratory. This is simply a one-to-one transformer. Its cost is so small that the investigator soon forgets his indebtedness to its silent work.

Sometimes all efforts fail unless a shielded room is used. Some very complex types are built, sold and advocated. They are usually too complicated, too small and too expensive. In new buildings a room can be completely shielded by incorporating copper screening in its walls, floors and ceilings. In my experience few such rooms are perfect.

A very serviceable inexpensive shielded cage may be built which is quite good for most uses. It consists of a wooden frame covered with copper or bronze mosquito-screening, which is grounded or earthed. Ideally it is made sectional to allow rapid dismantling. Its floor is raised and of course the screening must be continuous across the under surface of the floor as everywhere else (fig. 2.21). No great inventiveness is required to design a cheap but first-rate enclosure of this type that fits local needs.

Chapter 3

Towards understanding the EMG signal[1]

Carlo J. De Luca

O investigator, do not flatter yourself that you know the things nature performs for herself, but rejoice in knowing the purpose of those things designed by your own mind.

LEONARDO DA VINCI (MS. G47r)

THE electromyographic (EMG) signal is the electrical manifestation of the neuromuscular activation associated with a contracting muscle. It is an exceedingly complicated signal which is affected by the anatomical and physiological properties of muscles, the control scheme of the peripheral nervous system, as well as the characteristics of the instrumentation that is used to detect and observe it. Most of the relationships between the EMG signal and the properties of a contracting muscle which are presently employed have evolved serendipitously. The lack of a proper description of the EMG signal is probably the greatest single factor which has hampered the development of electromyography into a precise discipline.

In this chapter, a structured approach for interpreting the informational content of the EMG signal is presented. The mathematical model which is developed is based on current knowledge of the properties of contracting

---

[1] With the passage of time and the influx of computers, the now classic old Chapter 3 by Professor A. D. Moore has been "retired." Its place is taken by this new Chapter 3 by my former student Carlo J. De Luca, PhD., who is research associate in orthopaedic surgery at the Children's Hospital Medical Center (Boston); associate in orthopaedic surgery (anatomy) at Harvard Medical School; adjunct associate professor at Boston University; project director of the Liberty Mutual Insurance Company Research Center; and lecturer in mechanical engineering of the Massachusetts Institute of Technology (Boston).

normal muscles. These properties are discussed with particular emphasis on those aspects which are sufficiently understood to allow formulation. Most of the information presented pertains to human muscles, with occasional reference to mammalian muscles for supportive data. The extent to which the model contributes to the understanding of the EMG signal is restricted to the limited amount of physiological knowledge currently available. However, even in its present form, the modelling approach supplies an enlightening insight into the composition of the EMG signal.

The work discussed in this chapter was instigated by the pioneering efforts of Moore (1966). His attempt to synthesize the EMG signal provided inspiration for many in the search for understanding the EMG signal.

The Motor Unit Action Potential

Under normal conditions, an action potential propagating down a motoneuron activates all the branches of the motoneuron; these in turn activate all the muscle fibres of a motor unit (Krnjevic and Miledi, 1958; Paton and Waud, 1967). When the post-synaptic membrane of a muscle fibre is depolarized, the depolarization propagates in both directions along the fibre. The membrane depolarization, accompanied by a movement of ions, generates an electromagnetic field in the vicinity of the muscle fibres. A recording electrode located in this field will detect the potential or voltage (with respect to ground), whose time excursion is known as an action potential. A schematic representation of this situation is presented in figure 3.1. In the diagram, the integer n represents the total number of muscle fibres of one motor unit that are sufficiently near the recording electrode for their action potentials to be detected by the electrode. For indwelling needle electrodes, the muscle fibres of the motor unit must be less then 1.5 mm from the electrode (Buchthal *et al.*, 1957). For the sake of simplicity, only the muscle fibres from one motor unit are depicted. The action potentials associated with each muscle fibre are presented on the right side of figure 3.1. The individual muscle fibre action potentials represent the contribution that each active muscle fibre makes to the signal detected at the electrode site.

For technical reasons, the recording electrode is typically bipolar and the signal is amplified differentially. The shape of the observed action potential will depend on the orientation of the recording electrode contacts with respect to the active fibres. For simplicity, in figure 3.1 the recording electrode contacts are aligned parallel to the muscle fibres. With this arrangement, the observed action potentials of the muscle fibres will have a biphasic shape and

MOTOR UNIT ACTION POTENTIAL

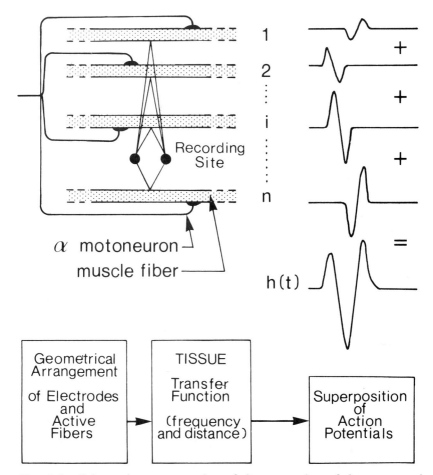

FIG. 3.1. Schematic representation of the generation of the motor unit action potential.

the sign of the phases will depend upon the direction from which the muscle membrane depolarization approaches the recording site (Geddes, 1972). To clarify the relative position of the neuromuscular junction of each muscle fibre and the recording site in figure 3.1, lines have been drawn between the nearest point on each muscle fibre and the contacts at the recording site. In the diagram, a depolarization approaching from the right side is reflected as a negative phase in the action potential and *vice versa*. Note that when the depolarization of the muscle fibre membranes reaches the point marked by the two lines, the corresponding muscle fibre action potential will have a zero inter-phasic value.

In human muscle tissue, the amplitude of the action potentials is dependent on the diameter of the muscle fibre, the distance between the active muscle fibre and the recording site, and the filtering properties of the electrode. The amplitude increases as $V = ka^{1.7}$, where a is the radius of the muscle fibre and k is a constant (Rosenfalck, 1969), and decreases approximately inversely proportional to the distance between the active fibre and the recording site (Buchthal *et al.*, 1957). The filtering properties of a bipolar electrode are a function of the size of the recording contacts (Geddes, 1972), the distance between the contacts (Lindström *et al.*, 1970) and the chemical properties of the metal-electrolyte interface (De Luca and Forrest, 1972).

The duration of the action potentials will be inversely related to the conduction velocity of the muscle fibre, which ranges from 2 to 6 msec. The relative time of initiation of each action potential is directly proportional to the difference in the length of the nerve branches and the distance the depolarizations must propagate along the muscle fibres before they approach the detectable range of the electrode. It is also inversely proportional to the conduction velocities of the nerve branch and the muscle fibre. The time delay caused by propagation along the muscle fibres is an order of magnitude greater than that caused by the nerve branch because of the much faster nerve conduction velocity (30 to 75 m/sec).

The shapes and, therefore, the frequency spectrum of the action potentials will be affected by the tissue between the muscle fibre and the recording site. The presence of this tissue creates a low-pass filtering effect whose bandwidth decreases as the distance increases (Lindström, 1970). This filtering effect of the tissue is much more pronounced for surface electrode recordings than for indwelling electrode recordings because indwelling electrodes are located closer to the active muscle fibres.

Thus far, muscle fibre action potentials have been considered as distinguishable individual events. However, since the depolarizations of the muscle fibre of one motor unit overlap in time, the resultant signal present at the recording site will constitute a spatial-temporal superposition of the contributions of the individual action potentials. The resultant signal is called the motor unit action potential (MUAP) and will be designated as h(t). A graphic representation of the superposition is shown on the right side of figure 3.1. This particular example presents a triphasic MUAP. The shape and the amplitude of the MUAP are dependent on the geometric arrangement of the active muscle fibres with respect to the electrode site as well as all the previously mentioned factors which affect the action potentials.

If muscle fibres belonging to other motor units in the detectable vicinity of the recording electrode are excited, their MUAPs will also be detected. However, the shape of each MUAP will generally vary due to the unique geometric arrangement of the fibres of each motor unit with respect to the recording site. MUAPs from different motor units may have similar amplitude and shape when the muscle fibres of each motor unit in the detectable vicinity of the electrode have a similar spatial arrangement. Even slight movements of indwelling electrodes will significantly alter the geometric arrangement and, consequently, the amplitude and shape of the MUAP.

Given the various factors that affect the shape of an observed MUAP, it is not surprising to find variations in the amplitude, number of phases and duration of MUAPs recorded by one electrode, and even larger variations if MUAPs are recorded with different electrodes. In normal muscle, the peak-to-peak amplitude of a MUAP recorded with indwelling electrodes (needle or wire) may range from a few microvolts to 5 mv, with a typical value of 500 μv. According to Buchthal *et al.* (1954a), the number of phases of MUAPs recorded with bipolar needle electrodes may range from one to four with the following distribution: 3% monophasic, 49% biphasic, 37% triphasic and 11% quadriphasic. MUAPs having more than four phases are rare in normal muscle tissue, but do appear in abnormal muscle tissue (Marinacci, 1968). The time duration of MUAPs may also vary greatly, ranging from less than 1 to 13 msec (Stålberg *et al.*, 1975; Basmajian and Cross, 1971).

It should be emphasized that the amplitude and shape of an observed MUAP are a function of the properties of the motor unit, muscle tissue and recording electrode properties. The filtering properties of the electrode (and possibly the cable connecting the electrode to the preamplifiers, as well as the preamplifiers themselves) can cause the observed MUAPs to have additional phases and/or longer durations. This is an inevitable behaviour of most filter networks.

Properties of Motor Unit Action Potential Trains

The electrical manifestation of a MUAP is accompanied by a twitch of the muscle fibres. The twitch response is delayed by approximately 2 to 3 msec. In order to sustain a muscle contraction, the motor units must be repeatedly activated. The resulting sequence of MUAPs is called a motor unit action potential train (MUAPT). The shape of the MUAPs within a MUAPT will remain constant if the geometric relationship between the electrode and the active muscle fibres remains constant, if the properties of the recording electrode do not change and if there are no significant biochemical changes in the muscle tissue. Biochemical changes within the muscle could affect the conduction velocity of the muscle fibre and filtering properties of the muscle tissue.

The muscle fibres of a motor unit are randomly distributed throughout a subsection of the normal muscle and are intermingled with fibres belonging to different motor units. Evidence for this anatomical arrangement in the rat and cat has been presented by Edström and Kugelberg (1968), Doyle and Mayer (1969) and Burke and Tsairis (1973). There is also indirect electromyographic evidence suggesting that a similar arrangement occurs in human muscles (Stålberg and Ekstedt, 1973; Stålberg *et al.*, 1976). The cross-sectional area of a motor unit territory ranges from 10 to 30 times the cross-sectional area of the muscle fibres of the motor unit (Buchthal *et al.*, 1959; Brandstater and Lambert, 1973). This admixture implies that any portion of the muscle may contain fibres belonging to 20 to 50 motor units. Therefore, a single MUAPT is observed when the fibres of only one motor unit in the vicinity of the electrode are active. Such a situation occurs only during a very weak muscle contraction. As the force output of a muscle increases, motor units having fibres in the vicinity of the electrode become activated and several MUAPTs will be detected simultaneously. This is the case even for highly selective electrodes used by Stålberg *et al.* (1975), which detect action potentials of single muscle fibres. As the number of simultaneously detected MUAPTs increases, it becomes more difficult to identify all the MUAPs of any particular MUAPT due to the increasing probability of overlap between MUAPs of different MUAPTs.

A variety of different techniques have been employed by numerous investigators for detecting and identifying MUAPTs. Indwelling needle and wire electrodes as well as surface electrodes with monopolar and bipolar configurations have been used. The MUAP separation has been performed visually and *via* computer. Reliable separation has been obtained for 3 to 5 MUAPTs in contractions up to 50% of the maximal value. For higher force levels the reported data is sparse. In this force range, the technical requirements become quite intricate. The following information presents some pertinent details describing the properties of MUAPTs.

Inter-Pulse Intervals

When MUAPTs can be properly identified, it is possible to measure the time between adjacent discharges of a motor unit, i.e., the inter-pulse interval (IPI). The IPI has been observed to be irregular and can be described as a random variable with characteristic statistical properties (De Luca and Forrest, 1973a).

The most general characterization of the IPI is a histogram, which is a discrete representation of the probability distribution function. The histogram should only be computed for relatively short durations of the MUAPT (less than 20 seconds). The shape of the IPI histogram, as reported by various investigators, is not consistent. Buchthal *et al.* (1954b), Leifer (1969), Clamann (1969) and others have reported the shape to have a Gaussian distribution. De Luca and Forrest (1973a, b), Person and Kudina (1972) and others have reported an asymmetric distribution with positive skewness. Person and Kudina stated that as the mean duration decreases, the IPI histogram becomes more symmetric, with a substantial decrease in the standard deviation. Gurfinkel' *et al.* (1964) noted that the IPI histogram also becomes more symmetric with a lower standard deviation when visual and/or audio feedback of the motor unit discharges are provided for the subject. The use of different muscles, erroneous discrimination of MUAPs and time-varying firing rates could contribute to the differences.

Two common parameters of the probability distribution function, or the histogram, are the mean and the standard deviation. These two parameters have been used to describe the IPIs. Tokyzane and Shimazu (1964) reported that it is possible to differentiate between two types of motor units (tonic and kinetic) by plotting the mean and standard deviation of the IPIs. They found that the standard deviation decreases rapidly as the firing rate increases from its threshold value and reaches a constant value at higher rates. The standard deviation of all motor units is approximately the same when they are initially recruited, but tonic motor units are recruited at much lower firing rates. Leifer (1969), Person and Kudina (1972), De Luca and Forrest (1973a) and Hannerz (1974) found no such distinction. They instead found a continuous range of motor unit behaviour between these two extremes. The rapid decrease of the standard deviation with increasing firing rate after recruitment is found in the data of Hannerz and of Person and Kudina. The latter investigators proposed that after-hyperpolarization is responsible for reducing the standard deviation, and thus increasing the regularity of the intervals.

Gurfinkel' *et al.* (1964) reported several influences on the standard deviation for the IPIs of individual motor units. They found a tendency for the standard deviation to decrease when normal subjects used surrogate means of control (audio or visual feedback) in addition to proprioception. In patients with disturbances of joint perception, the standard deviation was considerably reduced compared to a normal individual, but in patients with cerebellar disturbances, no differences were seen. Sato (1963) found that the coefficient of variation (standard deviation divided by the mean) for motor units from the dominant hand of right handed subjects tended to be lower than that of the left hand. Voluntary oscillations were more regular when performed with the right hand. This suggests that a lower coefficient of variation corresponds to greater capability of precision control.

Another statistical parameter of interest for describing the IPIs of a motor unit is their interdependence. The greatest amount of dependence (if any) should occur between adjacent intervals. Dependence may be tested by plotting the values of the adjacent IPIs against each other in the form of a scatter diagram. If the adjacent IPIs are independent and the random process is stationary (time invariant), the points on the scatter diagram will be randomly distributed in a fashion determined by the probability distribution function of the IPIs. In case of dependence, the points on the scatter diagram will have statistically dependent coordinates. An alternative test for dependence is serial correlation. If the average product of the adjacent IPIs is equal to the square root of the average of the IPIs, then the serial correlation is zero and the IPIs are linearly independent. Lesser values indicate a negative serial correlation and the tendency of short IPIs to be followed by long IPIs, and *vice versa*. If the IPI random process is not stationary, the above tests may indicate dependence, when none exists. Therefore, measurements for IPI dependence must be performed over sufficiently short time periods, to reduce time-varying effects.

Several authors have noted weak, negative correlations between adjacent IPIs of single motor units. Kranz and Baumgartner (1974) found some motor units that exhibited negative serial correlation, some weakly positive, and some with no significant correlation. It should be noted that in their procedure the subjects were provided with audio and visual feedback and were asked to maintain a constant firing rate. Clamann (1968) and Masland *et al.* (1969) found no significant serial correlation in most cases. Person and Kudina (1971, 1972) found negative serial correlation only for motor units firing at rates above 13 pulses per second. At these firing rates they found a constant, small standard deviation (5 msec) and symmetric IPI histograms. They attributed these results to the effect of after-hyperpolarization. De Luca and Forrest (1973a) used a *chi-square* test on the joint interval histogram for adjacent intervals. No dependence of any statistical significance was found.

Few authors report having made calculations from the IPI data of single motor units to test for higher order interval dependence. The 2nd through 10th order serial correlation coefficients computed by Kranz and Baumgartner (1974) were of lesser magnitude than the first order coefficients, and the *chi-square* test on the 3rd order joint interval histogram computed by De Luca and Forrest (1973a) revealed no dependence.

Recruitment and Firing Rate

The firing rate and recruitment parameters will be discussed simultaneously because there is a substantial interplay among them during a muscle contraction.

In a normal muscle, the motor unit does not discharge at constant intervals. Therefore, discharges of a motor unit must be measured in terms of an average firing rate, which is the reciprocal of the average IPI. However, for the firing rate values to be meaningful, they should be measured over a representative time interval of at least 1 second. Measurements made over shorter time intervals have lead to erroneous reports of uncommonly large firing rate values in excess of 60 pulses per second.

During a constant-force isometric contraction, Gilson and Mills (1941), Masland *et al.* (1969), De Luca and Forrest (1973a) and Grimby and Hannerz (1977) have indicated that a motor unit which is active at the beginning appears to remain active throughout a contraction. The issue of time-dependent recruitment during a constant-force isometric contraction is not yet resolved. Edwards and Lippold (1956), Vredenbregt and Rau (1973) and others have postulated that time-dependent recruitment should occur. De Luca and Forrest (1973a) were not able to verify the latter conjecture. The firing rate of motor units decreases monotonically during a sustained constant-force isometric contraction (Person and Kudina, 1972; De Luca and Forrest, 1973a). Hence, the sequential discharge of a motor unit is a time-dependent process. Gurfinkel' and Levik (1976) have presented evidence that the twitch tension of a motor unit increases during sustained stimulation. It is conceivable that these two phenomena complement each other to maintain a constant force level.

The force dependence of motor unit recruitment and firing rate during an isometric contraction has been studied by many investigators. The currently available information does not provide a consistent description of the mechanisms involved in force modulation. A frequently studied human muscle is the biceps brachii. Clamann (1970) found that the firing rate of motor units recruited at the lowest force levels was 7 to 12 pulses per second. The firing rate increased with increasing isometric force to a maximum of approximately 20 pulses per second. The minimal firing rate of a motor unit increased linearly with the threshold of recruitment. Almost no motor units fired above 20 pulses per second even near 100% maximal voluntary contraction (MVC), and no recruitment was observed above 75% MVC. Clamann (1970) also found that motor units near the muscle surface had higher thresholds of recruitment than those deep in the muscle. Leifer (1969) found that all motor units fired at approximately 11 to 16 pulses per second throughout the entire range of contraction force. After a motor unit was recruited, its firing rate increased slightly with increasing force and then remained constant at a preferred rate. He found that this preferred rate increased slightly with increasing threshold of recruitment. As the force level decreased, the firing rate decreased to 30 to 40% of the preferred rate before becoming inactive.

Also working with the biceps brachii, Gydikov and Kosarov (1974) found that all motor units had a firing rate of 6 to 10 pulses per second when they were recruited. Minimal recruitment occurred above 60% MVC. For some motor units, the firing rate increased to approximately 13 pulses per second and then remained constant with increasing force, whereas for other motor units, the firing rate increased linearly with force up to 100% MVC. The former were generally recruited at lower force levels than the latter. Based on their data, they proposed the existence of two types of motor units, tonic and kinetic. However, their small sample source (a total of 30 motor units from 15 subjects) limits the significance of their proposal. Clamann (1970) and Leifer (1969) never recorded two types of motor units in the biceps brachii; however, the firing rate characteristics found by these two investigators appear to differ slightly. In the rectus femoris muscle, Person and Kudina (1972) found that the low threshold motor units began firing at 5 to 11 pulses per second and reached 18 to 21 pulses per second at 45% MVC. They also found that the higher the recruitment threshold of the motor unit, the less the motor unit increased its firing rate with increasing force. At force levels up to 47% MVC the motor units with lower thresholds had higher firing rates.

Hannerz (1974) and Grimby and Hannerz (1977), working with the tibialis anterior and short toe extensors, found that the minimal firing rate of motor units recruited below 25% MVC was 7 to 12 pulses per second and the maximal firing rate was 35 pulses per second. For motor units recruited above 75% MVC, the minimal firing rate was 25 pulses per second and the maximal firing rate was 65 pulses per second in the tibialis anterior and 100 pulses per second in the short toe extensors. Thus, both the average firing rate and the initial firing rate at recruitment increased with force. They also found that all motor units recruited above 80% MVC discharged in bursts with pauses of 1 second or more at constant force levels. Interestingly, they also observed that during voluntary twitch contractions only motor units that would normally be activated at high force levels were recruited.

Milner-Brown *et al.* (1973a) have studied the activity of single motor units in the first dorsal interosseous muscle contracting at force levels below 50% MVC. They found that when recruited, motor units began firing at 8.4 ± 1.3 pulses per second and increased their firing rate 1.4 ± 0.6 pulses per second for each 100 g of force output, independent of the force at which each motor unit was recruited. They also found that a change in the force rate affected this result. At slow rates of increasing force (100 g per second), the firing rate had a tendency to reach a plateau, while at faster rates of increasing force (1000 g per second) motor units were recruited at lower force levels, but with higher initial firing rates. This difference was not apparent during decreasing voluntary force contraction.

Freund *et al.* (1975) also performed an extensive investigation of single motor unit activity in the first dorsal interosseous muscle. They found that all motor units, regardless of their recruitment threshold, began firing at approximately the same rate (6.8 ± 1.4 pulses per second). However, the lower threshold motor units increased their firing rates with increasing force much faster than the higher threshold units. The firing rates increased with force asymptotically to a maximum rate which also depended on recruitment threshold. These maximum rates varied from approximately 10 to 25 pulses per second for low to high threshold motor units respectively. However, none of the studied motor units was recruited above a force of 700 g. The rate of force increase tested by Freund *et al.* (1975) was slower than that tested by Milner-Brown *et al.* (1973a). This difference might account for some of the observed discrepancy.

The most consistent observation of motor unit behaviour reported in the literature concerns the order of recruitment as a function of size. Henneman *et al.* (1965) observed in decerebrate cats that the order of recruitment during a stretch reflex is from smallest to largest diameter motoneurons. Since the number of muscle fibres innervated by a motoneuron is proportional to its size, smaller motor units are therefore recruited first and larger motor units last. Freund *et al.* (1975), working with humans, measured the nerve conduction velocities and found that the slower conduction velocities, and thus the smaller nerves, were associated with the lower threshold motor units. By averaging the force output of the muscle as each pulse from a single motor unit occurred, Milner-Brown *et al.* (1973b) were able to determine the twitch tension of each motor unit. They found a linear relationship between twitch tension and recruitment force, indicating that the fractional increment in force ($\Delta F/F$) is constant. Their result also indicates that increasingly larger motor units are recruited as the force output of a muscle increases. Goldberg and Derfler (1977) have presented direct evidence indicating that in the masseter muscle, motor units with high recruitment thresholds tend to have larger amplitude MUAPs and twitches with greater peak tensions than motor units recruited at lower force levels.

Synchronization

All the above properties of the motor units have described the behaviour of individual motor units. Synchronization, the tendency for a motor unit to regularly discharge at or near the time that other motor units discharge, describes their behaviour with respect to each other. This includes, but is not limited to MUAPTs which are phase-locked or entrained. In a mathematical sense, synchronization can be defined as *dependence* between MUAPTs. Hence, cross-correlation is a sufficient, but not exclusive condition for synchronization. The latter technique has proven useful for detecting synchronization.

Evidence of the symptoms of synchronization has been reported by several authors. Lippold *et al.* (1957, 1970) found that the MUAPTs from different motor units tended to group at the rate of approximately 9 bursts per second. This grouping became more evident when the muscle became fatigued. Missiuro *et al.* (1962) and others have claimed to observe synchronization by noting the appearance of large periodic oscillations in the EMG signal as the muscle fatigued. Direct evidence was noted by Mori (1973), who observed that motor unit discharges in the soleus muscle synchronized during quiet stance in man. In a later study, Mori and Ishida (1976) demonstrated that the discharge of motor units would indeed become synchronized if the feedback from the muscle spindle in the muscle was sufficiently large.

Kranz and Baumgartner (1974) and Shiavi and Negin (1975) performed a cross-correlation analysis between the MUAPs of simultaneously recorded MUAPTs. They concluded that during non-fatiguing constant-force isometric contractions of the first dorsal interosseous, flexor digitorum profundus, extensor digitorum indicis and tibialis anterior, there was no significant cross-correlation. However, Buchthal and Madsen (1950) and Dietz *et al.* (1976), using the same technique, did find evidence of weak cross-correlation in normal muscles. The amount of the cross-correlation increased in diseased muscles. The degree of cross-correlation also increased as the amplitude of the physiological tremor increased.

The phenomenon of motor unit synchronization has not been analyzed and documented as fully as the other motor unit properties. Evidence has been presented that motor units tend to synchronize when the muscle is fatiguing, during physiological tremor, and in some disease states. However, no detailed description of the behaviour of synchronization as a function of measurable parameters such as force and time has been given. This has been mainly due to the limitations of the recording and analysis techniques. During relatively high force contractions, currently available electrodes detect too many MUAPTs simultaneously. With present techniques, the individual MUAPTs cannot be accurately identified and separated for analysis.

Summary

It is clear from the preceding information that at the present time there is no definitive explanation of motor unit behaviour during muscle contractions. The apparent contradictions in the reported data are likely due to the disparity in the too few studies that have been performed, and to the varying behaviour of different muscles. However, the following description emerges from the currently available information.

During a constant-force isometric contraction the firing rate decreases as a function of contraction-time. Motor units which are active at the beginning of a contraction remain active throughout the contraction. Time-dependent recruitment, which might be expected, has not been observed to occur in any significant fashion. However, there is some evidence that the twitch tension of motor units increases as a function of contraction time.

During force-varying isometric contractions the following interplay between firing rate and recruitment occurs. At the beginning of a contraction, recruitment is the dominant factor with the smallest motor units being recruited first. The firing rate at recruitment is unstable and has a minimal value of 5 to 6 pulses per second. Up to 30% MVC recruitment remains the dominant factor, with progressively larger motor units being recruited as the force increases. As a secondary factor, the firing rate also increases. For force levels ranging from 30 to 70% MVC, the dominant factor is the increase in the firing rate. Some recruitment of increasingly larger motor units also occurs, but plays a secondary role. At force levels above 70% MVC, the further increase in the firing rate continues to be the dominant factor. In most muscles, little (if any) recruitment occurs. In those cases where recruitment has been observed, the firing rate has been irregular and the amplitude of the MUAPs has been comparatively large, implying that the latter motor units should be among the largest in the muscle. Finally, the force rate of a muscle contraction may have a significant effect on the firing rate and recruitment interplay.

Synchronization of the discharges of motor units has been noted when a

muscle fatigues, during both constant-force and force-varying isometric contractions. Finally, the IPIs between adjacent MUAPs have been shown to be
independent in most cases, and to have a weak dependence in a few cases.

A Model for the Motor Unit Action Potential Train

Several investigators have attempted to formulate mathematical expres-
sions for the MUAPT (Bernshtein, 1967; Libkind, 1968; De Luca, 1968;
Coggshall and Bekey, 1970; Stern, 1971; Brody *et al.*, 1974; Gath, 1974; De
Luca, 1975). Of these investigators, only Libkind (1968), De Luca (1975) and
Gath (1974) have employed empirically derived information to construct the
model.

As discussed in the previous section, the MUAPT can be characterized by
its IPIs and the shape of the MUAP. The most commonly used parameters
of the EMG signal are the mean rectified value and the root-mean-squared
value. To obtain these parameters for the MUAPT, it is only necessary to
consider the average firing rate and the shape of the MUAP in the MUAPT.

Evidence has been presented indicating that the firing rate of a motor unit
is dependent on the time duration, force, and possibly the force rate of a
muscle contraction. Although it has yet to be clearly shown, it is conceivable
that it is also dependent on the velocity and acceleration of a muscle
contraction. For the purpose of our discussion, the firing rate will be considered to be only a function of time, t, and force, F, and will be denoted as
$\lambda(t,F)$. This restriction in the notation is adopted for convenience. However,
it should be clearly understood that the derivations which follow apply for
any general description of the firing rate. If future investigations reveal
definitive relationships between the firing rate and the force rate, velocity
and acceleration of a contraction, they can be readily incorporated into the
ensuing model with no loss of generality. A systematic way of obtaining a
mathematical expression for $\lambda(t,F)$ is to fit the IPI histogram with a probability distribution function, $p_x(x,t,F)$. The inverse of the mean value of
$p_x(x,t,F)$ will be the firing rate, or

$$\lambda(t,F) = \left[\int_{-\infty}^{\infty} x p_x(x,t,F) dx \right]^{-1}$$

Alternatively, a mathematical expression for $\lambda(t,F)$ could be obtained by
performing a regression analysis of the IPIs as a function of time and force.

On the other hand, it would be extremely difficult to give a unique
mathematical description of the MUAP because there are many possible
shapes. However, if a MUAPT is isolated and the MUAP can be identified,

it would be possible to make a piece-wise approximation of the shape. Refer to De Luca (1975) for additional information on the mathematical representations.

From a mathematical point of view it is convenient to decompose the MUAPT into a sequence of Dirac delta impulses, $\delta_i(t)$, which are passed through a filter (black box) whose impulse response is $h_i(t)$. If each Dirac delta impulse marks the time occurrence of a MUAP in a MUAPT, the output of the filter will be the MUAPT or $u_i(t)$. The integer i denotes a particular MUAPT. This decomposition, shown in figure 3.2, allows us to treat the two characteristics of the MUAPT separately.

The Dirac delta impulse train can be described by

$$\delta_i(t) = \sum_{k=1}^{n} \delta(t-t_k)$$

It follows that the MUAPT, $u_i(t)$, can be expressed as

$$u_i(t) = \sum_{k=1}^{n} h_i(t-t_k)$$

where $t_k = \sum_{l=1}^{k} x_l$ for $k, l = 1, 2, 3 \cdots, n$. In the above expressions, t is a real continuous random variable, t_k represents the time locations of the MUAPs. x represents the IPIs, n is the total number of IPIs in a MUAPT, and i, k, l are integers which denote specific events.

It is now possible to write the following expressions:

$$\text{Mean rectified value} = E\{|u_i(t,F)|\} = \int_0^\infty \lambda_i(\hat{t},F) \, |h_i(t-\hat{t})| \, d\hat{t}$$

$$\text{Mean-squared value} = MS\{u_i(t,F)\} = \int_0^\infty \lambda_i(\hat{t},F) \, h_i^2(t-\hat{t}) \, d\hat{t}$$

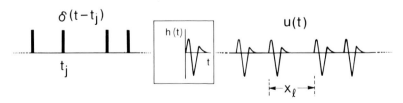

FIG. 3.2. Schematic model for the motor unit action potential train.

where $\hat{t}$ is a dummy variable and E is the mathematical symbol for the expectation or the mean. Although the above equations can be solved, the computation requires the execution of a convolution. De Luca (1975) has shown that since $\lambda(t,F)$ is slowly time varying, the above expressions can be greatly simplified to:

$$E\{|u_i(t,F)|\} = \overline{|h_i(t)|}\ \lambda_i(t,F)$$

$$MS\{u_i(t,F)\} = \overline{h_i{}^2(t)}\ \lambda_i(t,F)$$

This approximation introduces an error less than 0.001%. The bar denotes an integration from zero to infinity as a function of time. These mathematical computations are displayed in figure 3.3. It should be noted that the first term on the right side of the equations is now a scaling value and is

TERM	DIAGRAM	EXPRESSION
a) GENERALIZED FIRING RATE OF A TYPICAL MOTOR UNIT	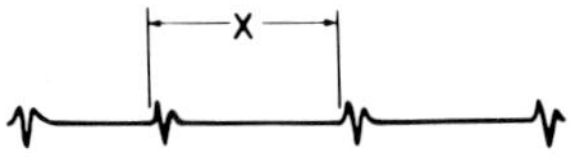 MOTOR UNIT ACTION POTENTIAL TRAIN	$\lambda(\tau,\varphi) = \dfrac{1}{E(X)}$
b) MOTOR UNIT ACTION POTENTIAL	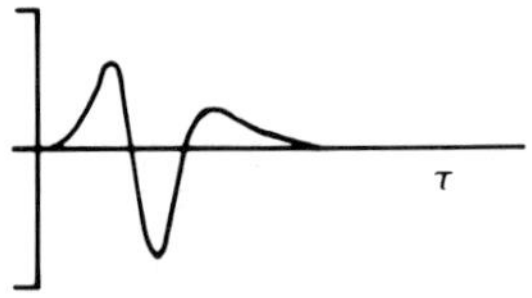	$h_i(\tau)$
c) AREA UNDER THE RECTIFIED MOTOR UNIT ACTION POTENTIAL		$\overline{|h_i(\tau)|} = \displaystyle\int_0^\infty |h_i(\tau)|\,d\tau$
d) AREA UNDER THE SQUARE OF A MOTOR UNIT ACTION POTENTIAL	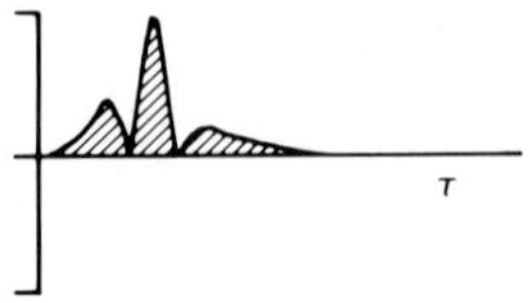	$\overline{h_i^2(\tau)} = \displaystyle\int_0^\infty h_i^2(\tau)\,d\tau$

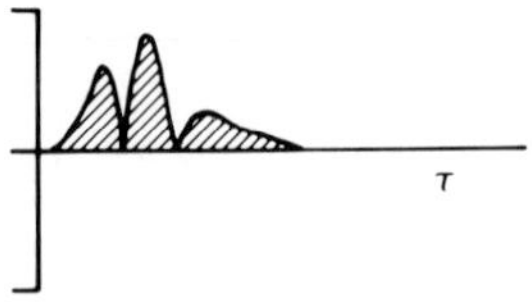

FIG. 3.3. Explanation of some of the terms in the expressions in the text.

independent of time. Hence, these MUAPT parameters are reduced to the expression of the firing rate multiplied by a scaling factor.

To compute the expression for the power density spectrum (frequency content) of a MUAPT it is necessary to consider additional statistics of the IPIs and the actual MUAP shape. The IPIs can be described as a real, continuous, random variable. Only minimal (if any) dependence exists among the IPIs of a particular MUAPT. Therefore, the MUAPT may be represented as a renewal pulse process. A renewal pulse process is one in which each IPI is independent of all the other IPIs.

The power density spectrum of a MUAPT was derived from the above formulation by Le Fever and De Luca (1976) and independently by Lago and Jones (1977). It can be expressed as:

$$S_{u_i}(\omega, t, F) = \frac{\lambda_i(t, F) \cdot \{1 - |M(j\omega, t, F)|^2\}}{1 - 2 \cdot \text{Real}\ \{M(j\omega, t, F)\} + |M(j\omega, t, F)|^2}\ \{|H_i(j\omega)|^2\} \qquad \text{for } \omega \neq 0$$

where

ω = the frequency in radians

$H_i(j)$ = the Fourier transform of $h_i(t)$

$M(j\omega, t, F)$ = the Fourier transform of the probability distribution function, $p_x(x, t, F)$, of the IPIs.

By representing $h_i(t)$ by a Fourier series, Le Fever and De Luca (1976) were able to show that in the frequency range of 0 to 40 Hz the power density spectrum is affected primarily by the IPI statistics. A noticeable peak appears in the power density spectrum at the frequency corresponding to the firing rate and progressively lower peaks at harmonics of the firing rate. The amplitude of the peaks increases as the IPIs become more regular. Beyond 40 Hz, the power density spectrum is essentially determined by the shape of $h_i(t)$.

Concepts of Normalization and Generalized Firing Rate

A generalized representation of the EMG signal must contain a formulation which allows a comparison of the signal between different muscles and individuals. This is not a problem in some contractions such as those involving ballistic movements. However, it is a requirement in isometric and anisometric contractions. The formulation for comparison may be obtained by normalizing the variables of the EMG signal with respect to their maximal measurable value in the particular experimental procedure. For example, in a constant-force isometric contraction, the time is normalized with respect to the duration that the individual can maintain the designated force level. The contraction force is normalized with respect to the force value of a MVC. The *normalized contraction-time* will be denoted by τ, and the *normalized force* by ϕ, and their maximal value is 1.

In the model, expressions for the parameters of the EMG signal are formed by a superposition of the equations of the MUAPT derived in the previous section. Such an approach requires that the mathematical relationship of the firing rates of all the individual MUAPTs be known. It is difficult to obtain such information from the EMG signal. To overcome this barrier, De Luca (1968) introduced the concept of the *generalized firing rate,* and defined it as the mean value of the firing rates of the MUAPTs detected during a contraction. For a detailed description of the calculation of the generalized firing rate, refer to De Luca and Forrest (1973a). For constant-force isometric contractions the generalized firing rate can be expressed as:

$$\lambda(\tau,\phi) \; = \; \frac{1000}{\beta(\tau,\phi)\Gamma[1 \, + \, 1/\kappa(\tau,\phi)] \, + \, \alpha} \text{ pulses per second}$$

$$k(\tau,\phi) \; = \; 1.16 \, - \, 0.19\tau \, + \, 0.18\phi$$
$$\beta(\tau,\phi) \; = \; \exp\,(4.60 \, + \, 0.67\tau \, - \, 1.16\phi) \text{ msec}$$
$$\alpha \; = \; 3.9 \text{ msec}$$

$$\text{for } 0 < \tau < 1, \; 0 < \phi < 1$$

The above values are valid for the middle fibres of the deltoid muscle. It is conceivable that other relationships may exist for other muscles. The above equation is plotted in figure 3.4.

A Model for the EMG Signal

De Luca (1968) modelled the EMG signal, $m(t,F)$, as a linear, spatial and temporal summation of the MUAPTs detected by the electrode. Biro and Partridge (1971) obtained empirical evidence which justified this approach. The modelling approach was later expanded by De Luca and van Dyk (1975) and Meijers *et al.* (1976). A schematic representation of the model is shown in figure 3.5. The integer s represents the total number of MUAPTs which contribute to the potential field at the recording site. Each of the MUAPTs can be modelled according to the approaches presented in figures 3.1 and 3.2. The superposition at the recording site forms the physiological EMG signal, $m_p(t,F)$. This signal is not observable. When the signal is detected, an electrical noise, $n(t)$, is introduced. The detected signal will also be affected by the filtering properties of the recording electrode, $r(t)$, and possibly other instrumentation. The resulting signal, $m(t,F)$, is the observable EMG signal.

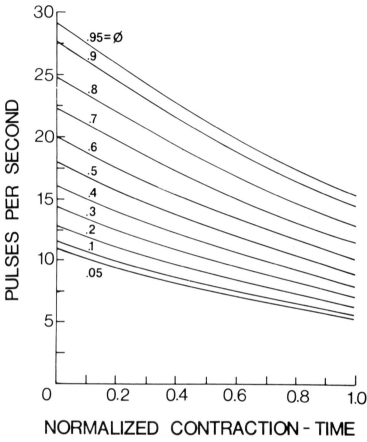

Fɪɢ. 3.4. Generalized firing rate of motor unit action potential trains as a function of normalized contraction-time at various normalized constant-force levels. The force was normalized with respect to the maximal isometric contraction.

The location of the recording site with respect to the active motor units determines the shapes of $h(t)$, as described at the beginning of this chapter.

From this concept it is possible to derive expressions for the mean rectified value, the root-mean-squared value, and the variance of the rectified EMG signal. The expressions are presented in figure 3.6. Their derivation can be found in the article by De Luca and van Dyk (1975). In figure 3.6, each of the terms of the expressions are associated with five physiological correlates which affect the properties of the EMG signal.

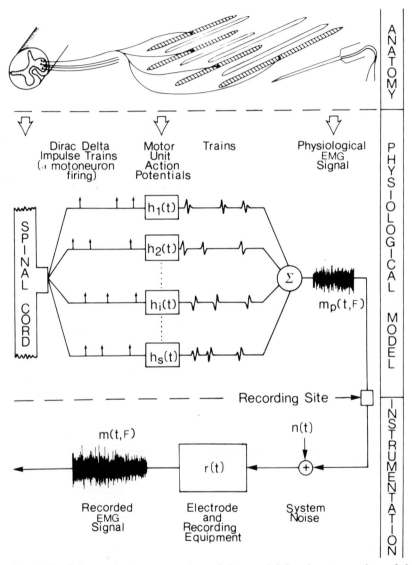

Fig. 3.5. Schematic representation of the model for the generation of the EMG signal.

In the equation of the mean rectified value, the term $J(t,F)$ is a non-positive term which accounts for the cancellation in the signal due to the superposition of opposite phases of the MUAPTs. In a sense, the *superposition term* represents the EMG activity which is generated by the muscle, but is not available in

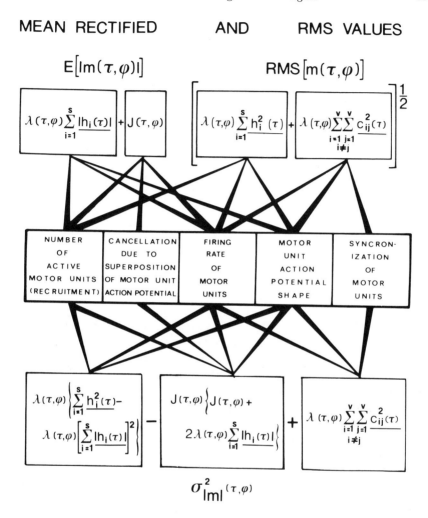

MEAN RECTIFIED   AND   RMS VALUES

$$E\bigl[|m(\tau,\varphi)|\bigr]$$

$$RMS\bigl[m(\tau,\varphi)\bigr]$$

$$\lambda(\tau,\varphi)\sum_{i=1}^{s}|h_i(\tau)| + J(\tau,\varphi)$$

$$\left[\lambda(\tau,\varphi)\sum_{i=1}^{s}h_i^2(\tau) + \lambda(\tau,\varphi)\sum_{i=1}^{v}\sum_{j=1}^{v}c_{ij}^2(\tau)\right]^{\frac{1}{2}}$$

NUMBER OF ACTIVE MOTOR UNITS (RECRUITMENT)	CANCELLATION DUE TO SUPERPOSITION OF MOTOR UNIT ACTION POTENTIAL	FIRING RATE OF MOTOR UNITS	MOTOR UNIT ACTION POTENTIAL SHAPE	SYNCRON-IZATION OF MOTOR UNITS

$$\lambda(\tau,\varphi)\left\{\sum_{i=1}^{s}h_i^2(\tau) - \lambda(\tau,\varphi)\left[\sum_{i=1}^{s}|h_i(\tau)|\right]^2\right\} - J(\tau,\varphi)\left\{J(\tau,\varphi) + 2\lambda(\tau,\varphi)\sum_{i=1}^{s}|h_i(\tau)|\right\} + \lambda(\tau,\varphi)\sum_{i=1}^{v}\sum_{j=1}^{v}c_{ij}^2(\tau)$$

$$\sigma^2_{|m|}(\tau,\varphi)$$

VARIANCE OF THE RECTIFIED SIGNAL

FIG. 3.6. Theoretical expressions for parameters of the EMG signal and their relation to physiological correlates of a contracting muscle.

the observed EMG signal. The expression for the mean rectified value confirms that this parameter of the EMG signal is dependent on the number and firing rates of the MUAPTs detected by the electrode, the area of the MUAPs and the amount of cancellation occurring from the superposition of the MUAPTs.

The integral of the mean rectified value is a commonly used parameter in electromyography. By definition, it will be dependent on the same physiological correlates as the mean rectified value. It is often used to obtain a relationship between the EMG signal and the force output of the muscle. A linear relationship has been reported often. Considering all the physiological correlates involved, the non-linearity of the motor unit behaviour, and the visco-elastic properties of muscle tissue, a linear relationship is unlikely. A non-linear relationship is more plausible.

The root-mean-squared value is also dependent on the number and firing rates of the MUAPTs and the area of the MUAPTs, but is not affected by the cancellation due to the MUAPT superposition. However, it is affected by the cross-correlation (due to synchronization) between the MUAPTs, represented by the $c_{ij}^2(\tau)$ terms. In the corresponding equation in figure 3.6, the integer v denotes the number of MUAPTs that are cross-correlated. Note that any two MUAPTs can still be synchronized even if their cross-correlation term is zero, since the lack of cross-correlation is not sufficient to prove independence. Under this condition, their synchronization has no effect on the root-mean-squared value.

The expression for the variance of the rectified signal is more complicated, containing all the terms which are present in the previous two parameters. Therefore, it reflects the combined effect of all the physiological correlates. This parameter represents the ac power of the rectified EMG signal and should prove to be useful in analyzing the EMG signal. However, it has been used sparingly. Most of the past investigations have dealt with the dc level of the EMG signal.

The approach used thus far has been directed at relating the measurable parameters of the EMG signal to the behaviour of the individual MUAPTs. However, when the recording electrode detects a large number of MUAPTs (greater than 15), such as would typically be the case for a surface electrode, the law of large numbers can be invoked to consider a simpler, more limited approach. In such cases, the EMG signal can be effectively represented as a signal with a Gaussian distributed amplitude.

A Test for the Model

According to the model, the power density spectrum of the EMG signal can be formed by the summation of the power density spectrum of each of the MUAPTs if they are independent. By considering the spectral analysis of a MUAPT discussed previously, it can be shown that if the EMG signal contains several independent MUAPTs with MUAPs of approximately the same amplitude, but different firing rates, the region below 40 Hz is relatively smooth. The peaks and valleys which exist in this region of the individual power density spectra will be smoothed by cancellation. Such is the case in most EMG signal spectra. However, in some cases De Luca (1968) and Hogan (1976) have noted a large peak that occurs between 8 and 20 Hz, with no significant muscle tremor. The model predicts this situation under the following two conditions: if the EMG signal contains a predominance of regularly firing MUAPTs with somewhat similar firing rates, or if the EMG signal is dominated by a high amplitude MUAPT. In fact, Hogan (1976) has shown that as successively more motor units are detected by the recording electrode during increasing force level contractions, the amplitude of the peak in the 8 to 20 Hz region diminishes with respect to the remainder of the spectrum. Above 40 Hz the shape of the power density spectrum is determined by the shapes of the MUAPs of the constituent MUAPTs. The amplitude of the power density spectrum increases as the number of MUAPTs and their firing rates increase.

Before testing the remainder of the parameters derived by the model, it is necessary to comment that the neuromuscular system is an extraordinary actuator. It is capable of generating and modulating force under a wide variety of static (isometric) and dynamic (velocity, acceleration) conditions. It has been shown that the behaviour of the MUAPTs varies for different types of contractions. However, it is possible to test the model for the EMG signal recorded during constant-force isometric contractions.

In a recent study performed by Stulen and De Luca, EMG signals were simultaneously recorded differentially with bipolar surface and needle electrodes while 11 subjects performed sustained constant-force isometric contractions at 25, 50 and 75% MVC. The empirical values of the parameters corresponding to those derived previously were calculated and compared. Let us consider the empirical root-mean-squared parameter which is plotted in figure 3.7. The solid lines represent the average value for the 11 subjects. The vertical lines indicate one standard deviation about the average. For convenience, the magnitude of the values has been normalized with respect to the largest value of the average.

Note that the amplitude of the root-mean-squared parameter increases as a function of time when the EMG signal is detected with surface electrodes and decreases when detected with indwelling electrodes. Why? These signals were recorded simultaneously from the same area of the same muscles. To explain this apparent paradox we must turn our attention to the model.

It is possible to solve the equation for the root-mean-squared parameter in figure 3.7, with the following restrictions: (1) no recruitment occurs during a constant-force contraction, (2) the areas of the MUAPs do not change and

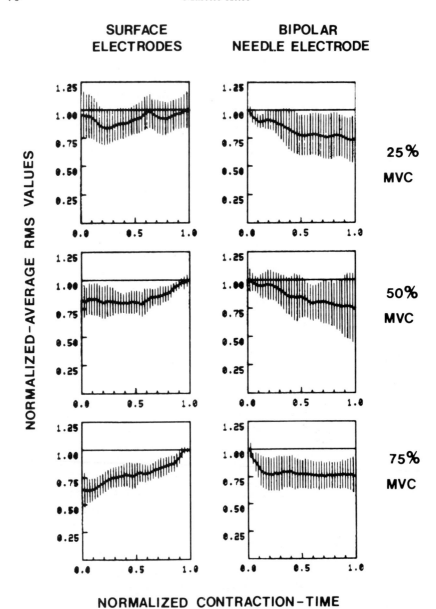

SURFACE
ELECTRODES

BIPOLAR
NEEDLE ELECTRODE

NORMALIZED-AVERAGE RMS VALUES

NORMALIZED CONTRACTION-TIME

FIG. 3.7. Average of the normalized root-mean-squared values from all subjects, plotted as a function of contraction time. Both the amplitude and time duration are normalized to their respective maxima. The vertical lines indicate one standard deviation about the average.

(3) the MUAPTs are not cross-correlated. With these assumptions, the root-mean-squared parameter is directly proportional to the square root of the generalized firing rate. In fact, if the generalized firing rate of figure 3.4 is normalized, it provides an exceptionally good fit to the mean value of the root-mean-squared parameter of the EMG signal recorded with indwelling electrodes at 25 and 50% MVC, but not at 75% MVC. It appears that the decrease in the EMG signal from contractions executed at less than 50% MVC is due to the decrease in the firing rates of the motor units, and that recruitment and synchronization do not play a significant role. But at 75% MVC, other physiological correlates affect the EMG signal. Earlier in this chapter, observations were presented indicating that in general, there is no recruitment over a 70% MVC. It appears that synchronization is the likely candidate to explain the different behaviour of the root-mean-squared curves at 75% MVC. This indication is also implied in the behaviour of the empirical mean rectified parameter and the variance of the rectified signal. However, it is necessary to emphasize that this result does not provide direct evidence of synchronization.

During a muscle contraction maintained at a constant force, if the firing rate of the motor units decreases and there is no significant recruitment, a complementary mechanism must occur to maintain the constant force output. One possible mechanism is the potentiation of twitch tension of the motor units as a contraction progresses. Recently, evidence for potentiation of twitch tension caused by sustained repetitive stimulation has been presented by Gurfinkel' and Levik (1976) *in situ* in the human forearm flexors and by Burke *et al.* (1976) *in vivo* in the cat gastrocnemius.

Now, let us consider the EMG signals recorded with surface electrodes. Why is the root-mean-squared value increasing during all three force levels when the firing rate is decreasing? The behaviour of the firing rate is not affected by the type of electrodes used to record the signal. One possible explanation is as follows. During a sustained contraction, the conduction velocity along the muscle fibers decreases (Stålberg, 1966; Lindström *et al.*, 1970). As a result, the time duration of the MUAPs increases (De Luca and Forrest, 1973a). This change in the shape of the MUAPs is reflected in the power density spectrum as a shift toward the lower frequencies which has been documented by Kadefors *et al.* (1968) and others. Hence, more signal energy passes through the tissue between the active fibers and the surface electrodes. Lindström *et al.* (1970) have made theoretical calculations which show that the muscle tissue and differential electrodes act as low-pass filters. As the distance between the active fibres and the electrodes increases, the bandwidth of the tissue-filter decreases. The increasing effect on the signal due to the tissue and electrode filtering overrides the simultaneously decreasing effect of the firing rate. The filtering effect is not seen in the signal recorded with the indwelling electrodes because the active fibres are much closer to the recording electrode.

Hence, the apparent paradox in the behaviour of the EMG signal recorded with surface and indwelling electrodes is resolved by considering the recording arrangement.

It has been shown that the modelling approach is in agreement with the empirical results that could be tested. The model has also helped to resolve some ambiguities as well as to give insight into the information contained in the EMG signal. The limited discussion and arguments based on the describable known MUAPT behaviour that have been presented are not sufficient to establish the generality of the model. However, as additional information describing the behaviour of MUAPTs becomes available (especially the force dependence), the model should prove to be more useful and revealing.

In closing, the author of this chapter wishes to express his appreciation to Dr. J. V. Basmajian for acting as the energetic catalyst that persuaded me to unify my thoughts on electromyography. Also, I am very grateful for the assitance provided by Messrs. Foster B. Stulen and Ronald S. Le Fever and Ms. Nancy W. Vignone during the preparation of the manuscript.

Chapter 4

Muscular tone, fatigue and neural influences

Most neurophysiologists now agree that electromyography shows conclusively the complete relaxation of normal human striated muscle at rest (Clemmesen, 1951; Basmajian, 1952; Ralston and Libet, 1953). In other words, by relaxing a muscle, a normal human being can abolish neuromuscular activity in it. This does not mean that there is no "tone" (or "tonus") in skeletal muscle, as some enthusiasts have claimed. It does mean, however, that the usual definition of "tone" should be modified to state that the general tone of a muscle is determined both by the passive elasticity or turgor of muscular (and fibrous) tissues and by the active (though not continuous) contraction of muscle in response to the reaction of the nervous system to stimuli. Thus, at complete rest, a muscle has not lost its tone even though there is no neuromuscular activity in it (Basmajian, 1957b).

In the clinical appreciation of tone, the more important of the above two elements is the reactivity of the nervous system. One can hardly palpate a normal limb without causing such a reaction. Therefore, the clinician soon learns to evaluate the level of "tone" and it may seem of little consequence to him that the muscle he is feeling is, in fact, capable of complete neuro-muscular inactivity. In spite of this, he would be surprised to learn that an experienced subject can simulate hypotonia or even atonia of lower motor neuron disease and successfully deceive—if only for a brief period—the most astute physician.

During the course of various electromyographic studies on spastic patients and spastic rabbits, I was impressed with the relative ease with which most spastic muscles also can be completely (though only temporarily) relaxed (fig. 4.1). Magoun and Rhines (1947) and Hoefer (1952) and others have also noted and commented on this, and it has been demonstrated by Kenney and Heaberlin (1962) in spastic children lying quietly, and by Holt (1966) also. The speed wth which voluntary relaxation can occur is quite impressive: Miyashita *et al.* (1972) found the mean values for relaxation reaction time in the biceps brachii of normal health adults to be the same as contraction reaction time.

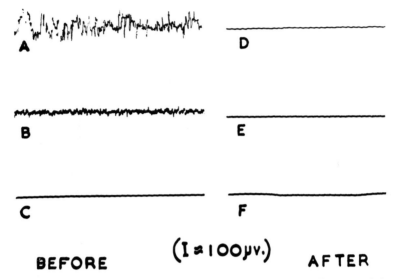

Fig. 4.1. Tracings *A*, *B* and *C* showing varying degrees of emg activity at rest in the spastic quadriceps of three different patients with severe spasticity. Many subjects can be quieted down to the "emg silence" of *C* without drugs. (Tracings *D*, *E* and *F* show the results of intravenous chlorpromazine in the same three patients.) (From Basmajian and Szatmari, 1955a.)

In thousands of electromyograms on normal human muscles, there has been complete and almost instantaneous relaxation when the subject has been ordered to relax. However, a small number of normal subjects do have great difficulty in relaxing quickly. In no normal muscle at complete rest has there been any sign of neuromuscular activity, even with multiple electrodes. As Stolov (1966) puts it: "We can therefore conclude that no alpha motor neuron discharge is present in normal muscle at rest, but may be present during a stretch that is rapid enough to initiate a reflex response."

"Complete rest" requires some qualification. A normal person does not completely relax all his muscles at once. Reacting to multiple interoceptive and exteroceptive stimuli, various groups of muscles show rising and falling amounts of activity. Iris Balshan Goldstein (1962) showed that 16 muscle groups measured at rest were related to a general muscular tension: these were mostly in the limbs and neck. Only frontalis and left sternomastoid were unrelated. Anxious subjects were not greatly different from non-anxious ones—except for the very anxious types of persons who showed marked reaction to new stimuli. These findings occurred in both women and men. Goldstein (1965) showed later that hysterics and certain other neurotics have very little increase in muscular tension over normal persons. Smith (1973) disagreed on the basis of widely separated electrodes on the forehead ("frontalis") which pick up EMG from most of the head and neck and hence reflect tension reasonably well. On the other hand, Alexander (1975) tended to agree with the earliest findings; his research revealed poor generalization between the forehead muscles and the whole body. Berger and Hadley (1975) found that the emg activity of observers taken from their arms and lips rose or fell according to whether they watched others ("models") arm-wrestle or stutter. A group in Los Angeles (deVries *et al.*, 1977) found that the flexors of the forearm are "probably best related to the factor of 'general resting muscular tension'." Schloon *et al.* (1976) found the newborns showed most of their resting activity in the region of the chin and neck.

During sleep, tonus of most head and neck muscles falls off, but, according to Jacobson, Kales, Lehmann and Hoedemaker (1964) "trunk and limb muscles exhibit stable levels of tonic activity throughout the night." Tauber *et al.* (1977) have recently confirmed this statement in an elegant study.

Leavitt and Beasley (1964), in accepting the absence of emg potentials at rest as true relaxation, wondered whether such emg silence could be maintained while the muscles being recorded were passively stretched. Quite surprisingly, they obtained absolute silence in both flexors and extensors of almost all subjects whose knee joints were flexed and extended passively, regardless of whether this was done slowly or rapidly. Although they actually were more interested in reciprocal inhibition of antagonists during active movements (see p. 93), Bierman and Ralston (1965) reported similar findings. Obviously, subjects could relax consciously during a state when normal myotatic reflexes would be expected to occur.

In a series of experiments with normal rabbits we found that if they were handled gently and firmly and the limbs placed and held in relaxed positions, the muscle being examined (generally the gastrocnemius or soleus) showed a rapid reduction of activity to *nil*. An experienced handler could get a rabbit to relax in a matter of seconds but any noise or other stimuli easily induced activity in the muscles.

Muscular "tone" is a useful concept if we keep in mind that at rest a muscle relaxes rapidly and completely. This has now been common knowledge among neurophysiologists for more than a decade. To repeat, tone is a function of the nervous system controlling muscle, but it also results from the natural elasticity of the of the muscular and fibrous tissue. The normal "feel" of the muscle is determined by its normal tissue turgor and its immediate reflex response to palpation. If one keeps one's hands off a resting normal muscle, it shows no more neuromuscular activity than one with its nerve cut. In fact, it shows *less* because the fibres of denervated muscle engage in many

fine random contractions invisible through the skin but detected by electromyography as "fibrillation potentials." The muscles in lower motor neuron
denervation actually exhibit very fine invisible contractions while normal
resting muscles exhibit complete neuromuscular silence. (These fibrillations
are not to be confused with fasciculation, the coarse contractions of motor
units visible through the skin and also often called fibrillations by older
neurologists.) To add confusion, deVries *et al.* (1976) concluded erroneously
that their pickup of 0 to 20 μv potentials from surface electrodes—"tissue
noise"—over resting muscles proves that those muscles were the source.
Electrical energy displayed by surface electrodes may and does come from
many sources and these micropotentials which DeVries *et al.* emphasize are
not motor unit potentials from the resting muscle.

Where did the false concept of continuous neuromuscular activity during
rest originate? Chiefly it seems to be a wide-spread misinterpretation of
Sherrington's *postural tonus* (fig. 4.2). There is no denying that any muscle
that is helping to hold the subject upright shows various degrees of activity.
On the other hand, our group, among a number of others, has shown that
not all the muscles of the leg need to be active in the upright position. That
is, the human upright position allows many of the limb muscles to relax
completely. Yet these muscles will immediately respond to any change
endangering the loss of balance (see p. 180). The neural basis of much of the
interoceptive stimuli that lead to rapid responses is unquestionably the
gamma-loop system (Granit, 1964).

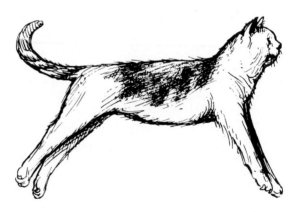

FIG. 4.2. Postural tone of decerebrate cat. (From Pollack and Davis, 1930.)

Spasticity

In another set of experiments we found that the spastic limb muscles of most human beings with lesions of the central nervous system could be relaxed completely (fig. 4.1). In the remaining spastic subjects the activity could be materially reduced, but, under the experimental conditions, complete relaxation was not obtained—probably due to the presence of considerable environmental stimuli (e.g., a well-lighted room, the activity of the investigators, apprehensiveness). In all spastic subjects, very slight stimuli (even conversation) causes immediate electromyographic activity. Similarly, the muscles of spastic rabbits can be completely relaxed, as we noted above, but much more careful effort is required on the part of the handler. As with spastic human subjects, the slightest stimuli causes marked activity which takes some considerable soothing to abolish.

Shimazu, Hongo, Kubota and Narabayashi (1962) confirmed our conclusion with the spastic and rigid muscles of patients with Parkinsonism. They found no spike discharges in resting muscles, but there was a greatly exaggerated stretch reflex.

The findings described for spastic human beings and rabbits at rest therefore are not surprising. While the increased tone is simply an overactive reflex contraction, the limbs can be relaxed completely, albeit with greater "effort" than that required by normal subjects.

An incidental, perhaps inevitable, result of our electromyographic studies of human spasticity and the effects of chlorpromazine (Basmajian and Szatmari, 1955a, b), seems to have been the general adoption of electromyography for studying the effect of the newer relaxant drugs and other types of therapy (Brennan, 1959). In view of what has been already discussed, it hardly seems necessary to warn against a naive acceptance of "electromyographic" evidence if the only criterion used is the amount of the activity "at rest." Unfortunately some workers have already published such data. In our recent studies of the effects on spasticity of new drugs, we use reflex responses to controlled constant stimuli (see, for example, Basmajian and Super, 1973).

Fatigue

Since its inception, electromyography has been used by some workers in the investigation of fatigue. I shall observe, at once, the traditional and necessary warning that fatigue is a complex phenomenon and perhaps a complex of numerous phenomena. The fatigue of strenuous effort is probably quite different from the weariness felt after a long day's routine sedentary work. Undoubtedly the following types exist: emotional fatigue, central nervous system fatigue, "general" fatigue and peripheral neuromuscular fatigue of special kinds.

Seyffarth (1940) showed that the increasing fatigue of prolonged voluntary periodical contraction of muscles of the forearm is accompanied by reduction of potentials. This is exaggerated by ischemia caused by a tourniquet. There is a diminution and variation in amplitude of the size of the motor unit potential (fig. 4.3). Loofbourrow's observation (1948) appears to agree. He concluded that reduction in mechanical response with fatigue during indirect supramaximal tetanization (in cats under anesthesia) is accompanied by a corresponding decrease in amplitude of emg potentials (fig. 4.4). An increased amplitude is always obtained by increasing the mechanical tension through stronger stimuli. Lindqvist (1959) of Helsinki also agrees with Seyffarth that progressive fatigue is accompanied by a decrease in the amplitude of motor unit potentials (fig. 4.5).

Lundervold (1951) found that after repeated contractions of hand muscles to the point of being "completely tired out," the accessory muscles involved showed no serious emg alteration. Indeed, more and more potentials were recruited. Muscles not considered essential to a particular movement were recruited and hyperactive (fig. 4.6). After resting the limb because of extreme subjective fatigue, tremors appeared.

In contrast, Merton (1954) has demonstrated a peripheral fatigue which appears not to be due to neuromuscular blocking. He showed that the blood supply is the significant factor even in contractions of a single small muscle. Even in extreme fatigue under the special experimental conditions used by Merton, action potentials evoked by motor nerve stimulation were not diminished (fig. 4.7). Recovery from this type of fatigue does not take place if the circulation is kept arrested, further underlining Merton's assertion. Curiously, the peripheral ischemic fatigue resulting from high-frequency electrical stimulation appears to be related much more to the total number of evoked potentials than the frequency of stimuli (Marsden *et al.*, 1976).

Christenson (1962) hypothesizes that myohemoglobin plays a very important rôle in preventing fatigue from acute activity of brief duration. On the other hand, it is used up in longer exercise resulting in the symptoms, objective signs and positive tests that indicate fatigue. Repeated periods of great effort not exceeding ½ minute with equal rest periods did not cause fatigue.

In the past few years our experience has shown that the fatigue experienced in heavily loaded upper limbs is not accompanied by any significant muscular activity. Yet "fatigue" becomes unbearable. This particular type is apparently due to the painful strain on the ligaments and capsules. This concept will be

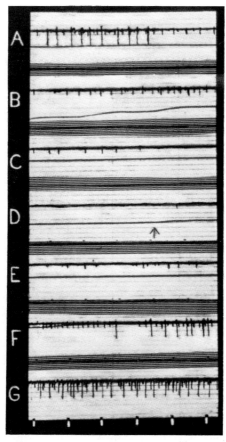

Fig. 4.3. Fatigue experiment. *A*, Single maximum contraction 30 minutes after the blood pressure cuff is inflated to obstruct circulation to forearm. The larger units have a slow frequency, though there is maximum effort. *B* to *G*, The subject is trying all the time (40 seconds) to maintain maximum load. *B*, ½ second after *A*. Though there is maximum effort, the contraction develops very slowly and is painful. Only the small unit is left. *C*, 10 seconds after *B*—the unit disappears in spite of the higher tension (produced by the muscles above the cuff). *D*, 18 seconds after the beginning of maximum contraction at *B*. Next, the cuff is relaxed. *E*, 9 seconds after *D*—the small unit appears. *F*, 13 seconds after *D*—the larger unit appears. *G*, 22 seconds after *D*—several units are present, but the frequency of the larger unit is slow (13 per second), though there is maximum effort. (From Seyffarth, 1940.)

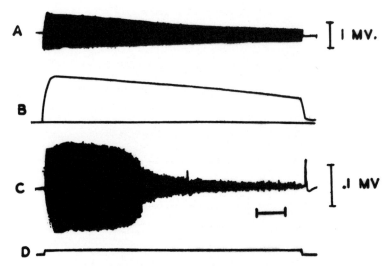

Fɪɢ. 4.4. Effects of fatigue in an isotonic contraction during indirect stimulation. Stimulus 4V, 38 per second. After-load, 20 g. Time axis, 2 seconds. *A*, EMG taken with bipolar needle electrode (tips 1 mm apart); *D*, stimulus signal. The amplitude of the "gross" EMG, *A*, closely parallels the contraction height, *B*. The sudden decline in amplitude of the electrical record of a small group of fibres, *C*, is interpreted as the result of the "dropping out" of a few fibres, whose loss is not apparent in the "gross" EMG. (From Loofbourrow, 1948.)

enlarged upon below under the heading "Muscles Spared When Ligaments Suffice" (p. 164).

During sustained contraction recorded with surface electrodes over flexor digitorum superficialis in the forearm, Eason (1960) found that the amplitude of the integrated EMG increases progressively with time in both active and passive muscle. He suggested that additional motor units are progressively recruited to compensate for the loss in contractility due to impairment of fatigued units. The action potentials summate with those of already active units to more than offset the drop of amplitude of the impaired units. Eason also reported that surface EMGs progressively increased in amplitude with continuous or repeated contractions *not* associated with fatigue. The rate of increase was proportional to the magnitude of contraction. However, Wright *et al.* (1976) assert that emg changes as reflected in an arm muscle provide a relatively insensitive indicator of general fatigue, and with this well-documented view it is difficult to argue.

Part of the findings of Missiuro, Kirschner and Kozlowski (1962a, b) of Warsaw were rather similar. Integrated EMGs from surface electrodes over

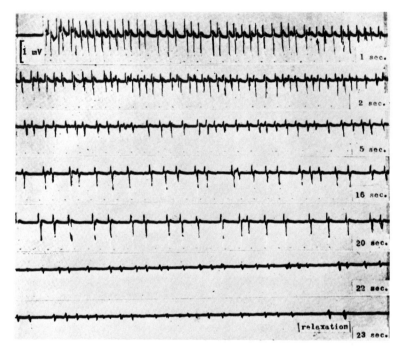

FIG. 4.5. Fatigue EMG. The train of single oscillations of a slightly irregular motor unit potential. The shape of the impulse is rather stable. The decrease of amplitude approaches the noise level. (Partly paralyzed extensor digitorum, bipolar needle electrode.) (From Lindqvist, 1959.)

biceps and triceps grow steadily during intense physical exertion. The EMG in the final stage of exertion shows evidence of synchronization of potentials. Missiuro *et al.* suggest two mechanisms of fatigue during effort: one, peripheral, produced by intense exertion; and the other, central occurring with prolonged low-intensity work. In the latter case, there are few emg signs of peripheral changes, i.e., even up to ultimate stages of fatigue when the subject consciously reduces his effort and therefore the output of potentials.

Zhukov and Zakharyants (1959) of Leningrad studied progressive fatigue in biceps and triceps put under continuous supporting activity against a load. At a certain stage of fatigue the supporting or lifting of the load appeared to be subjective. When the effort is maintained, typical changes appear in the EMG. The amplitude of potentials rises and the integrated level of voltage rises; synchronization of potentials appear, as reported also by Missiuro *et al.* (1962a, b) and Currier (1969).

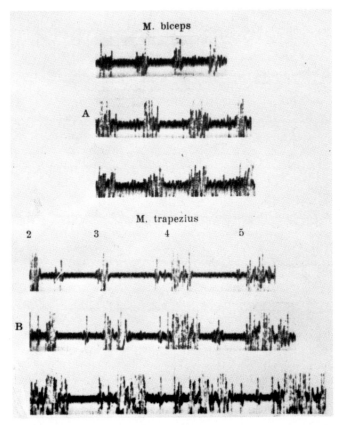

Fig. 4.6. Fatigue EMG recorded from needle electrode. *A*, biceps **brachii**, while the subject is repeatedly striking a typewriter key with the right index finger at maximal speed. The curves are led off at intervals of 1 minute and four strokes are recorded each time. *B*, trapezius, while the subject is continuously striking the keys at maximal speed with the four fingers. The curves are recorded at intervals of 1 minute. (From Lundervold, 1951.)

The Scottish physicians Lenman and Potter (1966) found that electrical changes associated with fatigue to be more prominent in patients with rheumatoid diseases and in the myopathic disorders. Yet the mean slope of the voltage-tension regression curves do not differ from that of healthy persons.

Studying the fatigue in rectus femoris of normal young men made to perform the Harvard step test, Sloan (1965) found little or no change in electrical activity in the absence of local fatigue. When more challenging

A   B

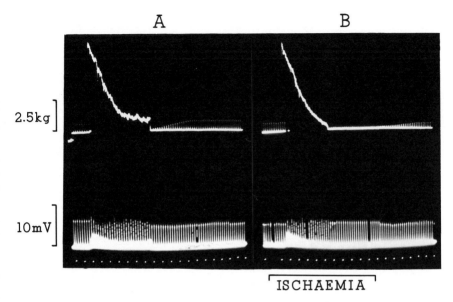

2.5kg

10mV

ISCHAEMIA

Fig. 4.7. Fatigue: mechanical myogram and EMG. *A*, maximal voluntary effort persisted in until severely fatigured. A series of single twitches (evoked by nerve shocks) precede and follow. *Lower tracing*, the corresponding emg action potentials. Time markers, ½ minute. *B*, the same but with the circulation occluded for the period indicated by a line (labelled ischaemia) beneath the record. (From Merton, 1954.)

exercise was imposed and fatigue occurred, the EMG amplitude increased with no consistent change in frequency. He could demonstrate no synchronization of potentials reported by others.

Using computerized methods of auto- and cross-correlation of potentials from different parts of one muscle during fatigue experiments, Person and Mishin (1964) confirmed beyond any doubt that synchronization of motor units *does* occur. More recently, Herberts and Kadefors (1976) showed that subjective fatigue appears constantly in the supraspinatus in overhead work by welders. Spectral analysis of the EMG signal revealed a significant decline in its frequency content, meaning an increase in the average duration of the active motor units in that muscle. Further, Viitasala and Komi (1977) found with isometric contraction of the rectus femoris maintained at 60% of maximum that the power density curve shifts toward lower frequencies. The mean power frequency decreased linearly as a function of fatigue time. They proposed that in addition to a natural recruitment of new motor units the fatigue includes a marked reduction in the conduction velocity of action potentials along the used muscle fibres.

Sato, Hayami and Sato (1965) of Tokyo were particularly interested in the differences in the fatigability of two-joint and one-joint muscles. They found little difference in emg changes in two-joint and one-joint muscles that lie superficial. But during fatigue the emg spectrum fell much more in the gastrocnemius than it did in soleus which lies deep to it. Deep one-joint muscles appear to be more difficult to fatigue than an overlying two-joint muscle. Sato (1966) further showed that the increasing rate of the global EMG accompanying fatigue in the squatting posture (common among Japanese laborers) is greater in some muscles of the thigh than others, being greater in vastus lateralis compared with rectus femoris and vastus medialis. The degree of lowering in the frequency spectrum of the EMG was much smaller in the rectus femoris compared with the one-joint muscles.

Scherrer and his colleagues showed in a various emg studies (1957 *et seq.*) that different views of normal muscular fatigue result from experimenting with human subjects and with animals and, apparently, also with the type of experimental technique used. There is, they find, a difference between the fatigue of repeated maximal effort and that of continuous contraction. This is not to deny the findings of a number of groups, such as Poudrier and Knowlton (1964) who find that fatigue brought on by intense repetitive or sustained strong contraction is peripheral to the myoneural endplate.

Scherrer, Lefebvre and Bourguignon (1957) suggest that Merton's demonstration of a reduction in the mechanical response without a concomitant reduction in the electromyographic potentials during maximal contractions is quite different from their own finding of progressive change in the EMG with continued prolonged mechanical work. The potentials show increased amplitude and decreased frequency (Scherrer and Bourguignon, 1959). Mortimer, Magnusson and Petersén (1970) elaborate this further. It is their opinion that the shift in the emg frequency spectrum during fatigue is in large part caused by a decrease in the conduction velocity of the muscle fibres, not just the synchronization of the firing of motor units. Stephens and Taylor (1972) find evidence in human physiological studies for the idea that in a maximal voluntary contraction, neuromuscular junction fatigue is important at first, superseded by contractile element fatigue. The former is believed to be most marked in high-threshold motor units while the latter affects low-threshold units. The influence of circulation must be important (Myers and Sullivan, 1968). Kuroda, Klissouras and Milsum (1970), bioengineers at McGill University, proved this to be the case, finding that the O_2 consumption rate is a concave upward function of EMG and force.

In recent years DeVries (1968) has attempted to develop a method for evaluating muscular fatigue from plots of integrated EMG as a function of time during isometric contractions at different strengths. Other investigators generally have not followed his lead.

Under the direction of Professors Monod and Scherrer, Phuon-Monich (1963) confirmed that there is an augmentation of electrical activity in muscle during fatigue caused by intermittent static work. The results were similar to those found with other forms of local work. Larsson, Linderholm and Ringqvist (1965) found that after intense sustained contractions and, to a lesser extent, after rhythmical contractions, polyphasic potentials increased in number. This was reversible. As in other studies, their action potentials decreased in duration but there was no change in amplitude.

In effect, the character of the electrical activity during voluntary exercise is a reflection of the EMG, or *vice versa*. Phenomena appear during the fatigue of prolonged work which are probably of spinal cord origin and still not well understood. Various authors have described various phenomena that occur with the progressive fatigue of continuous activity. These include synchronization of potentials (Lippold *et al.,* 1957), the rhythm of Piper (1912), and augmentation of the amplitude and duration of potentials, and an increase in polyphasic potentials. The specially interested reader should see the two long papers of Scherrer *et al.* (1957, 1960) for an excellent review of the opposed opinions on these matters.

VIBRATION FATIGUE. In an effort to determine the optimum seat-control configuration for pilots in helicopters where intense vibration of the operator's trunk and arms is a serious physiologic problem, Walter D. Brunohler (personal communication) undertook a series of studies using fine-wire EMG at Fort Rucker, Alabama. At the time of his first report (summer, 1973) he had analyzed the response of biceps brachii to heavy sine-wave vibrations applied to the simulated cyclic control handle and configuration of a VSTOL (vertical short take-off and landing) aircraft under varying conditions. From what I have seen of these interesting and important experiments, they undoubtedly will reveal factors of great scientific and practical usefulness in the understanding and prevention of a special type of fatigue.

In conclusion, it must be stated that, upon the available evidence, direct fatigue of human muscle fibres *per se* under normal conditions is insignificant and that the ordinary fatigue experienced by mortals is a much more complex phenomenon.

Control of Muscle Contraction

FEEDBACK. There is no doubt now that the state of contraction of a muscle is controlled by information "fed back" from it to the spinal cord centres.

Muscles Alive

Such feedback loops have been the subject of considerable basic research. (See Granit, 1964, for a brief but clear review.) Lippold, Redfearn and Vučo (1957) showed that a periodicity or modulation appears in normal motor unit potentials (from 8 to 14 per second) which is related to tremor, but is not normally visible to the naked eye. This modulation, increased by stretching a muscle or by the fatigue of effort and decreased by cooling is due to oscillation in the stretch reflex "servo loop" (fig. 4.8). Perhaps the tremors of various disorders are an exaggeration of this physiological periodicity or rhythmical activity of groups of motor units.

Eble (1961) has shown electromyographically that the muscles of the back

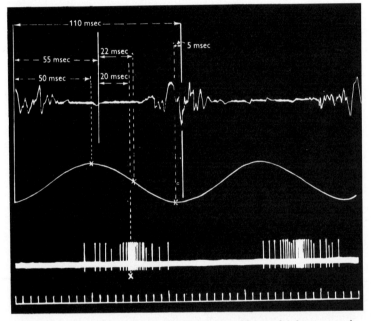

Fig. 4.8. Diagram of the relation (in human calf muscle) between electrical activity (*top*), mechanical record (*middle*) and spindle activity (whose time relations are calculated from a reflex time of 35 msec, which was the mean value of 10 determinations made during the same experiment). The action potential bursts are spaced at 110-msec intervals. The muscle is fully shortened at 50 msec later, which equals a phase lag of 170°. At the calculated time of greatest spindle discharge the velocity of lengthening of the muscle is at a maximum. Under these conditions, spindles therefore are sensitive to velocity (not tension or displacement). Displacement record shows an increase in muscle length as a downward deflection. (From Lippold, Redfearn and Vučo, 1957.)

(in acute spinal-rabbit preparations) function reflexly in various antagonistic and synergistic pairs. He concluded that the normal rôle of reciprocal innervation is to modify the excitability levels of appropriate neurons rather than to diminish activity in antagonist muscles which is, as we shall see below, normally non-existent.

TONIC NECK REFLEXES. For a discussion of reflex contraction in the neck which apparently increased the performance of upper limb movements see page 399.

MUSCULAR RESPONSE TO PASSIVE STRETCH. Although Granit and his associates (1956, 1957, 1958) suggest that in mammals there exist both tonic and phasic motor units just like those in invertebrates, this concept has not been generally accepted. Becker (1960) adduced electromyographic evidence that for the first time strongly supported this view as being valid—at least in certain muscles in the normal human being. Passive stretch of the long head of triceps brachii and soleus muscles—but not the short head of triceps and gastrocnemius—produced peculiar motor unit potentials which were unlike voluntary motor unit potentials. These potentials Becker has analyzed and he concludes that they come from special tonic motor units responding only to stretch. This preliminary work shows promise, but requires confirmation (see also p. 18).

Coordination, Antagonists and Synergy

One often sees the owlish statement that the brain does not order a muscle to contract but orders movements of a joint. As clever as it sounds, this statement is only true in part. Under certain circumstances the movement is, in fact, the result of contraction in only one or two muscles. This we have shown repeatedly by our various studies. For example, pronation of the forearm is usually produced by one muscle alone—pronator quadratus—unless added resistance is offered to the movement; then, more muscles are called upon (Basmajian and Travill, 1961). My colleagues and I have found this to be true in elbow flexion too, where brachialis alone often suffices, and in other movements. Therefore, it is wrong and misleading to believe that nature always calls upon groups of muscles to produce simple movements. On the other hand, there are complex movements (such as rotation of the scapula on the chest wall during elevation of the limb) which obviously call upon groups of cooperating muscles (see p. 191).

Antagonists, too, have been misrepresented in the normal functioning of muscles. The unfortunate and incorrect impression has been fostered by many physiologists and even more anatomists that during the movement of a joint in one direction muscles that move it in the opposite direction show some sort of antagonism. The truth of the matter, first proposed by Sherrington as "reciprocal inhibition" is that the so-called antagonist relaxes completely (Travill and Basmajian, 1961) except perhaps with one exception—at the end of a whip-like motion of a hinge joint. Here, apparently, the short sharp burst of activity in some antagonists occurs to prevent damage to the joint; this was first implied by Barnett and Harding (1955) and later supported by our own work (Basmajian, 1957, 1959) (see fig. 4.9) and that of Bierman and Ralston (1965). These investigators at the Biomechanics Laboratory of the University of California in San Francisco recorded the emg potentials in rectus femoris and biceps femoris while subjects had their knee moved passively and when they actively performed flexion and extension of the knee (fig. 4.10). When they turned their attention to what the antagonists are doing during active movements, they found that toward the end of such a movement, potentials occurred in the antagonist (fig. 4.10). They did not ascribe this to a stretch reflex as such, but they did consider the action as a regulatory one acting in proper timing through central feedback loops. They would agree that this brief terminal activity in antagonists serves a protective function to "avoid damage which such a force [in the prime mover] could produce."

Equally concerned with antagonist function are a group of French workers

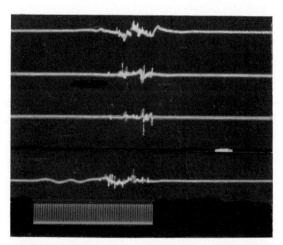

Fɪɢ. 4.9. Emg of elbow-flexors during rapid *extension* of elbow (0.1-msec time marker indicates duration). EMG tracings (*from above downwards*): biceps brachii long head; short head; brachialis; and brachioradialis. Burst of activity in the flexors at the end of extension.

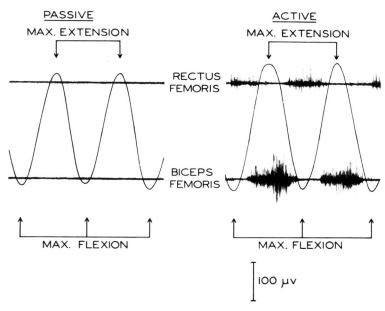

Fig. 4.10. EMG of rectus femoris and biceps femoris during passive and active flexion of the knee. (From Bierman and Ralston, 1965.)

(Goubel and Bouisset, 1967; Bouisset and Goubel, 1967; Lestienne and Bouisset, 1967; Goubel, Lestienne and Bouisset, 1968; Pertuzon and Lestienne, 1968; Lestienne and Goubel, 1969; Bertoz and Metral, 1970). In summary, they find a pattern of responses in which low unsustained activity occurs in antagonists at low speeds of voluntary flexion and extension of the elbow; at middle speeds there are successive activities in the agonist and antagonist, including common electrical silence; at high speed of flexion and extension there was partial overlapping of phasic activities in agonist and antagonist. They focus their attention not on the speed *per se*, but on the tension in the agonist and draw attention to the reflex activity during muscular recruitment especially in extension movements.

Patton and Mortensen (1970) also have studied the mechanical factors which affect agonist-antagonist interaction at the elbow joint. Extension of the elbow always causes more cocontraction of antagonists than does flexion. Increasing load increases cocontraction during both flexion and extension. Skilled subjects have reduced cocontraction. When voluntary cocontraction preceded a movement, there was marked reciprocal inhibition in the antagonist during an active movement. These findings assume an intermediate position in the continuing dialogue concerning the existence of cocontraction and probably come close to the truth in this complicated area of muscle physiology. Holt *et al.* (1969) found a reflex effect of antagonist contraction and head position on the responses of the agonist muscle which, for example, was augmented by prior strong contractions of the antagonist. Cohen (1970) goes even further in demonstrating that what is being done by the opposite limb affects the EMG of rhythmic movements in the studied limb in a varying manner. Generally it is agreed that voluntary slow movements in normal man do not cause stretch-reflex cocontraction of the antagonists, but rather, when it occurs, it occurs with rapid movements (Patton and Mortensen, 1971; Angel, 1975; Hallett *et al.*, 1975; de Sousa *et al.*, 1975; Morin *et al.*, 1976; Jacobs, 1976).

In effect, cocontraction of antagonists occurs to greater or lesser degree in some movements, in some people, at some ages and under some circumstances. With increasing age and training and at slower speeds, it tends to reduce to *nil.* When it occurs, it sometimes is due to reflexes and sometimes appears to be extravagant overflow.

The oft-used term *antagonist* should be replaced, in my opinion, by the companion word *synergist.* When "antagonists" act they really act just to prevent undesired movement, and their only important application as antagonists is in their acting against gravity. Because nervous coordination is so fine, there is no need for muscles to act in antagonism to others simultaneously. The rule, then, is for the "antagonist" to relax.

Wiesendanger *et al.* (1967) in a study on reaction-time at the elbow found that the muscular activity of a volitional reaction movement was short and usually showed reciprocal activity of the antagonist; in some cases there was reciprocal inhibition. Triceps activity in the position of the antagonist always was less marked than that of biceps as the antagonist.

One finds that the activity of muscles in the position of antagonists during a movement is a sign of nervous abnormality (e.g., the spasticity of paraplegia) or, in the case of fine movements requiring training, a sign of ineptitude. Indeed, the athlete's continued drill to perfect a skilled movement exhibits a large element of progressively more successful repression of undesired contractions. O'Connell has demonstrated this convincingly in her unpublished emg studies at Boston University. A group of physical education majors required to perform "head stands" while being studied electromyographically could be graded as to their actual experience by the amount of overflow of undesired activity in muscles that were only casually related to the exercise. (See also section under "Training," p. 105.)

Hirschberg and Dacso (1953), on the other hand, would seem to disagree with my opinion. In an early emg study, they appeared to conclude that simultaneous activity of agonists and antagonists is a common phenomenon, but unconsciously they come closer to my own position with their almost parenthetical statement that such activity is seen in " . . . strenuous motion or in tense experimental subjects." Furthermore, Lundervold's extensive experiments (1951) referred to on page 84, appear also to contradict Hirschberg and Dacso. Miles, Mortensen and Sullivan (1947) in an early study stated that potentials could be recorded from topographical antagonists, but the circumstances of their experiments were somewhat too specialized to make so sweeping a generalization today.

Dempster and Finerty (1947) in an early emg study set out to determine the influence of varying gravitational effects on the large number of muscles that may cross one joint—specifically, for 15 muscles that cross the wrist held in a horizontal position. Furthermore, they were concerned with the influence of torques or moments of force at the pivot. Finally, they employed rather esoteric calculations (of no interest to the general reader) to explain their findings.

For static support, the torque at the wrist produced by gravity must be balanced neatly by the torque of those muscles which are in an advantageous position, i.e., crossing above the horizontal level of the wrist pivot. However, Dempster and Finerty found that synergists were active as well and these were obviously not in a position to exert an antigravity torque. This activity in the synergists or stabilizers was about half that in the antigravity or main group (which they referred to as "agonists"). Muscles that were below the wrist pivot and therefore in no position to act against gravity showed activity too; this was one quarter as much as that in the agonists, according to Dempster and Finerty. They then unfortunately dubbed these muscles "antagonists." If indeed any true activity of this nature occurs—and refined emg techniques seem to deny it—the activity is not a matter of antagonism to the agonists, for gravity does not require help. Rather it must be due to secondary synergic and postural functions of the muscles of the wrist and fingers.

By rotating the horizontally held wrist (supination and pronation) different groups of muscles were brought to a superior position. Here they assumed the burden of the gravity torque; others were placed in less advantageous positions in which, however, they continued activity as synergists, but with reduced intensity.

Using as a model the act of prehension of the hand, Livingston, Paillard, Tournay and Fessard (1951) of Paris demonstrated the plasticity of synergists during voluntary movements. Thus, the interplay of activity of the flexors of the fingers and of the thumb with those of the forearm was shown during normal activity to vary significantly depending on the information of peripheral origin, e.g., position of joints, angle at which the synergists act, the nature of objects grasped, etc. More recently, Weathersby (1966) reported that there is considerable synergistic activity in certain forearm flexors during ordinary movements of the thumb.

Missiuro and Kozlowski (1961) illustrated the ultimate plasticity of synergists. In a study of rabbit muscle transplanted to the place of its "antagonist," they found the transplant took on the function of the anatomical and functional "antagonist." Obviously the nervous system is able to adapt readily to such changes.

We know that many contractions of any one particular muscle may be accompanied by synergistic activity in other muscles to steady the adjacent joints. Gellhorn (1947) thus demonstrated the rôle of far-removed synergists in movements of the wrist. While flexor carpi radialis was activated in very slight flexion of the wrist, triceps brachii became active with the increasing effort in the prime movers (the extensors of the wrist remaining relaxed meanwhile). Only with very strong static flexion of the wrist would activity—and that only occasionally—appear in the antagonists.

Gellhorn found three stages of recruitment of synergists, depending on the stress. In the first, the activity is confined to the agonist at the wrist. In the second, action potentials appear in the agonist and a muscle of the upper arm according to the following rule: biceps muscle becomes active with flexion of the supine wrist and with extension of the prone wrist, whereas the triceps becomes active with the reverse conditions (i.e., extension of the supine and flexion of the prone wrist). In the third stage, with excessive straining, some activity appears in antagonists as well but it is never equal to the activity of the prime mover and of the synergists. The exact significance of Gellhorn's patterns of recruitment are obscure but may be of fundamental importance. In any case, they stress the concept that "antagonists" are really only synergists.

Along the same line, experiments were done by Sills and Olsen (1958) in the hope of demonstrating activity in the unexercised arm while the opposite arm was exercised by normal subjects. There was, in these normal persons, little if any such "spread" to the opposite limb musculature unless extremely powerful movements were made. (See also our similar findings, p. 257 and the section, p. 102, under "Effects of Cross Exercise.") Their conclusions effectively demolish the basis for certain contralateral exercises that have been advocated for developing muscles, especially for an injured limb too painful or too immobilized to be moved itself.

Novel electromyographic studies of abnormalities in the plantar reflex response have fallen neatly into this general concept, too. The "up-going toe" of upper motor neuron lesions has been found by Landau and Clare (1959) to be the result of an exuberant overflow of activity to the great toe extensors;

even though the flexors continue to contract, the extensors overpower them (fig. 4.11).

In the very young normal child and especially premature babies, the same sort of phenomenon was demonstrated by Fényes, Gergely and Tóth (1960) with "flexion reflexes" observed electromyographically. Both agonists and antagonists contract in what they term a "co-reflex phenomenon." The same is true in spastic children with cerebral palsy during locomotion (Kenney and Heaberlin, 1962; Feldkamp *et al.*, 1976). There is an abrupt onset of the agonists and a rapid response of the antagonists with sufficient power to be obstructive. Under considerable resistance, normal children give the same response of exuberant (but wasteful or useless) overactivity of antagonists.

Rao (1965) has shown by EMG that, contrary to general opinion, reciprocal inhibition does not occur with the ankle jerk reflex. But he confirms its validity when voluntary actions are performed. Motor units in tibialis anterior act as briskly as those in gastrocnemius when the tendon of Achilles is tapped. He explains this reversal of normal inhibition in the "antagonist" as part of the positive supporting reaction in which the principle of reciprocal innervation is not applicable.

Agonist-antagonist interactions have been widely studied in the Soviet

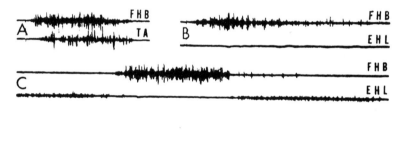

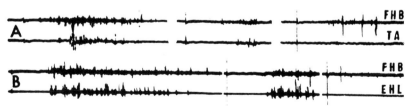

Fig. 4.11. Normal flexor response (*upper set of traces*) compared with the abnormal extensor response (*lower set*). EMGs from flexor hallucis brevis and extensor hallucis longus. (Composite of segments of two illustrations from Landau and Clare, 1959.)

Union (Baranov-Krylov, 1969; Person, 1965, 1969; Kozmyan, 1965) from the viewpoint of central motor controls. Person's work has been the most thorough and extensive. She showed that relaxation and tensing of an antagonist is learned and can be trained to increase or decrease. Kozmyan revealed that the latency of antagonist inhibition and agonist excitation varied most frequently during movements responding to non-rhythmic stimulation. With rhythmic repetitive movements, the latencies as well as dissociation of reciprocal inhibition diminished. Thus, inhibition of the antagonist muscles were to be expected in rhythmic activity with any element of supraspinal control or learning. Bratanova (1966) of Sofia found essentially the same thing with rhythmic activity of biceps and triceps brachii. In the "training" stages, coactivation was common apparently as the result of excitation radiation but later it was extinguished. Gatev (1967) also of Bulgaria, but working independently, found that as infants mature the excessive cocontraction typical of childhood diminishes progressively. This appears again to be the result of learned or patterned supraspinal control eliminating "undesirable" or "useless" cocontraction.

In a study of reflex reactivity of biceps and triceps in children at different developmental stages, the Polish investigator, Missiuro (1963), found a spread of electrical activity to other muscles of the same extremity. With increasing age this decreases so that in adult life it is minimal.

Vladimir Janda (1966, personal communication) of Prague has shown a significant linkage of emg activity in certain separate muscle groups, especially in children. During a strong effort in a particular muscle, he finds a high incidence of activity (in a predictable pattern) in far removed muscles of the same limb and trunk musculature. Hellebrandt and her colleagues have convincingly drawn our attention to a patterned spread of gross muscular activity to wider and wider areas during forceful effort or exercise stress (Hellebrandt and Waterland, 1962a, b; Waterland and Hellebrandt, 1964; Waterland and Munson, 1964a, b). Employing the Fukuda Stepping Test, Waterland and Shambes (1970) showed that the head-shoulder linkage of muscular activity was the key to body displacement and rotational directions when allowed to respond spontaneously.

In insects, simultaneous emg activity in antagonist muscles has been reported (Hoyle, 1964, in grasshoppers; Wilson, 1965, in cockroaches and locusts). These have no simple relationship and probably do not bear on the problem of synergy in mammals. The only possible connection is in the findings of Stuart, Eldred, Hemingway and Kawamura (1963) who showed that in shivering there are synchronous contractions in antagonistic muscles of mammals.

Silent Period and Reciprocal Inhibition

The "silent period" of muscle is the period of cessation of activity which occurs when a twitch contraction is superimposed on a voluntary effort (e.g., electrically or by a tendon tap). Its normal range (in adductor pollicis) is 87 to 151 msec (Higgins and Lieberman, 1968). It also occurs following the sudden release of a voluntarily innervated muscle provided a certain minimum rate of shortening is achieved; if shortening is not permitted or is slow, no silent period results (Struppler, 1975). Struppler has shown clearly that the silent period is not the result of reciprocal inhibition (autogenic inhibition); rather it is primarily due to the cessation of facilitatory impulses in the primary afferent fibres from the muscle spindles. At the end of the silent period, there is a rebound burst of EMG activity. A sudden stretch of a relaxed muscle does not recruit a monosynaptic (muscle-spindle) reflex in man (Marsden, Merton and Morton, 1976).

Garland and Angel (1971) showed that during a rapid voluntary movement, the agonist produces two distinct volleys of emg activity separated by a relative silence. When the active limb was unloaded during the movement, the second burst was significantly reduced. Apparently, the second burst is due to spinal reflexes. The mechanical properties of the muscle were shown by Agarwal and Gottlieb (1972) to have a significant influence on the duration of the silent period that follows the electrically induced H-wave in soleus muscle. The primary contraction in soleus (M-wave) coincides with reciprocal inhibition of the tibialis anterior. Yabe and Tamaki (1976) have shown the voluntary elbow extension is immediately preceded by a silent period in the unexercised contralateral agonist ("contralateral agonist silent period"), without any contralateral antagonist contraction occurring. Obviously this occurrence cannot be a unilateral type of reciprocal inhibition at the spinal level.

A different kind of silent period has been demonstrated in the muscles of mastication. Single stimulation of the teeth induces inhibition in actively contracting temporalis and masseter muscles with a 30 to 40 msec latency. The same type of inhibition occurs from sudden unload of the muscles (Ahlgren, 1969; Beaudreau *et al.*, 1969; Griffin and Munro, 1969). Munro and I (1971) more recently demonstrated that the inhibition in all the elevators of the mandible was almost synchronous; in most subjects there was a synchronous burst of activity in the chief depressor muscle (anterior belly of digastric) some 15 to 27 msec after initial tooth contact. Hannam (1972) has demonstrated that an enhancement of the masseteric reflex by voluntary contraction of the jaw-closing muscles may be due to autogenic factors, synergistic factors or both. Stimulation of the muscle spindles and facilitation caused by the voluntary activity before tooth contact must be involved.

Effects of Cross Exercise

The hypothesis that there is a transfer of activity to the contralateral limb during prescribed exercise on one side has been frequently postulated, but now it is being seriously questioned. Probably it is invalid except in very special circumstances. Gregg, Mastellone and Gersten (1957) of Denver, Colorado, found that overflow to the unexercised, contralateral muscles did not occur during simple non-resistive exercises or during isometric contractions of one biceps brachii. As the exercise stress increased, however, there was some "overflow" to the opposite triceps and, after even greater stress, to the biceps. Increasing fatigue played an important rôle in the "overflow" but was reversible, for after a rest of two minutes "overflow" would at first be absent. Moore (1975) found overflow activity to be between 10% and 20% of the maximal intensity of activity in the exercised limb. She believed that even this small amount of overflow gives sufficient justification for it to be used in maintaining muscular tone in immobilized limbs.

Samilson and Morris (1964) confirm the finding that in normal man activity of one upper limb is not accompanied by activity in the contralateral resting limb. However, in spastic children, there *is* such a spread. On the other hand Podivinský (1964) of Bratislava, Czechoslovakia finds a slight motor irradiation occurs from the strong contraction of finger flexors to the related muscles of the opposite limb ("crossed motor irradiation"). This perhaps is related to the findings of Hellebrandt and her colleagues regarding indirect learning, i.e., the improvement of strength in one limb by exercising the opposite limb (Hellebrandt and Waterland, 1962a, b). Its practical significance in ordinary life is unknown and appears to have been exaggerated since the days of Scripture *et al.* (1894). We have shown that at the finest levels of control in motor unit training the role of cross-training is not significant (Basmajian and Simard, 1966).

Further, the crossed reflex phenomenon described by Ikai (1956) of Tokyo is not really the same phenomenon as cross exercise. Ikai showed that the crossed reflex of limbs in spinal animals can be reproduced under certain conditions as a brief overflow of monosynaptic reflexes to the opposite limb.

Panin, Lindenauer, Weiss and Ebel (1961) seem to have delivered a serious blow to the concept of "cross exercise." In their extensive study they found that the spread of activity was minimal to insignificant. Insignificant potentials of low amplitude and frequency appeared in all non-exercised muscles in a widespread distribution in all four limbs. They appeared most in areas required for postural stabilization of the subject's body. Even then the amount of activity was so slight as not to constitute exercise effect.

Our own studies on quadriceps (p. 250) and those of Sills and Olsen (see above) largely confirm the conclusions of Gregg and his colleagues. We found in our studies of spastic patients (p. 83), however, that an exuberant overflow occurs to the opposite limb. Walshe (1923) and, more recently, Hopf *et al.* (1974) and Soto *et al.* (1974) have written about a similar phenomenon in hemiplegia. We must conclude that "cross education" is, at best, of dubious value in *normal* subjects.

Tremor

While general studies of tremor are common, primarily emg studies of the phenomenon are scattered. During a voluntary contraction, motor units composed of functionally identical muscle fibres start discharging at about several cycles per second and then accelerate up to 30 to 40 Hz as the strength of the contraction increases (Lippold, 1971). Physiological tremor (about 10 Hz) results from many stimuli including the pulse (Marsden *et al.*, 1969a; Dietz *et al.*, 1976) and external impulses; it is influenced by factors such as age, peak frequencies falling after the age of 50 (Marsden *et al.*, 1969b).

Parkinsonian tremor has a frequency of 4 to 5 Hz and exhibits alternating action of agonists and antagonists. However, clonus, a sign of upper motor neuron disease, which generally was thought to be alternating action of agonists and antagonists through reciprocal stretch reflexes, has been shown by Cook (1967) not to exhibit this alternating emg activity.

Shahani and Young (1976) showed that when the "tremor-at-rest" of Parkinson's disease is suppressed by voluntary activity, typical physiological tremor (8 to 12 Hz) appears. They also discriminate between this physiological tremor where the activity alternates in antagonists and essential-familial-senile tremor where a frequency of 5 Hz occurs synchronously in antagonists. Still other types of tremors occur in specific neurological diseases. Small-amplitude tremors may be determined mostly by viscoelastic components while larger displacements have a greater neuromuscular component (Stiles, 1976; Elble and Randall, 1977). Mary Schlapp (1973) of Edinburgh, in a fascinating emg study of violinists, showed that the voluntary *vibrato* has strong similarities to physiological tremor—but it is slower.

Spontaneous Muscle Cramps

Muscles in cramp have been studied electromyographically with needle electrodes and modern equipment by Denny-Brown and his colleagues (1948) and by Norris, Gasteiger and Chatfield (1957). Such cramps occur rather frequently in apparently normal people as localized, involuntary, sustained contractions, which are sudden and very painful. They occur in the calves of swimmers (particularly early in the swimming season) and during sleep ("night cramps") particularly in pregnant women. Perhaps they are related to the cramps that occur as a symptom of a variety of diseases, but this has been contested.

Norris *et al.* (1957) studied cramps in a series of subjects in whom they were produced by an ingenious technique which grew out of their observations. Cramps could be brought on by getting "normal," cramp-prone youths to make a voluntary effort while a large muscle under study (e.g., the biceps brachii) was in a shortened position. The action potentials that they recorded through fine indwelling wire electrodes were those of normal motor units and therefore were initiated by motor impulses from the central nervous system. Reflexes and other superimposed manoeuvres altered and even initiated the cramps, thus supporting their conclusions.

Disuse Atrophy

Although in an early report Buchthal and Clemmesen (1941) gave a substantial account of emg changes in atrophy including disuse atrophy, no extensive literature exists on this variant of normal EMG. Fudema, Fizzell and Nelson (1961) studied the question using external fixation apparatus on the hind limb of rats. They found a continuing decrease of electrical output from tibialis anterior through the period of immobilization. This reflected the reduction in size of the muscle and, apparently, the muscle-fibre membrane area. No spontaneous fibrillation potentials occurred and the shape of motor units remained normal, indicating that the myoneural junction is not implicated in disuse atrophy.

Cooper (1972) reported that the disuse atrophy caused by immobilization of a limb in cats produces a reversible increase of twitch contraction and relaxation times along with a reduction of tension in both a single twitch and a tetanus. No comparable emg results have appeared in the literature. Cortisone treatment in rats which produces similar muscular weakness reveals no evidence of emg abnormality (Prabhu and Oester, 1971). Intracellular recordings of muscle cells with disuse atrophy show no change of membrane characteristics (Brooks, 1970).

Effects of Overheating and Cooling

Edelwejn (1964) found in a series of experiments on rabbits subjected to overheating that there is a statistically significant increase in polyphasic potentials (from 11 to 31%) and shortening of time of single polyphasic potentials. These fell from 9.5 msec at 37°C to 7.1 msec at 41°C. No changes were found in amplitude. Edelwejn proposes that the cause is due to changes in impulse transmission in the muscle fibres with some unexplained disturbance of integration in the muscle fibres of the motor unit.

Under controlled hypothermia, Serra, Pasanisi and Natale (1963) have found a progressive fall of 50 μv in mean amplitude and 0.5 msec in duration of potentials during ordinary contractions. Similar findings were obtained when electrically stimulated contractions were studied.

Wolf and Letbetter (1975) found that a specific cutaneous cooling over a muscle (gastrocnemius of decerebrate cats) cause inhibition of motor units without actually changing intramuscular temperature.

Effects of Smoking

Serra and Lambiase (1957) of Naples have demonstrated that cigarette smoking causes changes in the action potentials. There is a decrease in the frequency of single motor unit potentials at maximum effort with spontaneous monophasic and diphasic spikes appearing in apparently relaxed muscles. On voluntary activity there are changes in the shape of the potentials, many becoming polyphasic. Serra and Lambiase suggested that these changes result from a light carbon monoxide poisoning plus nicotinic effect at the myoneural junction and on the muscle fibre itself.

Rao and Rindani (1962) and Rao (1963) have also studied the effect of smoking on the EMG. In the second study, Rao found a delayed second spike appeared in the composite wave from muscles stimulated electrically through their motor nerves. He ascribes this second spike to the effect of nicotine on some of the neuromuscular junctions but also admits that the cause may be primarily in the muscle fibres themselves.

Training

Professor Mitolo of the University of Bari in Italy has shown (1956, 1957) that progressive physical training of a specific muscle produces a gradual increase in the average duration of its potentials (with a progressive diminution of their average frequency) and a gradual "regularization" of the response. This last might be expected to be a function of training in athletics, i.e., with advanced training there is greater and greater efficiency and specificity of response.

The Russian physiologist, Person (1958), studied the electrical activity of the biceps and triceps brachii while subjects were trained in certain types of work (e.g., chopping and filing). Before the training, the rhythmical flexion and extension of the elbow were effected by exuberant, apparently wasteful, activity of the antagonist which is overcome by the greater activity of the agonist. With training, there is a progressive inhibition of the antagonist during the movements of flexion and extension until, with advanced training, the inhibition becomes complete. O'Connell's work, referred to on p. 96, agrees, as does that of Kamon and Gormley (1968) in principle. Hobart and his colleagues (see, for example, Hobart *et al.*, 1975) also confirm these views in their studies of changes that occur in motor functions during the acquisition of novel throwing tasks. Similar findings have been reported by Lloyd and Voor (1973) for the acquisition of a special skill during competition.

Motor Learning and Control

There is mounting evidence that motor learning and control are not a process of accretion but depend on patterning of inhibition in motor neurons. Electromyographic studies in health and disease indicate that the acquisition of skills occurs through selective inhibition of unnecessary muscular activity rather than the activation of additional motor units.

As noted in other sections, almost all resting muscles throughout the bodies of adults, both human and general mammalian, fall to a level of neuromuscular silence. This total relaxation occurs unless the muscles are needed to be tensed for a posture or movement or unless the person suffers from uncontrolled apprehension or neurotic and neurological disturbances. With this in mind, MacConaill and I (1969) enunciated the principle that there should be a minimal expenditure of energy consistent with the ends to be achieved. This self-evident principle embraces two laws: (1) *The Law of minimal spurt action*—no more muscle fibres are brought into action than are both necessary and sufficient to stabilize or move a bone against gravity or other resistant forces, and none are used insofar as gravity can supply the motive force for movement; (2) *The law of minimal shunt action*—only such muscle fibres are used as are necessary and sufficient to ensure that the transarticular force directed toward a joint is equal to the weight of the stabilized or moving part together with such additional centripetal force as may be required because of the velocity of that part when it is in motion.

Control of Movement

The neurophysiological literature is encrusted with the barnacle that "the brain does not order a muscle to contract but orders movement of a joint."

Recently, Phillips (1975) made a concerted effort to dispel this myth which has stultified research on the learning of motor behavior. In fact, the best movements are performed with an economy of muscular movements dependent upon impulses being sent to only one or two muscles or even a localized area of one muscle. What the brain has "learned" is patterning of these actions by means of a progressive inhibition of the inefficient mass responses that were natural to the child. Some movements are extremely economical in the well-trained person. For example, most of us are fairly well-trained in turning our hand over through pronation and supination of the forearm; in this learned act our nervous system calls upon only one or two muscles to produce the movements. Fortunately, the normally plastic human brain quickly adapts to shifts of function; otherwise tendon-transfer operations would be useless.

Physiologists and even some kinesiologists do not appreciate that each and every muscle has several (sometime many) component parts which are recruited in different functions at different times. Many investigations with intramuscular electrodes in many thousands of muscles lead me to believe that this local activity is patterned by progressive inhibition of motoneurons until an acceptable performance is achieved. Our studies of elbow flexion and thenar muscles, which show the interplay of motor unit functions dedicated to specific postures and movements, clearly indicate that the positioning of limbs is predetermined by sets of motor units which are permitted to act for that position. The same appears to be true for well-learned movements.

I believe that a mosaic of spinal motoneurons is dedicated to the learned response of a specific posture or movement of a joint through space. The ultimately superior performance of a skilled movement depends on the reproducibility of the ideal, an economically spare mosaic of motoneuronal activity (Basmajian, 1977). With different objects in mind, Payton *et al.* (1976) put it slightly differently: they found no statistically significant difference between prelearning and postlearning of a simple task in regard in the emg activity, movement time and range of movement. They concluded that all the prime movers that are going to contribute to the final learned act take part even before the skill is learned; thus as motor learning takes place, there is a marked reduction of activity only in the auxiliary muscles while the prime movers neither gain nor lose (Payton *et al.*, 1976).

When I first described the precision possible in controlling single motoneurons, I believed (as did many others) that this type of control was the building block of motor performance. Given visual and auditory cues through electronic amplification and feedback, subjects could be quickly trained to consciously activate single motoneurons with great precision. But conscious activation of single motoneurons in the single-motor-unit training paradigm depends on the same principles as the learning of any other novel task, that is, progressive (and sometimes rapid) inhibition of the motoneuronal activity that adds no useful function in producing a desired motor response (Smith, Basmajian and Vanderstoep, 1974).

Training, whether it is the unconcious process of the child learning simple social motor responses or the preparation for a specific skilled act (such as those of a musician or athlete), is a progressive inhibition of many muscles that flood into play when one first attempts to produce the required response. The athlete's continued drill to perfect a skilled movement exhibits a large element of progressively more successful repression of undesired contractions. Among others, O'Connell (1958) has demonstrated this convincingly. A group of physical education majors required to perform "head stands" while being studied electromyographically could be graded as to their actual experience by the amount of overflow of undesired activity in muscles that were only casually related to the exercise.

The young animal has enormous amounts of overactivity and reactive contractions in muscles that are serving no directed purpose in producing the desired movement or posture. Among others, Janda and Stará (1965) demonstrated in children a high incidence of mass responses in a predictable pattern even in muscles that are far removed from those which produce a required movement. As children mature this overactivity disappears and is absent in normal adults. It reappears in adults under psychological stress, but people can be trained to inhibit it to varying degrees. In patients with diseases and injuries of the central nervous system, the normal inhibition pattern is lacking; then mass responses from local interoceptive and exteroceptive bombardments of the motoneurons result in an exaggerated mass response described as spasticity.

The Moscow investigators led by Yusevich (see, for example, Okhnyanskaya *et al.*, 1974) attribute normal motor hyperactivity in infants and children to synkinesis or synergies of suprasegmental origin, pointing out the fact that they normally disappear by the time a person is adult.

The patterning of the inhibition would seem to come in part from obscure processes in diffuse centers of the cerebral cortex; since inhibition is a central feature, one must consider the possibility that brainstem centres and perhaps the cerebellum are critically important in the imprinting of the learning. It is too simplistic to consider a schema where an impulse is started at a tiny area of the cerebral cortex and is thence passed directly along a facilitatory path to a desired set of motoneurons. The motor learning process probably employs a neuronal network with the "main" pathway for motor activation being almost a small part of the whole.

Proprioceptive Effects

Gellhorn (1960) has described electromyographic studies which disclose the effects of central proprioceptive influences on movements elicited by the electrical stimulation of the motor cortex. Movements so produced are strongly reinforced by proprioceptive impulses which also determine, by and large, the type of movement that results. He showed, for example, that the contraction of triceps and flexor carpi muscles when stimulated through the cerebral cortex is greater if the elbow is at 45° than if it is at 110° or 160°. Furthermore, a cortical stimulus that is below threshold when a muscle is slack may become effective when the muscle is put on the stretch.

Electromyography of the Fetus and Newborn

Until the 1950s, no reliable information was available in regard to the earliest muscle potentials during fetal development. The characteristics of the earliest potentials are of great interest both in embryology and in the related fields of electromyography and neurology. Particularly important is the relationship of the time and innervation to the time of appearance of the earliest potentials. Our laboratory (then at Queen's University) developed a program employing special techniques with which to examine a series of living vertebrate fetuses.

The details of our methods and the detailed results have been published (Lewis and Basmajian, 1959; Ranney and Basmajian, 1960). A large number of rabbit fetuses and a limited number of goat fetuses were studied.

In the rabbits (which have a gestation period of 32 days) all fetuses aged 18 days or more showed electromyographic potentials. At 17 days, only some of the fetuses had emg activity. This is the earliest fetal age in which we first observed visible movements. At 16 days, only one of 16 fetuses exhibited true muscle potentials. None of the younger fetuses did so.

The emg potentials of these fetuses ranged from as low as 13 μv to as high as 250 μv (fig. 4.12). Their durations were too long to classify them with the short-duration, small potentials which are known as fibrillation potentials and are diagnostic of denervation in post-natal life. Therefore, it has not been possible to state that the potential we found indicated a lack of innervation of the fetal muscles.

A few authors have described early visible movements in mammalian fetuses and embryos but these have not been intrinsic or spontaneous; rather, they have been in response to prodding or electrical stimulation. Straus and Weddell (1940) found contractions of the forelimb in fetal rats stimulated electrically in the latter half of the 15th day (of a 21-day gestational period).

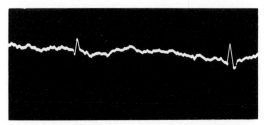

Fig. 4.12. Two emg potentials from shoulder region of a 17-day-old rabbit fetus (proved not to be ECG). Calibration (*at right*), 100 μv and 10 msec, peak to peak. (Ranney and Basmajian, 1960.)

This is in general agreement with the finding of Windle *et al.* (1935). Windle (1940), furthermore, states that the site of earliest activity is in the lower cervical region; it was nearby, in the shoulder region, that we picked up our earliest muscle potentials. Apparently, the onset of spontaneous activity requires a further degree of maturation beyond the stage at which muscles respond to an external stimulus.

Boëthius and Knutsson (1970) found in fetal chicks that an increase in membrane potentials of individual muscle fibres coincides with the transformation of the majority of the cells from myotubes to myocytes.

Marinacci (1959), in a report on the EMG of prematurely born infants, concluded that at the 6th month of intra-uterine life (relatively much older than our rabbit fetuses) about 20% of the muscle fibres have still to be innervated. At the time of birth, 5% apparently have not yet received their nerve supply. At the end of the 4th post-natal month, practically all the muscle fibres have been innervated. The delayed innervation is largely in the lower extremities especially in the intrinsic muscles of the feet.

Eng (1976) of Washington revealed the occurrence of spontaneous potentials—quite similar, if not exactly the same as, fibrillations—in almost half of 19 normal but premature infants. These disappeared in subsequent months. Normal newborns also had similar findings in a third of cases studied extensively and intensively.

Prior to innervation, primitive muscle fibres theoretically should possess an inherent tendency to spontaneous fibrillation and related electrical activity. Marinacci found that muscle fibres do fibrillate in premature infants corresponding to a stage in intrauterine life when, he believes, they might not all be innervated. Our own extended studies of fetuses in goats and rabbits (Ranney and Basmajian, 1960) failed to reveal any of the signs of spontaneous pre-innervation potentials. Botelho and Steinberg (1965) confirmed this in the canine fetus. We can offer no real explanation for this discrepancy; perhaps the finding of spontaneous fibrillation potentials in premature infants by Marinacci and by Eng does not necessarily prove their existence in normal fetuses *in utero*. Finally there may be a species difference. In fetal sheep, Ängg-ård and Ottoson (1963) found that skeletal muscles could be made to contract at the 50th day, considerably before the time at which myelination (and therefore normal functioning) of the axons occurs.

During prenatal development the fetal sheep shows marked changes in neuromuscular function as studied by motor nerve stimulation. Änggård and Ottoson found marked changes from the early stage (50th day) to more advanced stages when spontaneous movements are common (100th day). Though not strictly an emg study, this work bears upon our present concern. It clearly indicates that, in mammalian embryos, speed of conduction of motor axons is related to the amount of progressive myelination. However, the excitable properties of the axons is independent. A key factor in neuro-muscular function is the clear establishment of motor endings on the muscle fibres. In fetal sheep this occurs after about 50 days.

In normal newborn babies, Schulte and Schwenzel (1965) of Göttengen found irregular spontaneous bursts of normal motor unit activity in upper and lower limb muscles. These were sometimes reciprocal and sometimes strictly alternating between antagonistic muscles. In the upper limb the flexors were preferred. With hypertonic newborns, there was often constant tonic activity in certain muscle groups. This was widespread in the more severe cases.

EMG in Normal Fishes, Reptiles and Birds

Biologists have taken a new interest in the possibilities of using EMG for studying neuromuscular functions in intact vertebrates other than mammals. This is not the place for an exhaustive review of what has been accomplished. However, a few interesting illustrations are called for.

Ballintijn and Hughes (1965a, b) have investigated the muscular basis of the respiratory pumps in the trout and the dogfish. In brief, they find that the muscles of the mouth and gills of the trout may be divided into two main groups according to whether they are active or passive during the expansion or contraction phase of the pumps. The protractor hyoideus (geniohyoideus) is active only during the contraction phase. Moreover, there are differences depending on the depth of ventilation. Shallow ventilation is maintained by one group while deeper ventilation calls upon another group. Only during strong ventilation does the dilator operculi play a rôle in abduction of the gills.

Gans and Hughes (1967) extended this work to the tortoise, showing that in all phases of the respiratory cycle muscles were active. The increase of pressure in the peritoneal cavity is accompanied by activity in the transversus abdominis and the pectorales muscle which draws the shoulder girdle back into the shell. During the opposite phase (when increase in volume of the cavity leads to the reduction of intrapulmonary pressure below atmospheric) the obliquus abdominis and serratus major muscles act. In the leopard frog, the intrinsic laryngeal muscles have been studied by Schmidt (1972). The posterior constrictors do not tense the vocal folds; rather, they merely oppose them through lever action, Other details of glottic function are given in Schmidt's fascinating paper.

Ballintijn and Hughes showed that there are variations in the pattern in different individuals and in the same individual at different times. They also find that the pattern in dogfish is different from that in the trout. In dogfish, electrical activity takes place almost synchronously among all the muscles when the two cavities through which water is passed are decreasing in volume. During normal resting ventilation no electrical activity was recorded in the hypobranchial muscles. They became active during hyperventilation and when biting. During swimming of certain sharks, the whole branchial region remains in a relaxed condition and water enters the mouth. The amount is regulated by the adductor mandibulae (Ballintijn and Hughes, 1965b).

In the dragonet, *Callionymus lyra*, Hughes and Ballintijn (1968) found that the chief ejector muscles are the adductor mandibulae, protractor hyoideus and hyohyoideus. Generally, during the expansion phase of the respiratory cycle, levator hyomandibulae and sternohyoideus are active. In the carp, Ballintijn (1960) found a general principle for free-swimming teleosts: the lateral expansions and contractions of the buccal and opercular cavities are always synchronous with, or may precede, the lateral expansion. The levator operculi of the carp does not participate in the opercular pump; rather it is an abductor of the lower jaw.

Other emg work on fish is now in progress in Leiden, Netherlands by J. W. M. Osse (personal communication) with the European perch, and in Tokyo by Kaseda and Nomura (1973) with the carp. No doubt these types of studies will multiply rapidly but they are not the object of this book and must be chronicled elsewhere. Now at the University of Michigan, Carl Gans has studied the locomotor pattern of snakes. Fowl have been studied fairly extensively because of the occurrence of hereditary muscular dystrophy in chickens. Bekoff (1976) used EMG to study the changes in motor behavior of leg muscles of chick embryos while Engelhart *et al.* (1976) studied tissue-cultured chick skeletal muscles. There appear to be no limits!

Hungarian scientists Czéh and Székely (1971) recorded potentials from four pairs of homologous muscles of normal and grafted supernumerary forelimbs in ambystoma. The major contribution of their beautiful work is the light it sheds on coordination of rhythmic patterns between grafted limbs and normal limbs.

One now can expect a widespread increase in the use of EMG among biologists. With the marvellous improvement and practicality now available in telemetering devices, biologists have been provided with an excellent tool for dynamic studies of normal function.

Electromyography of Insects

Beránek and Novotný (1959), of Prague, Czechoslovakia, have recorded EMGs from the limb muscles of cockroaches. The records of voluntary movements reproduced in their paper resemble closely the complex interference pattern of the mammalian electromyogram except that the duration of individual spikes is somewhat shorter. Single motor unit potentials appear to consist of a slightly more complex wave-form. According to Beránek and Novotný, there is an isolated rhythmical discharge of such single motor units during "absolute motor rest" for long periods of time; this, of course, suggests that the insect does not relax completely at rest. [These workers also went on to demonstrate in denervated muscles of the insect spontaneous potentials which do not resemble fibrillation potentials in denervated mammalian muscles. Since a discussion of denervation is not called for here, specially interested readers should consult the original paper, which is in English.]

The neuromotor mechanisms during flight of the honey bee were the object of electromyographic experiments of Bastian (1972), and Kammer and Heinrich (1972), and those of several dipteran species were studied by Wyman (1970). Crickets, too, have been studied; Bentley and Kutsch (1966) found that the muscle action potentials during stridulation show that specific thoracic wing muscle groups are involved during calling and aggressive song pulses while others produce the "tick" of courtship song and still others the soft pulse phase of the song. The upstroke of flight evokes the same motor response as the closing stroke of stridulation. Bentley and Hoy (1970) have further elaborated on this work in postembryonic crickets, and Hustert (1975) studied the abdominal ventilation in locusts. The work of Hoyle and his group with locusts and other animals deserves mention although it is outside the concerns of this book (see, for example, Hoyle and Willows, 1973).

In Denmark, Eric Gettrup (1966) is studying the integrative processes within the pterothoracic ganglia of locusts, using records of sensory and motor events. Variation of wing twisting during flight is controlled by motor unit activity that is influenced by impulses from sensilla found on both hindwing and forewing. Baker (1972) has elaborated a sophisticated method for obtaining myopotentials from a locust during free turning in flight. One would hope that greatly increasing and widespread use will be made of such techniques for the study of the muscular activity involved in flying, hopping and walking in many different species of insects.

Chapter 5

Conscious control and training of motor units and motor neurons

S TUDIES of neuromuscular and spinal-cord function have been growing increasingly complex in recent years without offering clearer answers to many fundamental problems. Especially confusing and fragmentary are theories on the influence of various cortical and subcortical areas on spinal motor neurons and motor units in man. It was therefore refreshing to be able to develop and advocate a technique that not only proved to be quite simple but also promised to reveal considerable fundamental information. Ironically, the technique was only a modification of ordinary electromyography. This modification consists of regarding electromyographic potentials not for their own intrinsic value but as the direct mirroring of the activity of spinal motor neurons. Thus the group of muscle fibres in a motor unit is considered only as a convenient transducer that reveals the function of the nerve cell.

Perhaps the ultimate irony is that in their classic paper establishing the modern era of electromyography in 1929, Adrian and Bronk suggested that " . . . The electrical responses in the individual muscle fibres should give just as accurate a measure of the nerve fibre frequency as the record made from the nerve itself." Even earlier, Gasser and Newcomer (1921) had shown that "the electromyogram is a fairly accurate copy of the electroneurogram." Perhaps as a reflection of the general turning away from man as an experimental animal in favour of more exotic beasts and preparations, no real use of these early conclusions has been made until recently. In fact, the implications in Gasser and Newcomer's work did not lead to any systematic use of electromyography for studying the behaviour of individual spinal motor neurons in any species even though the action potential of a motor unit picked up by direct electromyography reflects the activity of its spinal motor neuron.

No great progress was made until 1928–29 when Adrian and Bronk published two classic papers on the impulses in single fibres of motor nerves in experimental animals and man. Their method consisted of cutting through all but one of the active fibres of various nerves and recording the action currents from that one fibre. They also succeeded in making records directly from the muscles supplied by such nerves. Somewhat incidentally, Adrian and Bronk introduced the use of concentric needle electrodes with which the activity of muscle fibres in normal human muscles could be recorded. Meanwhile Sherrington (1929) and his colleagues had crystallized their definition of a motor unit as "an individual motor nerve together with the bunch of muscle-fibres it activates." (Universally, later workers have also included in their definition the cell body of the neuron from which the nerve fibre arises.)

Although in subsequent years the concentric needle electrode was seized upon for extensive use, until the Second World War only a handful of papers appeared on the characteristics of action potentials from single motor units in voluntary contraction. In 1934, Olive Smith reported her observations on individual motor unit potentials, their general behaviour and their frequencies. She showed that normally there is no proper or inherent rhythm acting as a limiting factor in the activity of muscle fibres; rather, the muscle fibres in a normal motor unit simply respond to each impulse they receive. Confirming earlier work of Denny-Brown (1929) she set at rest the false hypothesis of Forbes (1922) that the muscle fibres or motor units were fatiguable at the frequencies they were called upon to reproduce by their nerve impulses.

Forbes had also suggested that normal sustained contraction requires rotation of activity among quickly fatiguing muscle fibres. Smith proved that such a rotation need not occur and that an increase in contraction of a whole muscle involves both increase in frequency of impulses in the individual unit and an accession of new units which are independent in their rhythms. The frequencies ranged from 5 to 7 per second to 19 to 20 per second, although "highly irregular discharge may occur at threshold both during the onset of a contraction and during the last part of relaxation." Finally, she proved that tonic contraction of motor units in normal mammalian skeletal muscle fibre, the existence of which was widely debated, does not exist. Two generations later, there are people in muscle research still not aware of her definitive studies.

Lindsley (1935), working in the same physiology laboratory as Smith, determined the ultimate range of motor unit frequencies during normal voluntary contractions. Although others must have been aware of the phenomenon, he seems to have been the first to emphasize that at rest "subjects can relax a muscle so completely that . . . no active units are found. Relaxation sometimes requires conscious effort and in some cases special training."

In none of his subjects was "the complete relaxation of a muscle difficult." Since then, this finding has been confirmed and refined by hundreds of investigators, using much more sophisticated apparatus and techniques than those available in the early 30's.

Lindsley also reported that individual motor units usually began to respond regularly at frequencies of 5 to 10 per second during the weakest voluntary contractions possible and some could be fired as slow as 3 per second. The upper limit of frequencies was usually about 20 to 30 per second but occasionally was as high as 50 per second. Earlier, Adrian and Bronk (1928, 1929) had found the same upper limit of about 50 per second for the nerve impulses in single fibres of the phrenic nerve and from the diaphragm of the same preparations.

Gilson and Mills (1940, 1941), recording from single motor units under voluntary control, reported that discrete, slight and brief voluntary efforts may call upon only a single potential (i.e., a single twitch) of a motor unit being recorded. Twenty years later, Harrison and Mortensen (1962) showed that by means of surface and needle electrodes action potentials of single motor units could be identified and followed during slight voluntary contractions in tibialis anterior. Subjects provided with auditory and visual cues could produce "single, double and quadruple contractions of single motor units . . . " and in one case, " . . . the subject was able to demonstrate predetermined patterns of contraction in four of the six isolated motor units."

Using special indwelling fine-wire electrodes (p. 32), I had no difficulty in confirming these findings (Basmajian, 1963), and on this basis I was able to elaborate techniques for studying the fine control of the spinal motor neurons, especially their training, and the effects of volition. Later, my colleagues and I further developed and described our system of testing and of motor unit training. We demonstrated the existence of a very fine conscious control of pathways to single spinal motor neurons (Basmajian, Baeza and Fabrigar, 1965). Not only can human subjects fire single neurons with no overflow (or perhaps more correctly, with an active suppression or inhibition of neighbours), but they can also produce deliberate changes in the rate of firing. Most persons can do this if they are provided with aural (and visual) cues from their muscles. Many investigators have documented the qualitative and quantitative aspects (for example: Simard, 1969; Zappalá, 1970; Gray, 1971a, b; Török and Hammond, 1971; Clendenin and Szumski, 1971; Harrison and Koch, 1972; and others, some of whom are cited elsewhere in this chapter).

Following the implantation of fine-wire electrodes and routine testing, a subject needs only to be given general instructions. He is asked to make contractions of the muscle under study while listening to and seeing the motor unit potentials on the monitors (fig 5.1). A period of up to 15 minutes is sufficient to familiarize him with the response of the apparatus to a range of movements and postures. [Kahn (1971) has shown that in limited circumstances and with on-line computer analysis to help, surface electrodes can prove useful and even preferable for special projects.]

Subjects are invariably amazed at the responsiveness of the loudspeaker and cathode-ray tube to their slightest efforts, and they accept these as a new form of "proprioception" without difficulty. It is not necessary for subjects to have any knowledge of electromyography. After getting a general explanation they need only to concentrate their attention on the obvious response of the electromyograph.

With encouragement and guidance, even the most naive subject is soon able to maintain various levels of activity in a muscle on the sensory basis

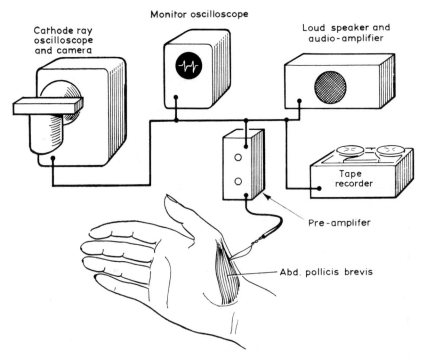

Fig. 5.1. Diagram of arrangement of monitors and recording apparatus for motor unit training. (From Basmajian, 1963b.)

provided by the monitors. Indeed, most of the procedures he carries out involve such gentle contractions that his only awareness of them is through the apparatus. Following a period of orientation, the subject can be put through a series of tests for many hours.

Several basic tests are employed. Since people show a considerable difference in their responses, adoption of a set routine earlier proved to be impossible. In general, however, they were required to perform a series of tasks. The first is to isolate and maintain the regular firing of a single motor unit from among the 100 or so a person can recruit and display with the technique described. When he has learned to suppress all the neighbouring motor units completely, he is asked to put the unit under control through a series of tricks including speeding up its rate of firing, slowing it down, turning it "off" and "on" in various set patterns and in response to commands. More elaborate techniques now used are really only controlled versions of the original methods (Basmajian and Samson, 1973). Johnson (1976) has tested and fashioned methods that meet statistical requirements more adequately.

After acquiring good control of the first motor unit, a subject is asked to isolate a second with which he then learns the same tricks; then a third, and so on. His next task is to recruit, unerringly and in isolation, the several units over which he has gained the best control.

Many subjects then can be tested at greater length on any special skills revealed in the earlier part of their testing (for example, either an especially fine control of, or an ability to play tricks with, a single unit). Finally, the best performers can be tested on their ability to maintain the activity of specific motor-unit potentials in the absence of either one or both of the visual and auditory feedbacks. That is, the monitors are turned off and the subject must try to maintain or recall a well-learned unit without the artificial "proprioception" provided earlier.

Lloyd and Leibrecht (1971) and Samson (1971) independently showed that the SMU training fulfills the requirements of the learning paradigm. The feedback methodology is not critical; thus, a highly artificial indication of successful training is satisfactory to a considerable degree. Leibrecht *et al.* (1973) went on to show that direct EMG feedback substantially improved initial learning. The nature and amount of learning, including the ability to use proprioceptive cues in controlling an SMU, were not affected; neither was the retention of learning.

Ladd, Jonsson and Lindegren (1972) investigated the learning process involved in the fine neuromotor control of single motor unit training which, of course, also embodies inhibition of motor activity. They employed trained units in five different muscles in 25 subjects. Voluntary inhibition, they found, is a conceptual type of response showing independence of the motor component; it generalizes and transfers positively from one muscle to another. However, the voluntary contractions of an individual unit is a specific perceptual motor type of response; the motor component of the response is essential and the learned response does not generalize or transfer from one muscle to another. Middaugh (1976) reported that subjects are not relying on peripheral factors in learning, i.e., only limited learning of peripheral sensory information and discrimination occurs. The finding by Vogt (1975) that there is little correlation between self-estimation of success and gross emg levels of contraction in the forearm supports Middaugh's finding for motor units.

Any skeletal muscle may be selected. The ones we have used most often are the abductor pollicis brevis, tibialis anterior, biceps brachii and the extensors of the forearm. However, we have easily trained units in buccinator (Basmajian and Newton, 1973) and in back muscles; Sussman *et al.* (1972) have trained units in the larynx while Gray (1971a) has trained them in the sphincter ani!

Ability to Isolate Motor Units

Almost all subjects are able to produce well-isolated contractions of at least one motor unit, turning it off and on without any interference from neighbouring units. Only a few people fail completely to perform this basic trick. Analysis of poor and very poor performers reveals no common characteristic that separates them from better performers.

Many people are able to isolate and master one or two units readily; some can isolate and master three units, four units and even six units or more (fig. 5.2). This last level of control is of the highest order, for the subject must be able to give an instant response to an order to produce contractions of a specified unit without interfering activity of neighbours; he also must be able to turn the unit "off" and "on" at will. The ultimate ability of human subjects was demonstrated by Kato and Tanji (1972a) who found that within 30 minutes their subjects could voluntarily isolate 73% of 286 motor units appearing on the oscilloscope during voluntary contractions.

Control of Firing-Rates and Special Rhythms

Once a person has gained control of a spinal motor neuron, it is possible for him to learn to vary its rate of firing. This rate can be deliberately changed in immediate response to a command. The lowest limit of the range

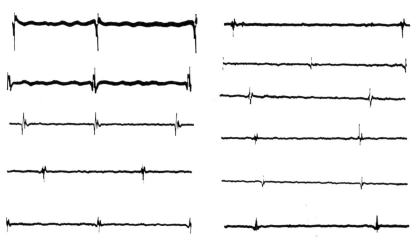

F𝒊𝗀. 5.2. Eleven different motor units isolated by a subject in quick succession in his abductor pollicis brevis. (From Basmajian, Baeza and Fabrigar, 1965.)

of frequencies is zero, i.e., one can start from neuromuscular silence and then give single isolated contractions at regular rates as low as one per second and at increasingly faster rates. When the more able subjects are asked to produce special repetitive rhythms and imitations of drum beats, almost all are successful (some strikingly so) in producing subtle shades and coloring of internal rhythms. When tape-recorded and replayed, these rhythms provide striking proof of the fitness of the control.

Units of low threshold have a wide (7- to 20-spike-per-second) firing frequency range and tend to be located deep in the muscle, according to Clamann (1970) who used human biceps brachii and bipolar fine-wire electrodes. The higher the tension threshold before a unit is recruited, the higher is the lowest firing frequency, the narrower is its frequency range, and the more superficially is it located. Approaching the threshold values in a slightly different way, Petajan and Philip (1969) reported *intervals* rather than frequencies. For limb muscles the threshold intervals were 90 ± 19 msec, but for the facial muscles they were only 40 ± 16 msec. These investigators also defined and recorded "onset intervals," the longest regular interval at minimal effort; in the limbs it was 132 ± 32 msec, and in the face, 86 ± 29 msec with the distal limb muscles having the longest onset intervals.

Individual motor units appear to have upper limits to their rates beyond which they cannot be fired in isolation; that is, overflow occurs and neighbours are recruited. These maximum frequencies range from 9 to 25 per second (when the maximum rates are carefully recorded with an electronic digital spike-counter). Almost all lie in the range of 9 to 16 per second. However, one must not infer that individual motor units are restricted to these rates when many units are recruited. Indeed, the upper limit of 50 per second generally accepted for human muscle is probably correct, with perhaps some slightly higher rates in other species.

Reliance on Visual or Aural Feedback

Some persons can be trained to gain control of isolated motor units to a level where, with both visual and aural cues shut off, they can recall any one of three favorite units on command and, in any sequence. They can keep such units firing without any conscious awareness other than the assurance (after the fact) that they have succeeded. In spite of considerable introspection, they cannot explain their success except to state they "thought about" a motor unit as they had seen and heard it previously. This type of training probably underlies ordinary motor skills.

Variables Which Might Affect Performance

Tanji and Kato (1971) found that cortical motor potential related to the discharge of a single motor unit is about the same size as that related to the contraction of whole muscles (e.g., as in key-pressing). This led to their obvious conclusion that cerebral mechanisms are involved in an important manner in conscious isolation of individual motor units; they later consolidated these views with more specific tests (Kato and Tanji, 1972b; Tanji and Kato, 1973a,b). However, my associates, McLeod and Thysell (1973) do not agree; their studies of evoked EEG potentials reveal no true response in the sensorimotor areas that can be related to single motor unit activity. Intensive research is in progress to resolve the question.

We find no personal characteristics that reveal reasons for the quality of performance (Basmajian, Baeza and Fabrigar, 1965). The best performers are found at different ages, among both sexes, and among both the manually skilled and unskilled, the educated and uneducated, and the bright and the dull personalities. Some "nervous" persons do not perform well—but neither do some very calm persons.

Carlsöö and Edfeldt (1963) also investigated the voluntary control over individual motor units. They concluded that: "Proprioception can be assisted greatly by exeroceptive auxiliary stimuli in achieving motor precision." Nevertheless, Wagman, Pierce and Burger (1965), using both our technique and a technique of recording devised by Pierce and Wagman (1964), emphasize the rôle of proprioception. They stress their finding that subjects believe that certain positions of a joint must be either held or imagined for success in activating desired motor units in isolation.

Over several years, we have conducted investigations of the various factors which affect motor unit training and control (Simard and Basmajian, 1967; Basmajian and Simard, 1967; Simard, Basmajian and Janda, 1968). We find that moving a neighbouring joint while a motor unit is firing is a distracting influence but most subjects can keep right on doing it in spite of the distraction. We tend to agree with Wagman and his colleagues who believe that subjects require our form of motor unit training before they can fire isolated specific motor units with the limb or joints in varying positions. Their subjects reported that "activation depended on recall of the original position and contraction effort necessary for activation." This apparently is a form of proprioceptive memory and almost certainly is integrated in the spinal cord.

Our observations were based on trained units in the tibialis anterior of 32 young adults. They showed that motor unit activity under conscious control can be easily maintained despite the distraction produced by voluntary movements elsewhere in the body (head and neck, upper limbs and contralateral limb). The control of isolation and the control of the easiest and fastest frequencies of discharge of a single motor unit were not affected by those movements (fig. 5.3).

Turning to the effect of movements of the same limb, we found that in some persons a motor unit can be trained to remain active in isolation at different positions of a "proximal" (i.e., hip or knee), "crossed" (ankle), and "distal" joint of a limb (fig. 5.4). This is a step beyond Wagman, Pierce and Burger (1965) who observed that a small change in position brings different motor units into action. Consequently they noted the important influence of the sense of position on the motor response. Our later investigation showed that in order to maintain or recall a motor unit at different positions, the subject must keep the motor unit active during the performance of the movements; and, therefore, preliminary training is undeniably necessary.

The control of the maintenance of a single motor unit activity during "proximal," "crossed" and "distal" joint movements in the same limb has been proved here to be possible providing that the technique of assistance offered by the trainer is adequate. The control over the discharge of a motor unit during proximal and distal joint movements requires a great concentration

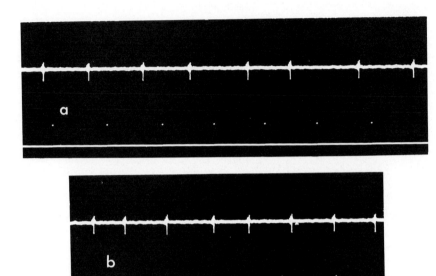

F$_{\text{IG}}$. 5.3. Sample EMG of (*a*) the easiest, and (*b*) the fastest frequency of discharge of a motor unit in the right tibialis anterior during movements of the contralateral limb (time mark: 10-msec intervals). (From Basmajian and Simard, 1967.)

on the motor activity. But when one considers the same control during a "crossed" joint movement, there are even greater difficulties for obvious reasons.

The observation that trained motor units can be activated at different positions of a joint is related to the work of Boyd and Roberts (1953). They suggested that there are slowly adaptive end organs of proprioception, which are active during movements of a limb. They observed that the common sustained discharge of the end organs in movements lasted for several seconds after attainment of a new position. This might explain why a trained single motor unit's activity can be maintained during movements.

Lloyd and Shurley (1976) studied the hypodynamic effects of sensory isolation on single motor units recorded through fine-wire electrodes in 40 normal subjects. A light panel indicated the trial onset, correct, and incorrect response. Isolation condition was produced by an air-fluidized, ceramic-bead bed in a light and sound attenuating chamber. A relearning session followed the initial session after a 2-week interim rest. Subjects were randomly assigned to the isolation or non-isolation condition for both sessions. The hypodynamic

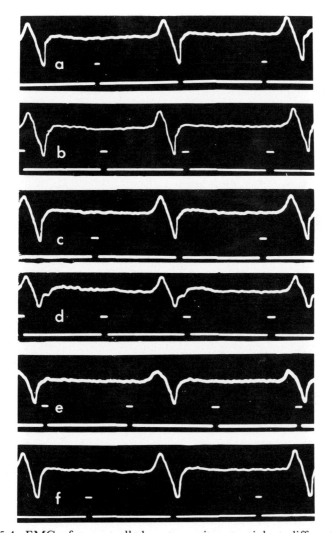

FIG. 5.4. EMG of a controlled motor unit potential at different "held" positions of the right lower limb: *a*, neutral; *b*, in lateral rotation at the hip; *c*, in medial rotation at the hip; *d*, dorsiflexion of the ankle; *e*, plantarflexion; *f*, toes extended. (Calibration: 100 μv, 100-msec intervals.) (From Basmajian and Simard, 1967.)

effects of sensory isolation increased the speed of learning to isolate and control a SMU. The results suggested that subjects were better able to attend to the relatively weak proprioceptive information provided by the SMU through the reduction of the amount and/or variety of competing stimuli.

The Level of Activity of Synergistic Muscles

The problem of what happens to the synergistic muscles at the "hold" position or during movements of a limb has been taken into consideration only in a preliminary way. The level of activity appears to be individualistic. Active inhibition of synergists is learned only after training of the motor unit in the prime mover is well established. Basmajian and Simard (1967) clearly demonstrated that as the subject focusses his attention on feedback from a single motor unit in one muscle (tibialis anterior) the surrounding muscles become progressively relaxed to the point of complete silence when isolation of the SMU is complete. Only such motor units in a limb as are needed to maintain its particular posture are still active. The process of "active inhibition," probably the more dramatic element of motor unit training, is thus achieved.

Influence of Manual Skills

Although our earlier studies had failed to reveal any correlation between the abilities of subjects to isolate individual motor units and the variable of athletic or musical ability, a systematic study (Scully and Basmajian, 1969) cast some light on the matter. We used the base time required to train motor units in one of the hand muscles as the criterion. To our surprise, the time required to train most of the manually-skilled subjects was *above* the median.

Henderson's (1952) work offers an explanation: the constant repetition of a specific motor skill increases the probability of its correct recurrence by the learning and consolidation of an optimal anticipatory tension. Perhaps this depends on an increase in the background activity of the gamma motoneurons regulating the sensitivity of the muscle spindles used in performing the skill. Wilkins (1964) postulated that the acquisition of a new motor skill leads to the learning of a certain "position memory" for it. If anticipatory tensions and position memory, or both, are learned, spinal mechanisms may be acting temporarily to block the initial learning of a new skill. Perhaps some neuromuscular pathways acquire a habit of responding in certain ways and then that habit must be broken so that a new skill may be learned. The "unstructured" nature of learning a motor unit skill would make this mechanism even more likely (Basmajian, 1972).

Influence of Age And Sex

Although the training of fine control of individual units is complicated when children are involved, we found it possible in children even below the age of 6 years (Fruhling, Basmajian and Simard, 1969). Simard and Ladd (1969) and Simard (1969) have further documented the factors involved.

Zappalá (1970) found only minor sex differences in the ability to isolate SMUs; males showed some superiority. In a different type of experiment, Harrison and Koch (1972) and Petajan and Jarcho (1975) found the opposite, but again the sex differences were not impressive.

Influence of Competing Electrical Stimulation

Any changes in the action potentials of trained motor units as a result of electrical stimulation of the motor nerve supplying the whole muscle must reflect neurophysiologic changes of the single neuron supplying the motor unit. Therefore, we investigated the influence of causing strong contractions in a muscle to compete with a discrete SMU in it which was being driven consciously (Scully and Basmajian, 1969). Each of a series of subjects sat with his forearm resting comfortably on a table top. The stimulator cathode was applied to the region of the ulnar nerve above the elbow. The effective stimuli were 0.1-msec square-wave pulses of 70 to 100 volts, delivered at a frequency of 90 per minute. Because stimuli of this order are not maximal, all axons in the ulnar nerve were not shocked and slight variation must have existed in axons actually stimulated by each successive shock.

Contrary to expectation, when the massive contraction of a muscle was superimposed on the contraction of only one of its motor units, the regular conscious firing of that motor unit was not significantly changed. Our experiments leave little if any doubt that well-trained motor units are not blocked in most persons. Even the coinciding of the motor unit potential with elements of the electrically induced massive contraction would not abolish the motor unit potential.

Influence of Cold

Brief cutaneous applications of ice over the biceps brachii in which an isolated motor unit had been trained elicited facilitation of both background activity and spontaneous activation of the trained SMUs (Clendenin and Szumski, 1971). Using a special electronic cooling device (Wolf and Basmajian, 1973), we confirmed this finding (Wolf, Letbetter and Basmajian, 1976). Seventeen subjects discharging SMUs at a comfortable resting frequency (5.2 ± 0.9 Hz) tended to get an inhibitory response in the initial minute of cooling. Most subjects (13 of 18) who held SMU discharges to 0.5 Hz first got an increase and then a significant decrease. Apparently the central excitatory state is the mediator of these local motor reactions to cutaneous cooling.

Effects of Handedness and Retesting

When a large number of subjects were studied on two occasions using a different hand each time, Powers (1969) found that they always isolated a unit more quickly in the second hand. Isolation was twice as rapid when the second hand was the preferred (dominant) hand; it was almost five times as rapid when the second hand was the non-preferred one. The time required to control a previously isolated unit was shortened significantly only when the preferred hand was the second hand. However, in a test-retest situation with much fewer subjects, Harrison and Koch (1972) found no significant improvement from test to retest.

Influence of Disease States

While we have found that partially paralyzed people can learn single unit controls quite easily, the factor of spasticity introduces considerable difficulty. In clinical studies, my colleagues and I have learned to overcome these difficulties by carefully training the patient to relax spastic muscles. Parkinsonian rigidity seems to be a different matter. Petajan and Jarcho (1975) reported that patients with Parkinson's disease are unable to adjust the firing rate of motor units that initiate contraction from zero to higher rates. Although the frequency modulation is not normal, motor units recruit in an orderly fashion. Levodopa treatment restores normal control of single motor units.

Reaction Time Studies

A number of investigators have used trained single motor units for psychological testing of reaction times. Thus, Sutton and Kimm (1969, 1970) and Kimm and Sutton (1973) have shown stable differences in the RT in triceps and biceps brachii and a showing of RT following the intake of alcohol. Generally, they concluded that single motor-unit spike RTs were slower than gross EMG and lever-press RTs. But Thysell (1969) disagrees, finding them to be comparable and rather like those of Luschei *et al.* (1967). Further, Vanderstoep (1971) questions the finding of inherent differences between muscles when the RT paradigm is used with triceps, biceps, the first dorsal interosseus and the abductor pollicis brevis. Zernicke and Waterland (1972), on the other hand, were able to show differences between the two heads of biceps brachii. The short head contains motor units that are easier to control than those in the long head. They relate this to various morphological and functional requirements of the two heads (e.g., the density of muscle spindles is greater in the short head). The willful fractionization of control between two heads of the same muscle, not entirely unexpected in view of the fineness of willful controls involved in single motor unit control, once more underlines the discrete nature of controls over the spinal motoneurons.

Practical Applications

Many applications are emerging for the use of motor unit training, e.g., in the control of myoelectric prostheses and orthoses, in neurological studies and in psychology. The growth of the field of "biofeedback" from this work is the subject of the next chapter and a separate book (Basmajian, 1978).

Chapter 6

More on EMG and myoelectric biofeedback

I n the third edition of this book (1974), the progress made in the new field of biofeedback in the 7 years that followed the second edition (1967) was the subject of amazement. In the middle of the 60's, there was little of consequence except for our single motor unit training experiments (Chapter 5) and the related operant conditioning research of Hefferline and his colleagues (described below). Biofeedback, especially with EMG, has grown immensely since then. It has become a widespread topic of concern to investigators and clinicians along with biofeedback of other psychobiological phenomena over which human subjects can acquire some control (e.g., heart rate, blood pressure and some waves of the EEG).

In this chapter, the present state of emg biofeedback (or better, *myoelectric biofeedback*) is summarized without becoming involved in the controversial aspects of general biofeedback applications; whether or not yogis, meditators and ordinary people can do certain satisfying tricks with alpha waves in the EEG must be left to other writers. Here we are concerned with skeletal ("voluntary") muscle and the EMG of a much grosser nature than single motor unit potentials—although the feedback principles controlling them are common to both.

Following confirmation of my studies of the single motor unit principles (outlined in Chapter 5), Green *et al.* (1969, 1970) rapidly extended biofeedback work into the clinical investigation of the effects of feedback relaxation. They combined this with other forms of electronic feedback and applied the results to a variety of general and local tension states believed to be the cause of pathological physiology. Simultaneously Gaarder (1971) was exploring practical means to control relaxation in patients with feedback devices.

Hoping to determine whether an ability to produce electromyographic patterns accurately reflects the ability to achieve specific muscle tensions,

131

Rummel (1974) studied a long series of normal subjects. She was amazed to find no statistically significant correlation. Schwartz, Fair *et al.* (1976a, b) on the other hand revealed patterns of covert activity in facial muscles that could be graded and correlated to states of affective imagery and mood. Earlier, Smith (1973) had found a positive correlation between personality traits of anxiety and EMG from the region of the forehead. This finding disagreed with the earlier work (p. 80) of Iris Balshan Goldstein (1962) but it must be remembered that Smith's forehead electrodes often pick up from a wide area (down to the clavicles). Similar findings were reported for the muscles of the jaw by Thomas *et al.* (1973) in explaining temporomandibular joint syndrome. Chapman (1974) showed that EMGs from the forehead reflected even the fact that the subjects were not alone but were in an audience (i.e., in a social-facilitation setting). Biofeedback appears to be superior to verbal feedback in inducing relaxation, at least in the research models used by Kinsman *et al.* (1975) and Coursey (1975); but Alexander (1975) disagreed on the basis of his research.

Psychophysiological Mechanisms

The question continues to arise: is biofeedback training based on volition or is it operant conditioning? Hefferline and Perera (1963) in their continuing search for the effect of proprioception in behaviour showed that subjects could be conditioned to respond to covert twitches in a thumb muscle (displayed by EMG). After the emg feedback was eliminated, the response often persisted. By coincidence the muscle used (abductor pollicis brevis) was the same as the one used in my early experiments on single motor unit training (Basmajian, 1963). Instead of asking the subject to shape the behaviour of the EMG within the target muscle, Hefferline and Perera conditioned him to press a key using another muscle. Their system was based on the operant conditioning paradigm. Fetz and Finocchio (1971) were able to condition awake monkeys to give bursts of cortical cell activity with and without simultaneous suppression of emg activity in specifically targeted arm muscles. Operant conditioning methodologies proved sufficient to bring about the correlated response.

In man, Germana (1969) demonstrated quite adequately and not surprisingly that conditioning may be employed in modifying electromyographic responses; perhaps more important, his work has tended to support "cardiac-somatic coupling" with which Obrist and various colleagues have been concerned (see Obrist, 1968). Cohen and Johnson (1971) found a high correlation between heart rate and muscular activity, supporting Obrist's theoretical position. Subtle changes in muscular activity did change heart rate both when subjects were intentionally modifying muscular activity as well as when spontaneous changes were occurring.

Refining his techniques, Cohen (1973) recently showed a relationship only in subject groups that had a moderately high emg output from skin electrodes over the muscles of the chin; lower emg outputs seemed unrelated to heart rate changes; thus the "cardiac-somatic coupling" is not absolute and mechanisms must exist in the central nervous system for separating cardiac and peripheral motor responses. Other autonomic functions have been linked with covert motor responses; thus Simpson and Climan (1971) have shown that there is some apparent effect of muscular activity on the pupil size during an "imagery" task in which subjects generated images in response to words.

General Relaxation

In the 1920's and 1930's, Edmund Jacobson of Chicago became the enthusiastic proponent of a clinical form of emg monitoring of his patients' progress during relaxation training. Limited by the apparatus available at the time, Jacobson developed methods of electrical measurement of the muscular state of tension and employed his measurements to induce progressive somatic relaxation for a variety of psychoneurotic syndromes (Jacobson, 1929, 1933). Fortunately he has seen the exuberant revival of his life's work along modern lines some 40 years later.

At the Menninger Clinic, Green *et al.* (1969), using a modification of the single motor unit training technique, suggested forcefully that emg biofeedback training would be useful in many states. They proposed specific equipment needs, as did Gaarder (1971). Mathews and Gelder (1969) studied the effect of relaxation training with phobic patients, showing the EMG (among other parameters) was altered during relaxation, and concluding that relaxation is in some way associated with a controlled decrease in "arousal level" with retention of consciousness. Paul (1969) compared hypnotic suggestion and brief relaxation training, showing the superiority of the latter in reducing subjective tension and distress. Wilson and Wilson (1970), while agreeing that muscle tension could be manipulated by feedback and conditioning, were much less sure of the desirable effects of relaxation. Whatmore and Kohli (1968) and Budzynski and Stoyva (1969) also contributed to the literature of emg biofeedback in relationship to general clinical disorders.

Specific Relaxation

Dixon and Dickel (1967) first drew attention to the idea that the EMG indicates that tension headache is accompanied by "fatigue contracture" of traumatized cranio-cervical muscles. While the idea of "fatigue contracture" is in itself unacceptable, and while emg feedback was not advocated as the therapeutic approach, nevertheless the linkage of demonstrable muscular over-activity in neck muscles and headache was significant. Shortly after, Jacobs and Felton in Los Angeles (1969) had actually employed visual emg feedback to facilitate muscle relaxation in patients with neck injuries as well as normal persons. Subjects were given ten 15-minute trials and clearly demonstrated the usefulness of the feedback. Although the injured group showed greater difficulty in relaxing, their ultimate success was as great—complete relaxation of the involved muscles.

Tension Headaches

At the University of Colorado Medical Center, Budzynski and Stoyva (1969) with various associates have concentrated on techniques and equipment for deep relaxation using emg feedback. In particular they have employed emg feedback for treating patients with severe tension headache (Budzynski *et al.*, 1970). The same workers have applied similar techniques in studies of various neurotic states. Controversies still simmer over whether biofeedback is necessary in the relaxation process (Cox *et al.* 1975).

While most American investigator-therapists have used the forehead placement of electrodes, the Polish physician, Pózniak-Patewicz (1975) successfully applied emg biofeedback relaxation to "cephalalgic" spasms of other head and neck muscles found to be related to headaches. Such muscles were found to be silent in non-headache subjects.

Chronic Anxiety

Chronic anxiety is often reflected in over-activity in the general body musculature. Townsend *et al.* (1975) compared treatment of chronic anxiety with emg biofeedback to treatment with group therapy in a control group. Significant improvements resulted as they did in a study by Canter *et al.* (1975) in which they compared biofeedback with Jacobsonian progressive relaxation. While the latter was effective, the biofeedback approach proved superior in reducing both muscular tension and chronic anxiety.

Chronic asthma, a condition notoriously aggravated by anxiety, was significantly improved (Davis *et al.*, 1973; Scherr *et al.*, 1975) using deep muscle relaxation mediated by biofeedback.

Temporomandibular Joint Syndrome

Dental specialists are increasingly enthusiastic about the new treatment of this common and distressingly painful jaw pain caused by over-active use of muscles that are normally relaxed or only lightly contracted. Myoelectric biofeedback training involves making the patient aware of hyperactivity in the masseter muscle and then training local relaxation of the muscle (Carlsson *et al.*, 1975).

Speech Apparatus and EMG Feedback

As noted in the earlier editions of this book, Hardyck *et al.* (1966) were among the first to modify the lessons of single motor unit training to applied biofeedback of useful function. Using feedback from surface EMG of the laryngeal muscles during silent reading, they were able to accelerate the reading skills of slow readers. Simultaneously McGuigan and various associates at Hollins College, Virginia were studying the covert oral language behaviour (as measured by surface EMG of chin muscles). McGuigan (1966) first showed that psychotic patients gave an emg response prior to reports of auditory hallucinations. Later McGuigan (1970) showed that the silent performance of language tasks by normal subjects is accompanied by emg responses; indeed it may be beneficial (McGuigan and Rodier, 1968). Inouye and Shimizu (1970) examined the hypothesis that verbal hallucination is an expression of so-called "inner speech." In nine schizophrenics, they found that the experience of verbal hallucination was accompanied about half the time by an increase in emg discharges from wire electrodes in the "speech musculature" (cricothyroid, sternohyoid, orbicularis oris and corrugator supercilii). The emg discharge was delayed by about 1.5 seconds and then it coincided with the duration of the verbal hallucination. The "louder" the hallucination, the more likely that emg activity would occur.

The Czech investigators Baštecký *et al.* (1968a, b) using delayed auditory speech feedback and EMG of mimic muscles (primarily mentalis at the chin) found that schizophrenic patients could be differentiated from normal subjects. This area of research, now in its infancy, requires a great deal of investigation. Thus, Sussman *et al.* (1972) have shown that individual units in the laryngeal muscles can be trained. The same group (Hanson *et al.*, 1971; MacNeilage *et al.*, 1972; MacNeilage and Szabo, 1972; and MacNeilage, 1973) have systematically exposed mechanisms of fine control of the laryngeal function which should have far reaching use.

Stuttering has been the special concern of Barry Guitar (1975). He taught stutterers to reduce resting emg activity in the lips and in the larynx with varying success. But the success of individual patients in specific speech situations was quite dramatic.

Targeted Muscle Retraining

While general muscle relaxation training found early use in the late 1960's, our demonstration of the exquisite controls that biofeedback could elicit in the 1950's and 1960's was turned into practical use for the neurologically handicapped only in the 1970's. My colleagues Ladd and Simard (1972), building on our earlier work together on single motor unit training, have trained and studied congenitally malformed children with the aim of using the limited sources of muscle power for myoelectric and other types of artificial limbs and orthoses. Payton and Kelley (1972) have explored the factors controlling biceps brachii and deltoid during performance of skilled tasks in a way that lends itself to feedback training.

Recently we have shown that with myoelectric biofeedback through fine-wire electrodes in the lips and buccinator muscles of the cheek, clarinet players can quickly revise the localized activities in bizarre ways without losing the ability to perform (Basmajian and Newton, 1973). We have also shown that trumpet and trombone players have different natural patterns, that these vary with proficiency and that they can be altered with emg feedback (Basmajian and White, 1973; White and Basmajian, 1973).

Rehabilitation

Practical approaches with practical biofeedback instruments are becoming a reality with commercial equipment being marketed. Some devices are shabby but several are of reasonable quality and price. Our own *Mini-trainer* which has been under development, testing and validation since 1972 is a small battery-operated instrument suitable for physical therapy and related applications. Yet it really has not been necessary to have specialized miniaturized devices for retraining of muscles: standard emg equipment obviously is serviceable. Thus, Booker *et al.* (1969) demonstrated retraining methods for patients with various neuromuscular conditions, and Johnson and Garton (1973) have succeeded with hemiplegic patients in retraining functions of the upper and lower limbs where other methods proved inadequate. What has been surprising to many people is the ease with which ordinary patients "take to" the feedback signals and learn to manipulate them by acquiring more precise control over the muscles requiring training or recruitment.

This book is not the place for details of how emg biofeedback may be used in rehabilitation. The topic has been covered thoroughly in my companion book, *Biofeedback: Principles and Practice for Clinicians* (1978). Additional details in journal-article form may be found for the following subjects:

Hemiplegia, cerebral palsy and spasticity (Amato *et al.*, 1973; Baykushev, 1973; Basmajian *et al.*, 1975; Wolpert and Wooldrige, 1975; Inglis *et al.*, 1976a, b; Nafpliotis, 1976; Teng *et al.*, 1976; de Girardi-Quirion, 1976; Brudny *et al.*, 1976; Grynbaum, *et al.*, 1976; Baker *et al.*, 1977).

Dystonias including spasmodic torticollis (Cleeland, 1973; Brudny, *et al.*, 1974; Vasilescu and Dieckman, 1975; Korein *et al.*, 1976; Farrar, 1976).

Pre- and postoperative retraining (Kukulka, Brown and Basmajian, 1975; Brown and Basmajian, 1977); and

Emphysema (Johnston and Lee, 1976).

Related Psychological Research

Since the valuable start given to it by the Montreal group in the early 1950's (see Malmo *et al.*, 1951; Malmo and Smith, 1955), a sort of electromyographic subculture has existed in the psychological literature. No great purpose would be served in reviewing that spotty literature for it contributes little of real value to the serious electromyographer; indeed some of it is picayune. Better for us to choose only those recent papers that bear upon the subject matter of this chapter. Some of the significant work was already cited in this and the previous chapter and the work of Goldstein described on page 80; below the more general papers will be reviewed briefly.

Following up a previous investigation of limb positioning with kinesthetic cues (Lloyd and Caldwell, 1965), Lloyd (1968) found no statistically significant relationship between position accuracy and the amount of contralateral activity as measured by EMG; but there was no doubt that such activity exists at a low level especially during passive movement of the ipsilateral limb. Lloyd concluded that a minimal level of activity was required for kinesthetic mediation of accurate limb position. It was this work that led Lloyd and his colleagues to study single motor unit responses (cited on pp. 119 and 124).

Wiesendanger *et al.* (1969) measured simple and complex reaction times with the EMG of biceps and triceps. While their chief concern was to find differences between normal persons and patients with parkinsonism—there were none in the simple tasks—they showed that the normal reciprocal inhibition of antagonists was modified in different ways, biceps activity always being present (see the general discussion of agonist-antagonist behaviour, p. 93).

Bartoshuk and Kaswick (1966) had shown earlier that general arousal level may not be necessary to produce emg gradients; instead selective facilitation may be sufficient.

The influence of environmental and emotional factors on emg activity is gaining widespread interest. A good example of this type of study is that of Lukas *et al.* (1970) who recorded the effect of sonic booms and noise from subsonic jet flyovers on skeletal muscle tension (in the trapezius muscle) as well as other parameters. The emg activity increased with sonic booms with lesser effect from the flyover noise.

Phasic changes in muscular and reflex activity during non-REM sleep were demonstrated in man and cats by Pivik and Dement (1970). The suppression of EMG from surface electrodes in the submental (chin) area was observed in all subjects during non-REM sleep, but occurred with the greatest frequency during sleep stages 2 and 4. The suppressions averaged a ¼-minute in duration and exhibited a higher frequency in the 10 minutes prior to the REM period than after. Larson and Foulkes (1969) confirmed that EMG suppression in chin and neck muscles heralds REM sleep onset. The amount of emg activity during non-REM sleep just prior to being awakened influences the recall frequency of dreams.

Pishkin and Shurley (1968) and Pishkin *et al.* (1968) demonstrated a positive correlation between emg responses and concept-identification performance which produces cognitive stress. About the same time, Aarons (1968) was exploring possible diurnal variations of myopotentials and word associations related to psychological orientation. Word-association tests revealed qualitative differences among responses before sleep, upon awakening and at noon. Some differences were related to psychological test variables (kinesthetic orientation, "need for change" and anxiety); the other influences were the time of the tests and, apparently, the intensity of emg response. EMG levels during sleep correlated highly with electroencephalographic sleep stages.

This brings us back to a group of studies on the effects of stress and anxiety on the EMG, first adequately investigated by Goldstein (see p. 80). Brandt and Fenz (1969) showed a peak of forehead EMG in conditions of induced mild stress, suggesting it might reflect inhibitory control. Incidentally, they questioned the specificity of the forehead source as the ideal one for such experiments—and well they might for we have shown with intramuscular fine-wire electrodes that the frontalis and corrugator supercilii are silent unless the face shows clear emotive responses (Vitti and Basmajian, 1973). Searching for a suitable muscle for stress-emg studies, Yemm (1969a, b) of Bristol, England concentrated on the masseter—not surprisingly for he is a dental scientist. He found an increase in masseter emg activity during the stress of cognitive manual-task performances in this postural muscle of the jaw. With patients who have temporomandibular dysfunction, the emg responses persisted abnormally long (Yemm, 1969c).

The use of muscles active in maintaining human posture has other advocates. Thus, Avni and Chaco (1972) used the EMG of supraspinatus muscle (which is described elsewhere in this book in its shoulder-posture role). Reasoning from our earlier work (Basmajian, 1961; Basmajian and Bazant, 1969) that drooping of the shoulder should influence supraspinatus activity, they studied a series of depressed patients. While normal controls showed normal antigravity reflex activity depressed patients all showed significant decrease while they were depressed but recovered the normal pattern on recovery from depression.

Notes on Technique

The use by Avni and Chaco and by Yemm of postural muscles (noted above) opposed to surface EMG of chin and forehead, raises the general question of appropriate methods for emg studies of tension. Unquestionably some of the techniques employed by investigators naive in EMG have been less than acceptable. Most EMG from the submental region would appear to reflect the frequency of swallowing—which of course may be a good criterion of tension. (For swallowing EMGs, see Chapter 17). As noted before, forehead emg work also may be questionable, although obvious facial mimicry often represents inner states, and so, in a distressed person, may be a satisfactory source of EMG.

It is true that *in the hands of experts* good surface EMG is quite adequate for tension studies. Bruno *et al.* (1970) and Kahn (1971) have even demonstrated its usefulness for precise identification of signals. But the factors affecting the reliability of surface EMG are many and appear to be ignored by many psychologists; they ought to read and re-read the paper by Grossman and Weiner (1966).

The foregoing facts mean that to be useful in biofeedback practice, integrated EMG from the forehead or frontal region need not come from frontalis muscle. Indeed, a wide source of myopotentials is much to be preferred as a reflection of general nervous tension. But we should know that (1) wide-source myopotentials are not "frontalis EMG" and (2) the numbers of "microvolts" produced on the meter of a commercial device or any other device simply indicate a microvolt reading at the input of the device. The intergated EMG from forehead surface electrodes generally reflects the total or global EMG of all sorts of repeated dynamic muscular activities down to about the first rib—along with some postural activity and nervous tension overactivity. The exact meter readouts can be taken with a grain of salt by the knowledgeable electromyographer at the same time as he is deliberately and wisely using them as (1) a rough indicator of progress in a clinical relaxation training program and (2) a visual placebo in reinforcing the patients' responses. Any higher level of reliance on such inflated numbers is self-deception (Basmajian, 1976).

Finally, a word on muscle re-education with myoelectric biofeedback. To consider it as anything other than a modified form of rehabilitation medicine is to invoke mysticism once more. Rehabilitation therapists have used the basic technique without electronics for many years; electromyography simply gives the experienced practitioner instant reinforcement and conditioning of patients' responses, and this reinforcement and conditioning have permitted dramatic acceleration or initiation of recovery in some neuromotor conditions in some patients.

Chapter 7

Nerve conduction velocity and residual latency

I NSEPARABLY wrapped up in function with the skeletal muscles are their motor nerves. The use of human nerve conduction velocity in studying abnormal states was introduced by Harvey and Masland (1941). Their technique which has been followed with minor variations by others consisted of recording action potentials from the hypothenar muscles following electrical stimulation of the ulnar nerve (figs. 7.1, *top, bottom*; 7.2). A simple and clear description of technique is given by Lundervold, Bruland and Stensrud (1965) and the entire field reviewed in book form by Smorto and Basmajian (1972). That book deals at great length with all aspects of nerve conduction tests, both motor and sensory, with an emphasis on clinical applications. Here in this chapter only a general consideration is given to the most significant features of normal electroneurography.

Hodes, Larrabee and German (1948) adapted this technique to the study of nerve conduction velocity as well as the action potentials of normal and abnormal nerves. Stimulating the ulnar nerve first at the elbow and then at the wrist while recording action potentials of the abductor digiti minimi, they determined the difference between the two latencies of response. This difference in latencies divided by the distance between the two points of stimulation yielded an accurate measure of the conduction velocity of the most rapidly conducting fibres in the nerve. Assuming the speed was constant, Hodes *et al.* calculated the time that should be taken by the impulse to reach the muscle, and found that this was always less than the measured latency of the muscle action potential. Therefore, there is a small "residual latency" caused by either a slower velocity in the finer terminal portions of the nerve, or a delay at the neuromusclar junction, or a combination of these two factors. Indeed, Trojaborg (1964) has shown that the conduction velocity falls in the distal

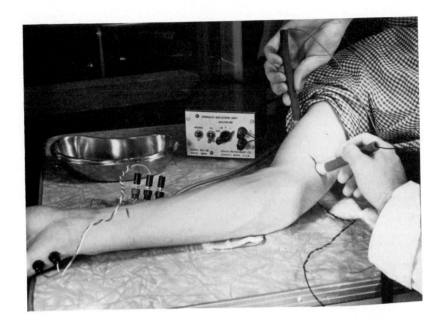

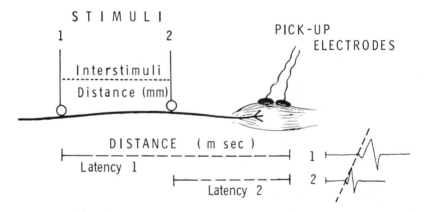

Fig. 7.1, *top*. Electromyographic electrodes on hypothenar muscles and stimulating cathode on ulnar nerve above the elbow. Black mark above wrist is site for other stimulation point. (From Low, Basmajian and Lyons, 1962.)

Fig. 7.1, *bottom*. Diagram showing method of computing conduction speed in a segment of nerve. (From Smorto and Basmajian, 1972.)

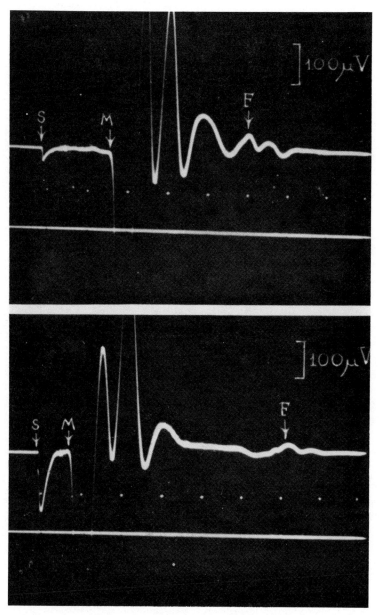

Fig. 7.2. Upper record, stimulus above elbow; Lower record, stimulus above wrist. Conduction time from stimulus (S) to the start of a muscle twitch (M) can be measured by comparing with the time signal (series of white dots at 5-msec intervals). (From Low, Basmajian and Lyons, 1962.)

143

part of the median and ulnar nerve trunks. Spiegel and Johnson (1962) hold the opposite view.

In a series of papers, Magladery, McDougal and Stoll (1950a, b, c) extensively analyzed patterns of electrical activity evoked by stimulation of mixed peripheral nerves in human limbs. Their results indicate a slowing of the impulse in distal portions of both motor and afferent nerve fibers, which accounts for a portion of the residual latency as described by Hodes *et al.*

Norris, Shock and Wagman (1953) determined ulnar nerve conduction velocities and "residual latencies" in 25 subjects ranging in age from 20 to 90 years. They found a decrease in velocity but no change in residual latency with increasing age.

Normal minimum values in average adults (rounded out, in metres per second) for various human nerves are: median and ulnar, 47 M per second; deep peroneal, 38 M per second; posterior tibial, 39 M per second (Schubert, 1963). The mean velocity for the radial nerve is 56 M per second according to Downie and Scott (1964), but it is 66 to 74 M per second according to Gassel and Diamantopoulos (1964). The mean velocity in the sciatic nerve is 51 M per second when abductor hallucis is recorded, and 56 M per second when a shorter length of nerve is studied, i.e., when gastrocnemius is used for recording (Gassel and Trojaborg, 1964). See Table 7.1.

Mayer (1963) gives the following motor conduction velocities for the age group 10 to 35 years, which I have rounded out to the closest whole number:

Median: wrist-elbow—59 ± 4, and elbow-axilla—66 ± 5 M per second

Ulnar: wrist-elbow—59 ± 4, and elbow-axilla—64 ± 3 M per second

Common peroneal: 50 ± 6 M per second

Posterior tibial: 46 ± 4 M per second

In a much older age group (51 to 80 years) the velocities were lower by about 4 or 5 M per second in almost all the above categories.

One always must apply "average normal" and "minimum" velocities with caution to individual cases. A wide fluctuation is possible during one session or from day to day as shown by our experiments, described below, and those of Christie and Coomes (1960). With variable physical factors playing a very real rôle, considerable error can creep into the results (Schubert, 1964).

LaFratta and Smith (1964) have shown that, on the average, conduction velocities are greater in women than in men although the overlap shown in their graph seems quite large. They also found a slight decline in velocity with the older age groups.

The use of these emg methods to determine slowing in nerves affected by injury and various neuropathies has been discussed by Johnson and Olsen (1960) and Bastron and Lambert (1960), and by others. In thorough articles,

TABLE 7.1*
Normal Readings for Motor Nerve Conduction Velocity (M/Sec)

	Mean	Range	S.D.	Source
Ulnar	55.1		6.4	Johnson & Olsen
	58.7	50.8–66.7	4.0	Abramson *et al.*
	59.1	49.1–65.5		Henriksen
	60.4	47.0–73.0	5.8	Thomas & Lambert
Median	56.1	46.8–68.4	4.5	Abramson *et al.*
	56.4	47.9–68.3	5.4	Jebsen
	58.5	53.0–64.3		Henriksen
Radial	58.4	45.4–82.5	6.7	Jebsen
	72.0		6.1	Gassel *et al.*
Tibial	46.2	37.4–58.9	3.3	Jebsen
	50.2		9.3	Johnson & Olsen
Peroneal	47.3	40.2–57.0	4.3	Jebsen
	50.1		7.2	Johnson & Olsen
	51.0		3.3	Thomas & Lambert
	51.5	45.6–56.3		Henriksen

* From Smorto and Basmajian (1972).

Thomas (1961) and Dunn *et al.* (1964) reviewed the progress made in the clinical applications of these techniques. The purpose of this chapter (and, indeed, this book) is to avoid purely clinical discussions while providing a solid foundation of information about normal neuromuscular function as revealed by this application of EMG. One does this reluctantly for the recent clinical literature on conduction velocity is profuse. (See *Clinical Electroneurography*, Smorto and Basmajian, 1972; a new edition, 1979, is in press.)

Our detailed study of normal conduction velocity and residual latency, carried out on volunteers over a period of hours, showed that conduction velocity in the motor fibres of the ulnar nerve between the elbow and the wrist fluctuates considerably, sometimes rather widely from time to time in any individual (Low, Basmajian and Lyons, 1962). Therefore, single clinical estimations must be viewed with caution. Residual latency, a function of the delay in the terminal neuromuscular apparatus, also shows a fluctuation with time. Nevertheless, the study of the residual latency offered a new tool in the investigation of neuromuscular function. It deserves wider investigation, but to date it has been neglected.

The most striking observation in individual experiments was the fluctuation in conduction velocity. In individual subjects, velocities measured 15 minutes apart fluctuated by as much as 8 M per seconds. Generally, the first two or three determinations (during the first 30 minutes of an experiment) tended to be higher than those that followed. Aside from this (inconstant) tendency for the velocity to decrease as the experiment progressed, values did not vary in a predictable way.

Because the temperature of the limbs was not controlled, temperature variations may account for some of the differences observed between individuals and for some of the differences from time to time in the same person. Such variations were considered and deliberately accepted from the start so that the results might be directly comparable to determinations made on patients in routine diagnostic clinics where elaborate temperature control is impractical.

Our mean control value for the conduction velocity in the ulnar nerve, which is 56.3 ± 4.2 M per second, agrees closely with values quoted in the literature, that given by Thomas, Sears and Gilliatt (1959) for the ulnar nerve being 56 ± 4.6 M per second.

No one had commented previously on the fluctuation in conduction velocity with the passage of time in normal subjects. Reasons for the irregular fluctuations are not evident in the literature. Undoubtedly, progressive cooling of the forearm occurs as it lies exposed during an experiment, but, as explained before, temperature changes can be only a part of the explanation.

Changes in blood flow, variations in metabolic processes of the nerve or nearby muscle, or changing influences from the central nervous system all must contribute to the fluctuations. An important implication of their occurrence is that we should be cautious in making only one or two determinations in a brief period to compare with "normal" or "abnormal" values. Johnson and Olsen (1960) state that their whole procedure takes " . . . five or ten minutes in a co-operative patient." None of the other writers reports exactly what time is involved in the determination of their velocities. Sources of error in routine "diagnostic" tests are many and complex. Gassel (1964) incriminates the following: recording of potentials arising from muscles at a distance from electrodes, anomalous innervation and spread of stimuli other than that over which the recording electrodes are placed. Simpson (1964) also warns against facile acceptance of results.

Nerve conduction velocities may be influenced by various conditions other than neurological disease and by pharmacological influences. For detailed reports (which would be out of place in this book) one should refer to the following: the effects of chronic alcoholic polyneuropathy—Mawdsley and Mayer (1965); the acute effects of ethanol—Low, Basmajian and Lyons (1962); effects of acute drug intoxication—Pihkanen, Harenko and Huhmar (1965), and of cold (de Jesus *et al.*, 1973).

Aging affects nerve conduction velocity, especially during the early years. From 1 month of age to 4 years, mean velocity rises from 33 m/sec to over 50 m/sec (Wagner and Buchthal, 1972). In adults, Trojaborg (1976) reported a drop of 2 m/sec in the musculocutaneous nerve for each decade increase in age. However, Nielson (1973) has shown that there is no significant difference due to the sex of the subject.

Residual Latency

The mean of our control residual latencies was 1.52 ± 0.21 msec. This is somewhat lower than the other three mean values of residual latency in adults reported in the literature. Hodes, Larrabee and German (1948), Norris, Shock and Wagman (1953), and Bolzani (1955) gave values of 2.2, 1.7 and 2.68 msec respectively.

Like the values for conduction velocity, the residual latencies determined in each person fluctuate throughout an experiment lasting several hours. They tend to drop slightly, probably for the same sort of reasons that cause variations in conduction velocity. Only one paper comments on any positive change in residual latencies during experiments. Hodes, Larrabee and German (1948) report that in regenerating nerves (ulnar and median) following suture, the residual latencies were greater than the values for the same nerves under normal conditions. At the same time, they found that nerve conduction velocity was slowed. In a study by Norris, Shock and Wagman (1953), performed on a large number of subjects from 50 to 90 years of age, the velocity of conduction in the ulnar nerve decreased with increasing age, but the residual latencies remained unchanged.

Before concluding this chapter, a more comprehensive definition of "residual latency" than that offered by other writers must be made by discussing its elements. An important component of the residual latency must be the time taken for the muscle to respond to depolarization of the endplate. The residual latency must also include a short time taken by the muscle action potential to pass beneath the proximal recording electrode because muscles have a slow conduction velocity ranging from 1.3 to 4.7 M per second, as shown by Ramsay (1960) and Eccles and O'Connor (1939). With surface electrodes, this factor is extremely small because they are placed over the area of innervation and they gather potentials from a relatively wide area.

Another component of the residual latency, the slowing of the impulse in the fine terminal fibres of the nerve, has been calculated by Eccles and O'Connor (1939) to be about 0.2 msec. There is a fourth and perhaps the most significant component: the time consumed by neuromuscular transmission or the "synaptic delay." Direct measurements of this delay give values of about 0.5 msec (Eccles and O'Connor, 1939). According to Nachmansohn (1959) and Fatt (1959), this delay must be due to the time taken by electrical events in the nerve ending causing the release of acetylcholine and the corresponding build-up of the postsynaptic potential. The transmission of acetylcholine across the 500 Å gap cannot be a factor in the delay since Eccles and O'Connor estimated that this process should take no more than 10 microsec.

Thus the latent period from the arrival of the impulse at the fine terminal nerve branches to the beginning of the muscle action potential is composed of the four factors discussed above. This time has been measured directly; in mammalian striated muscle it is approximately 0.85 msec (Eccles and O'Connor, 1939). The values for residual latency determined in our experiments are of the order of 1.5 msec; therefore, they must include factors in addition to the four previously mentioned.

According to some authors (Magladery and McDougal, 1950; Gilliatt and Thomas, 1959) there is a decrement in conduction velocity along a nerve in an extremity. Thus the distal portions conduct more slowly than proximal segments of the same nerve (Copack *et al.*, 1975). If this is true, , then the assumption that the velocity of the impulse from elbow to wrist is the same as that from wrist to muscle must introduce an error in calculation of the residual latency by the accepted method (i.e., subtracting the calculated latency of response in the hypothenar muscles applied at the elbow from the observed latency). This is mentioned by Magladery and McDougal and probably it contributes most of the remainder of the residual latency. In our series this portion would be about 0.67 msec.

Sensory Nerve Conduction

Although this subject is fully considered in our other book (Smorto and Basmajian, 1972), here it might be briefly reviewed as an introduction for some readers. Sensory conduction velocity is a relatively new field in electrodiagnosis. However, in the past few years many investigators have contributed information to this field. Direct recording of nerve potentials dates from 1949 when Dawson and Scott developed the first clinical approach to the study of the sensory nerve latencies in man. They showed that by summation of single responses, potentials may be recorded from the skin. They first applied the method to clinical diagnosis of peripheral nerve lesions.

The early experiments were concerned with sensory latencies obtained by applying electrical stimuli to the ulnar and median nerves at the wrist. The stimuli evoked afferent volleys of potentials in the sensory fibres of the nerve trunk. A pair of surface electrodes were placed over the course of the nerve at different levels of the arm and elbow. Dawson and Scott analysed these waves which were small, di- or triphasic and were less than of 50 μv in amplitude. The durations were 2 to 3 msec. This method elicited not only orthodromic ascending sensory waves, but also antidromic stimulation of the motor fibres of the same nerve.

Seven years later Dawson (1956) modified his technique in order to eliminate the antidromic stimulation; instead of applying the stimulus at the wrist (where both sensory and motor fibres are present) he stimulated the nerve at the fingers where no motor fibres are present. The sensory nerve potentials evoked in this way were recorded at the wrist. However, the evoked potentials were much smaller in amplitude (5 to 15 μv). Nevertheless, Dawson successfully recorded the small potentials in 14 normal subjects. (The sensory latency between the fingers and the wrist averaged 4.73 msec.) The motor latency from the wrist to hypothenar muscles or wrist to thenar muscle were about equal—5.12 msec. Details of methodologies and results are given in Smorto and Basmajian (1972; second edition, 1979).

Shagass (1961) described a method for recording cerebral evoked potentials after stimulation of major peripheral nerves of the body by using an averaging technique. In 1963 Liberson and Kim by using the mnemotron were able to record evoked potentials from the median, ulnar, radial, tibial and peroneal nerves. They also concluded that while the method of cerebral evoked potentials permits assertion of the continuity of a sensory nerve, a precise determination of the conduction velocity may be somewhat difficult.

Determination of conduction velocities in sensory fibres elcited by H and F reflex responses has held considerable interest but systematic study which early depended on the work of Liberson *et al.* (1966) has now expanded to involve many clinical investigators. The chief usefulness of the technique is in the presence of pathology; therefore a review of this field may be obtained from Smorto and Basmajian (1972; second edition, 1979).

Chapter *8*

Muscle mechanics

U NDER this chapter heading, we shall consider a number of miscellaneous electromyographic studies all concerned with the mechanics
of muscular action. The matters dealt with are of fundamental
importance though they have remained obscure until recent years.

Protective Muscular Responses

A number of reflex phenomena have been demonstrated with EMG.
Carlsöö and Johansson (1962) showed that when subjects fall to the ground
on the outstretched hand all the muscles which surround the elbow joint are
"strongly activated some tenths of seconds before the hand touches the
surface." Consequently the musculature is prepared to protect the joint. This
is partly a conditioned reflex and partly an unconditioned reflex arising from
tonic neck and labyrinthine reactions.

Independently, Watt and Jones (1966) found rather similar results in the
lower limb. EMG activity in gastrocnemius began 80 msec before and 40
msec after the landing impact of the foot. They suggested that there is "a
pre-programmed open-loop sequence of neuro-muscular activity virtually
unaided by myotatic feedback." Myotatic reflexes were found to play no
significant role in the deceleration for they came much to late.

Jones and Watt (1971) later showed by EMG that the human gastrocnemius has a stretch reflex (which they called the Functional Stretch Reflex)
elicited by a sharply applied and maintained dorsiflection at the ankle; it
occurs after a 120-msec delay. Gravitational forces elicit a "body jerk" in
muscles of the trunk and lower limbs whenever external force is removed
suddenly (Denslow and Gutensohn, 1967). O'Connell (1971) has demonstrated the effect of sensory deprivation on the ability of the human being to
achieve the erect posture when he is dropped vertically some moderate (but
to him unknown) distance when he lands on his feet.

More recently, Greenwood and Hopkins (1976a, b; 1977) showed that (1)
an initial peak of activity in many muscles arises from labyrinthine reflexes
and (2) a second peak in lower limb muscles comes after 200 msec as a

preparatory contraction to soften the landing and may or may not be efficient in doing so. Similar results have been reported for sudden arm drops (Wyrick and Duncan, 1974).

Taking another tack, Stein and Bawa (1976) studied the reflex responses of the human soleus muscle to small perturbations. They recorded EMG and force changes resulting from intramuscular stimulation of a small branch of the motor nerve while subjects maintained a steady voluntary contraction. In addition to the usual M-wave or twitch-contraction that follows such a stimulus, they revealed one or more later waves with a latency greater than 100 msec. They also found later waves that appeared to be reflex products of tension fluctuations, and after further analysis they concluded that supraspinal reflexes were involved.

Spurt and Shunt Muscles

By the application of mathematical analysis MacConaill (1946, 1949) has shown that skeletal muscles may act as "shunt" or "spurt" muscles. A shunt muscle is one that acts chiefly during rapid movement and along the long axis of the moving bone to provide centripetal force. On the other hand, spurt muscles are those that produce the acceleration along the curve of motion (fig. 8.1). However, there has been a lack of experimental data to confirm or disprove his theories. Examination of the findings of certain of our electromyographic experiments on the flexors of the elbow joint (performed with other aims in mind) appeared to confirm MacConaill's calculations and conclusions (Basmajian, 1959). However, a recent upsurge of interest in the theory has led to its critical re-examination and criticism by several workers. Thus, Stern (1971) has questioned its validity, at least as it applies to the elbow.

During slow flexion of the elbow, with or without a load of 2 pounds (about 1 kg), the brachioradialis was relatively quiescent in most of our subjects, while the biceps and the brachialis showed considerable activity. On the other hand, with quick flexion of the elbow the brachioradialis became very active in almost all the subjects.

During maintenance of flexed postures against the force of gravity, the biceps and the brachialis were almost always active while the brachioradialis was either inactive or only slightly active. Even the addition of the moderate load made little change in the activity of the brachioradialis.

During the slow extension there was some slight activity in all of the muscles acting against the force of gravity with or without the added load. With quick extension, there was a general increase of activity, that in the brachioradialis being most pronounced.

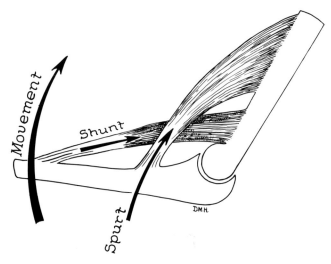

FIG. 8.1. Diagram of spurt and shunt muscles

On the basis of MacConaill's theory, the biceps and brachialis are chiefly spurt muscles at the elbow while the brachioradialis is chiefly a shunt muscle. In other words, except in complete extension, the former two muscles act mainly across the long axis of the forearm providing the acceleration along the curve of motion (fig. 8.1). The brachioradialis, on the other hand, remaining more or less parallel to the forearm throughout the range of motion, acts mainly along the long axis of the forearm to provide the centripetal or shunt force and the required stabilization at the elbow joint (MacConaill, 1949).

Obviously, when the joint is not moving, the muscular forces along the bone and through the joint are equal to the total load of the limb. We know from common experience, confirmed by electromyography, that with no added weight and the limb hanging free this force is minimal. The load here is the weight of the limb beyond the elbow joint and the ligaments alone are adequate to carry it. The addition of a weight held in the hand increases the muscular activity in the biceps and the brachioradialis.

During the very *slow* uniform flexion of the elbow joint, the shunt forces along the forearm will be approximately unchanging. Thus no great increase in the activity is required from the muscles for shunt or centripetal force. But if the flexion is rapid, a greater shunt or centripetal force *is* required. That force cannot be provided by the spurt muscles because they would impart a centrifugal acceleration along the tangent to the curve instead of producing a uniform rapid movement. Therefore the shunt muscles are called upon—indeed, must be called upon—for much greater activity. Our electromyographic findings confirm this. The brachioradialis, the typical shunt muscle, shows its greatest activity during uniform quick flexion of the elbow (fig. 8.2). During slow flexion and during the maintenance of flexed postures, it showed little or no activity even with a moderate load in the hand. Stern's (1971) experiments confirm our findings but his calculations of the torque

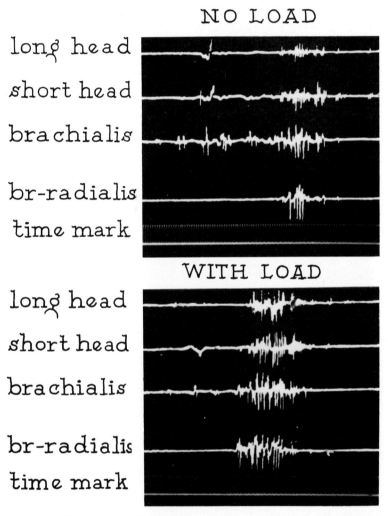

FIG. 8.2. EMG of main elbow flexors during quick elbow flexion. (Time mark intervals: 10 msec.)

forces involved across the elbow joint lead him to the conclusion that the findings can be explained without invoking the theory of "spurt" and "shunt." In fact, he suggests that brachioradialis is a poor candidate as a shunt muscle. If his calculations of required centripetal forces are correct, they are not of an order to recruit muscular activity, inert ligamentous structures being adequate. Elsewhere (p. 164) the competence of such structures is clearly demonstrated and the principle enunciated that ligaments rather than muscles are used to counteract linear transarticular distraction. Whether torque forces create special stresses is the unsettled question here.

Our experiments showed that during both slow and quick extension of the elbow all the flexor muscles show considerable activity in man. During slow extension the "letting-out" function against gravity is necessarily called upon. However, during quick extension there would seem to be a need for complete inhibition of the antagonists. Experimentally, this did not occur. Barnett and Harding (1955) concluded from similar findings with the biceps alone that the antagonists come into strong contraction at the end of a whip-like movement due to the stretch reflex. It would appear that this protects the joint which otherwise would be injured.

In our work in this area (Basmajian and Latif, 1957) we found that during extension, a short, sharp burst of activity occurred in all three muscles—biceps, brachialis and brachioradialis. The brachioradialis in general was more active than biceps or brachialis during quick extension, providing further confirmation of MacConaill's mathematical theory. The basic requirement for a shunt muscle is to provide centripetal force during rapid movement in the circular path and so the direction of movement (regardless of whether it is in the direction of flexion or extension) is of no consequence.

The Controversy of Cocontraction vs Ballistic Movements

A good review of the concepts of cocontraction of antagonists and of ballistic movements is presented by Hubbard (1960). It is obvious that he rejects the former in favour of the latter and, although some of the evidence he cites is dubious, I must admit that our own work (some of which is reported above) tends to confirm his thesis in regard to fast movements. However, I am dubious about his concepts in regard to slow movements.

Concontraction may be defined as the simultaneous contraction of both the agonists (or prime movers) and the antagonists, with a supremacy of the former producing the visible motion. Ballistic action may be defined as spurts of activity followed by relaxation during which the motion continues through the imparted momentum. When applied to fast movements, the ballistics concept is acceptable if not fully proved, but when applied to slow, controlled movements it is unacceptable and quite unproved. For a useful—though very partisan—discussion of these problems the reader should see Hubbard's chapter in *Science and Medicine of Exercise and Sports* (1960). This section might be ended, however, with the famous verdict used in the Scottish Law Courts of "not proven."

Two-Joint Muscles

A two-joint muscle is one that not only crosses two joints but is also known to have an important action on both. The best examples are found in the thigh, crossing the hip and knee joints—rectus femoris, the hamstrings, gracilis and sartorius; but the anatomist is soon reminded that such important muscles as gastrocnemius, biceps brachii and the long head of triceps also cross two joints. Moreover, the tendons of many muscles of the forearm and leg cross an even larger number of joints. Generally, however, there is little confusion about the significant actions and functions of this last group, and the unsolved problems are centered more on the functions of the simple two-joint muscles. For lack of exact knowledge about these functions, they are usually dealt with superficially and, at best, theoretically.

Markee and his associates (1955), basing their conclusions on dissections of human cadavers and nerve-stimulation studies in dogs, stated that two-joint muscles of the human thigh can act at one end without influencing the other end. This is an astonishing concept that appeared to run directly against the logical understanding of muscular action. The explanation they offered is that the middle of the muscle bellies may be moored in various ways and the pull on each end can then be from the middle. If this were true, the functional implications would be extremely interesting and important and the clinical applications would be obvious. Because our group was engaged in a systematic study of the integrated functions and control of skeletal muscle, it became necessary to test the above thesis. A series of normal male volunteers were examined electromyographically, using a row of three to five needle electrodes in each muscle examined (Basmajian, 1957a).

In the analysis of the activity in the proximal, middle and distal parts of the three muscles we may consider the significant movements of (1) the hip (i.e., proximal) joint and (2) the knee (i.e., distal) joint, disregarding whether these are flexion or extension.

In none of 21 muscles tested was there greater activity in the proximal part of the muscle when the proximal or hip joint was acted upon (figs. 8.3, 8.4).

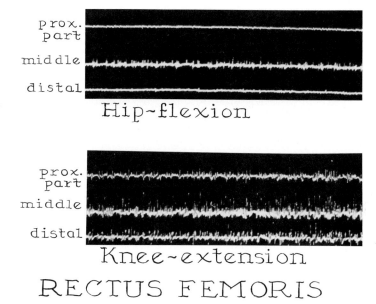

prox.
part

middle

distal

Hip-flexion

prox.
part

middle

distal

Knee-extension

RECTUS FEMORIS

FIG. 8.3. EMGs of three parts of one subject's rectus femoris during hip flexion and knee extension (with other joint "relaxed"). (From Basmajian, 1957a.)

Only one muscle in one subject, a semitendinosus, showed greater activity in the distal part when the distal joint (knee) was acted upon (the formula being middle > distal > proximal). However, this particular muscle showed the same formula with movements and postures of the hip, thus indicating that it has a relatively constant pattern regardless of the joint moved (Basmajian, 1957a). Miwa, Tanaka and Matoba (1963) confirmed these findings in a similar study.

The thesis put forward by Markee and his colleagues appeared at first to be attractive and important, but the electromyographic results showed that it is completely untenable in the case of normal human two-joint muscles. In fact, the evidence is overwhelmingly in favour of the orthodox view. These muscles pull directly from one end to the other simply because all parts of the muscle belly contract together, the greatest activity being at the middle of the belly. What has been said above is not true for muscle bellies in parallel or parallel heads of large muscles. For example, in another study we have demonstrated that the two heads of biceps brachii may act relatively independently (Basmajian and Latif, 1957).

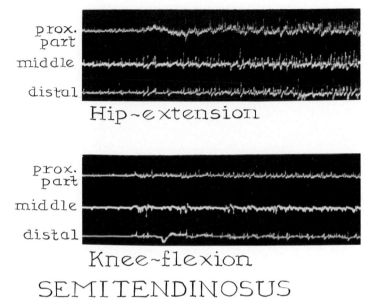

Hip~extension

Knee~flexion

SEMITENDINOSUS

Fig. 8.4. EMGs of three parts of one subject's semitendinosus during hip extension and knee flexion (with other joint "relaxed"). (From Basmajian, 1957a.)

Neurological Control of Two-Joint Muscles

To clarify the unresolved problem of how reciprocal innervation appears in a two-joint muscle that is being elongated (stretched) at one end and simultaneously shortened (activated) at the other, we studied the rectus femoris and medial hamstrings (Fujiwara and Basmajian, 1975). In 10 healthy volunteer adults, we inserted bipolar fine-wire electrodes into rectus femoris and the medial hamstrings, as well as the iliopsoas, vastus medialis and gluteus maximus. For iliopsoas, the fine-wire electrodes were situated in fleshy fibers near the hip joint, and for the other four muscles, the electrodes were in the middle of their bellies.

Each subject lay in a supine position and the leg was suspended in a sling with a balancing weight so that the hip and the knee were held passively flexed at right angles (fig. 8.5). The subject was ordered to maintain his limb in this position against new torque forces exerted by adding 5-kg weights. These forces were directed to make: (1) the hip or the knee to extend or flex as a monoarticular motion; (2) the hip and knee to extend simultaneously; (3) the hip to extend and the knee to flex; (4) the hip to flex and the knee to

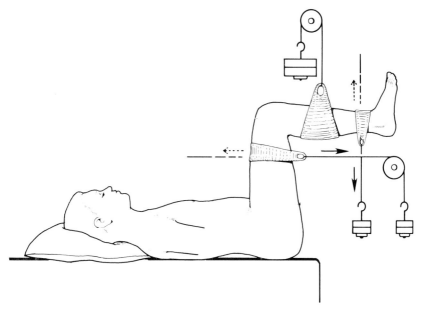

Fɪɢ. 8.5. Positioning of subject and forces. (From Fujiwara and Basmajian, 1975.)

extend; and (5) the hip and knee both to flex as a biarticular motion. Thus, the reactive contraction included both isometric and isotonic elements.

Monoarticular motions of the hip or the knee (fig. 8.6). Iliopsoas was active in hip flexion and knee extension. Rectus femoris was active in hip flexion and in knee extension, the activity being rather prominent in knee extension. Vastus medialis was active not only in knee extension but also in hip extension. Gluteus maximus was active in hip extension. More activity was observed from the medial hamstrings in knee flexion than in hip extension.

Biarticular motions (fig. 8.7). Iliopsoas showed activity in simultaneous hip and knee flexion, and hip flexion with knee extension. Vastus medialis showed activity only in the motions that included knee extension, and gluteus maximus in hip extension with knee flexion. During hip flexion with knee extension, we found maximum activity from rectus femoris in all subjects. Activity of rectus femoris was less in knee extension with hip flexion than in knee extension without hip motion, and no activity was observed in hip flexion with knee flexion. The other two-joint muscles, the medial hamstrings, showed maximum activity during hip extension with knee flexion; however, less activity occurred during knee flexion combined with hip flexion and no activity in hip extension combined with knee extension.

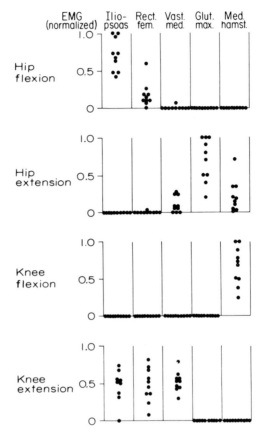

FIG. 8.6. EMGs in monoarticular motions (10 subjects). (From Fujiwara and Basmajian, 1975.)

The activity of two-joint muscles more or less influences both joints. If the other muscle works as a stabilizer of one joint, the actual kinetic effect would be limited on another joint. Monoarticular motions involving two-joint muscles are a result of a coordination with other muscles. Alone, a two-joint muscle cannot work as a one-joint muscle.

In countercurrent movements (fig. 8.8), rectus femoris and the medial hamstrings show maximum activity. However, in concurrent movements (fig. 8.9), the activity of the two-joint muscle does not follow a mathematical law. It is not the simple sum of the activity of two monoarticular motions. Two examples: (1) rectus femoris shows activity in hip flexion (fig. 8.8) but not in

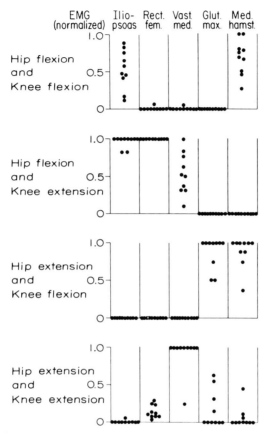

FIG. 8.7. EMGs in biarticular motion (10 subjects). (From Fujiwara and Basmajian, 1975.)

hip flexion accompanied by knee flexion (fig. 8.9), and (2) rectus femoris works as a knee extensor (fig. 8.9), but when hip extension accompanies knee extension, almost all rectus activity disappears (fig. 8.9).

In example (1), the activity of rectus femoris muscle is completely depressed to allow knee flexion (fig. 8.9). In (2), the activity of rectus femoris is depressed for hip extension (fig. 8.9), and the vastus muscles compensate for it to maintain or produce knee extension.

Depressed activity of rectus femoris in concurrent movements (fig. 8.9) is in response to a mechanical demand to flex the knee or to extend the hip. From the neurological standpoint, this may be called an antagonistic inhi-

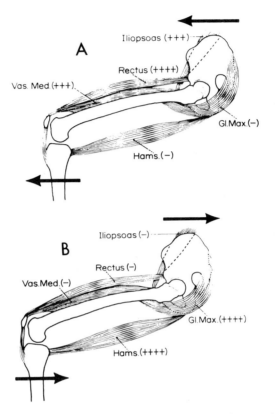

FIG. 8.8. *A.* Hip flexion and knee extension. Biarticular motion—counter-current movement (modified after Frost). *B.* Hip extension and knee flexion. Biarticular motion—countercurrent movement (modified after Frost). (From Fujiwara and Basmajian, 1975.)

bition of reciprocal innervation. Contraction of rectus femoris influences knee extension more than hip flexion. Like the rectus femoris, the medial hamstrings are also depressed in some concurrent movements and they are knee flexors more than hip extensors (fig. 8.9) (Fujiwara and Basmajian, 1975).

To summarize: The effect of contraction of two-joint muscles is never limited to one joint; whenever a two-joint muscle participates in a monoarticular motion its role shifts in close coordination with the other muscles. In biarticular concurrent motion the activity of the rectus femoris and the medial hamstrings is inhibited when they are antagonists, especially when motion of the knee is concerned.

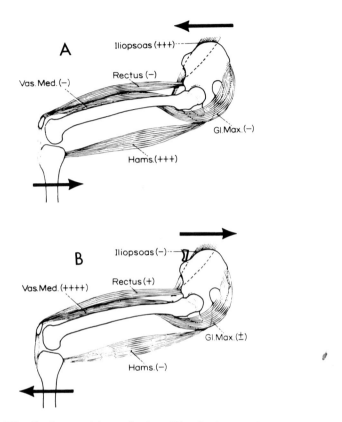

FIG. 8.9. *A.* Hip flexion and knee flexion. Biarticular motion—concurrent movement (modified after Frost). *B.* Hip extension and knee extension. Biarticular motion—concurrent movement (modified after Frost). (From Fujiwara and Basmajian, 1975.)

Related conclusions were drawn from a different approach by Yamashita (1975) of Kyoto University. He was concerned most with mechanisms that generate and transmit resultant leg extension forces by maximal isometric contractions in two directions while the hip and knee are kept at 90°. Single-joint muscles were most active when they crossed the joint where they limit extension force. Two-joint muscles were only moderately active then, even though the forces applied were great.

It may be argued that, in spite of the normal findings, occasions may arise when the proximal or distal part of a two-joint muscle does act independently. Such occasions must be very rare indeed. In fact, in upper motor neuron diseases the reverse is seen. In such cases the patient employs mass response of many neighbouring, and often unrelated, muscles. It is difficult to imagine his employing one isolated muscle, let alone the proximal or distal half of one.

Carlsöö and Molbech (1966) attempted to answer the underlying principle of two-joint muscles by comparing a number of them in the thigh both in free movements and during bicycling. In bicycling, during one revolution the effect of a muscle contraction can change from a flexing to an extending effect in the case of hamstrings and gastrocnemius. With the latter, during the knee extension of pedalling there is paradoxical activity, and Carlsöö and Molbech feel that this is part of a steered movement in a closed kinematic chain.

Finally, we must admit that under artificial experimental conditions, proximal and distal parts of two-joint muscles can be made to contract independently with relatively isolated effects. To accept this observation there is no need to invoke species differences and probably it would be reproducible in a human "preparation" if such were available.

Muscles Spared When Ligaments Suffice

Hardly any informed person would doubt that when gravity acts on the upper limb, and certainly when the limb carries a heavy load, muscles are the chief agents in preventing the distraction of the joints. Yet, as a result of many studies, I have concluded that this is a false belief. I originally stumbled on this idea by accident while working on the electromyography of shoulder muscles with Dr. F. J. Bazant (Basmajian and Bazant, 1959). I have also extended our observations to include the elbow region.

Essentially, the fundamental conclusions can be made that ligaments play a much greater part in supporting loads than is generally thought and, in most situations where traction is exerted across a joint, muscles play only a secondary rôle. A review of our experiments on the foot (discussed on p. 275), adds further confirmation to the idea that normally ligaments and not muscles maintain the integrity of joints.

Our broader studies in the shoulder and elbow region will be described in greater detail below (p. 196) and the full details have been published previously (Basmajian and Latif, 1957; Basmajian and Bazant, 1959; Basmajian, 1961). This electromyographic investigation dealt with supraspinatus, infraspinatus, deltoid, biceps and triceps muscles in a series of normal persons. In the case of the deltoid, the needle in the anterior fibres was 5 cm below the lateral end of the clavicle; that in the middle fibres was

5 cm below the lateral border of the acromion; and that in the posterior fibres was about 7.5 cm below the spine of the scapula. The electrodes in the supraspinatus and the infraspinatus were placed in or near the middle of their bellies. The electrodes in the biceps were placed in the middle of the muscle whereas those in the triceps were placed in the middle of its long head.

The subject was seated upright with his arm hanging in the relaxed neutral position (the forearm midway between pronation and supination). Two types of load were added to the subject's arm. The first of these was a load of about 7 kg (lead weights held in the hand to the limit of individual endurance which proved to be a variable factor). The other load, less precise but more effective, was a sudden heavy sustained downward pull by one of the observers on the subject's hanging arm. In five persons a longitudinal pull was applied to the arm which had been abducted to the horizontal plane and completely supported by another observer so that no abduction activity was required of the subject's muscles.

For studies of the elbow, 24 adults were studied with electrodes in both heads of biceps, in brachialis and in brachioradialis. In addition, the pronator teres muscles of eight other subjects were studied later and our findings published (Basmajian and Travill, 1961).

In all instances, electromyographs were made with the subject seated upright and the upper limb hanging straight downwards in a comfortable position. Thus considerable numbers of biceps muscles were studied, some with heavy and moderate loads and others with light loads. In making the electromyograms of the pronator teres, only a strong downward pull was used as the added load since experience had already shown the ineffectiveness of lesser loads.

Contrary to expectation, the vertically running muscles that cross the shoulder joint and the elbow joint are not active to prevent distraction of these joints by gravity (figs. 8.10, 8.11). Much more surprising is the fact that they do not spring into action when light, moderate or even heavy loads are added unless the subject voluntarily decides to flex his shoulder or his elbow and thus to support the weight in bent positions of these joints. Quite often, he may do this intermittently or, when uninstructed, from the very onset. But it must be clear that such muscular action is a voluntary action and not a reflex one.

Carlsöö and Guharay (1968) confirmed our findings of muscular inactivity in the heavily loaded shoulder and elbow joints—the muscles being biceps, triceps, brachialis and brachioradialis. In addition, they found that the temperature fell in these muscles, apparently because of a lower oxygen demand.

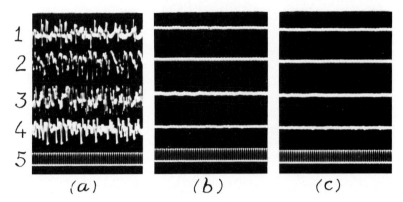

FIG. 8.10. EMGs, *a*, during abduction; *b*, unloaded arm hanging; and *c*, heavy downward pull applied to arm. Lines *1, 2* and *3*—anterior middle and posterior fibres of deltoid; line *4*—supraspinatus; line *5*—time marker; intervals 10 msec. (From Basmajian, 1961.)

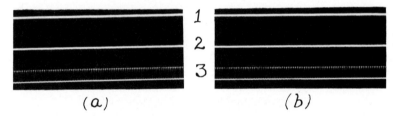

FIG. 8.11. EMGs of biceps (line *1*) and triceps (line *2*). *a*, hanging arm (unloaded), and *b*, heavily loaded. No change. (Line *3*, time marker, intervals 10 msec.) (From Basmajian, 1961.)

Even while the muscles were quiescent, our subjects rapidly felt local fatigue. What, then, is fatigue in the heavily-loaded limb? Normally, it would be thought of as "muscular fatigue" but we see now that this is incorrect. The "fatigue" that is experienced probably originates from the painful feeling of tension in the articular capsule and ligaments, not from overworked muscles. In fact, as we have seen, the muscles need not be working at all. Cain (1973) confirmed that feelings of fatigue in static contraction arise in part from structures outside the muscles.

An analogous situation occurs in the foot where we found, some years ago, that the muscles that are usually supposed to support the arches continuously were generally inactive in standing at rest (see p. 182). Independently Hicks (1954) showed by deduction that the plantar aponeurosis and plantar ligaments were the chief weight-bearers in this position. It would seem, then, that in the normal foot the fatigue of standing is not a muscular phenomenon.

The dual conclusion that articular ligaments suffice to prevent the downward distraction of joints in the upper limb and that fatigue is chiefly a form of pain in the ligaments appears to be of fundamental importance. It not only runs counter to "common sense" but it is of practical interest, for example, in explaining why dislocations by traction on normal limbs are rare. It should be noted especially that the capsule on the superior part of the shoulder joint including the coracohumeral ligament is extremely tight only when the arm hangs directly downward and the scapula is in its normal position. The special mechanism that includes this ligament together with the supraspinatus muscle and the normal slope of the glenoid cavity will be described elsewhere (p. 196). When the shoulder joint is abducted or flexed, however, the capsule is extremely loose and the shoulder joint depends for its integrity on the well-known "rotator-cuff" muscles.

All the experiments reported above finally led to one that has in turn led to new ideas. Elkus and I (1973) found that healthy subjects suspended by their hands from a trapeze can hold on for less than 3 minutes even when their fists are kept closed by a special gauntlet. Severe discomfort in the hands and forearms and uncontrollable slipping of the fingers (when no gauntlet was used) were the main causes for failure. Action potentials from a large number of muscles were unremarkable and all evidence pointed to a significant ligamentous force rather than muscles preventing articular distraction. Similar EMG studies by Tuttle and myself (1973) on apes (gorilla, chimpanzee and orangutan) confirm these findings. Further, Brantner and I (1975) found a clear-cut training effect in a remarkably increased endurance time of normal young human subjects; this was best explained by psychological adaptation.

Quite independent from us, Stener, Andesson and Petersén of Göteborg, Sweden were arriving at similar conclusions in regard to ligament sparing from a different type of experiment in cats and man. When Andersson and Stener (1959) greatly increased the tension in the medial ligament of the knee of the cat in specially designed experiments, no reflex muscular contractions appeared in the muscles of the thigh as would have been expected if the usual hypothesis of "ligamento-muscular protective reflexes" were valid (fig. 8.12). Furthermore, they showed convincingly that the absence of reflex motor effects was *not* due to absence of afferent discharges which were well registered from the articular nerves.

Petersén and Stener (1959) carried the above experiments forward to human subjects again using the medial ligaments of the knee. Their results were a complete vindication of the conclusions made in the animal experi-

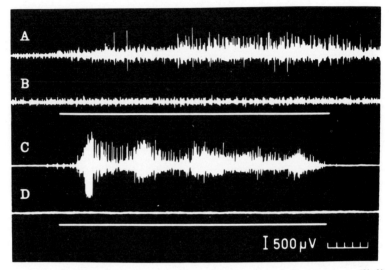

Fig. 8.12. EMGs of (decerebrate) cat from—*A* and *B*, vastus medialis; *C* and *D*, semitendinosus. *A* and *C* show reflex responses to stimuli other than stretch, to compare with the lack of response in *B* and *D* when the tendons were stretched by transverse loading (duration shown by straight horizontal white lines). (From Andersson and Stener, 1959.)

ments described previously. In addition their work suggests that if injured ligaments are pulled till pain results, muscles *do* show reflex contraction, but it the torn ligament is then anesthetized, they do not.

Following almost the same line of reasoning, deAndrade, Grant and Dixon (1965) distended human knees with non-irritating plasma (which emphasized the pressure phenomenon as opposed to pain). There was a definite and even marked inhibition of quadriceps contraction with a depression of motor unit activity. This is undoubtedly a reflex inhibition and helps further to explain the muscle weakness, atrophy and deformity that follow knee injury and disease. Czipott and Herpai (1971) reported that atrophy of the quadriceps that develops in connection with meniscal and ligamentous injuries is of neurogenic origin with disuse atrophy only secondary. However, Carlsöö and Norstrand (1968) could find no qualitative difference in muscle coordination in most of the thigh muscles examined by EMG when they compared patients with ruptured cruciate ligaments and normal subjects.

Freeman and Wyke (1966, 1967) obtained a definite and chronic drop in reflex postural tonus in cats by cutting the sensory nerve supply of the knee joint capsule. The mechanoreceptors in the joint are involved in reflex muscular activity to maintain posture in quadrupeds; undoubtedly the same mechanisms occur in man as well. Surely all the above observations are of great importance in orthopedics and surgery of joint injury; they deserve wide attention.

Becker (1960) analyzed the responses of several muscle groups to pulls applied to them, i.e., passive stretch. Distinctive responses were obtained in the soleus and long head of triceps brachii. In normal persons, these had a definite relationship with the phase of the stretch cycle, being about 50 μv initially and diminishing on repeated stretching. In patients with diseases characterized by hyperirritability of the skeletal muscles, the responses in these muscles were greatly exaggerated. Gastrocnemius and the lateral head of triceps were found to not react to repeated stretching. A few vague scattered discharges were all that could be obtained in the anterior tibial muscles, and these only in some of the abnormal cases. Becker feels that these responses indicate a special postural function of muscles that react to stretch.

Relation of EMG to Force or Tension

Isometrically contracting muscle most certainly shows a direct relationship between the mechanical tension and the integrated EMG. However, in the muscles of amputees, Inman, Ralston, Saunders, Feinstein and Wright (1951, 1952) found no direct quantitative relation when the muscles changed in length and they found no relationship between the muscle's inherent power and the EMG. With the studies involving rapid movement, they showed the mechanical tension lags (less than 0.1 second) behind the main burst of potentials. De Luca and Forrest (1973b) pointed out a further problem: in some movements a group of muscles contribute to the action and are not included in the total envelope of electrical activity.

Lippold and Bigland of University College, London, demonstrated in a fine series of papers (Lippold, 1952; Bigland *et al.*, 1953, 1954a, b) that, during a voluntary contraction, the tension is proportional to the measurable electrical activity both under isometric and—contrary to Inman *et al.* (above)—under *isotonic* conditions (figs. 8.13, 8.14). They also showed that the gradation of contraction is brought about mainly by motor unit recruitment. The maximum tension produced by maximum tetanic indirect stimulation only equalled that developed in maximum voluntary contraction. The maximum with tetanic stimulation in their experiments with human subjects occurred at frequencies of 35 to 40 per second. (In a series of unreported experiments with rabbits, Hugh Lawrence and I found essentially the same thing.) Finally, Bigland and Lippold (1954a) demonstrated that

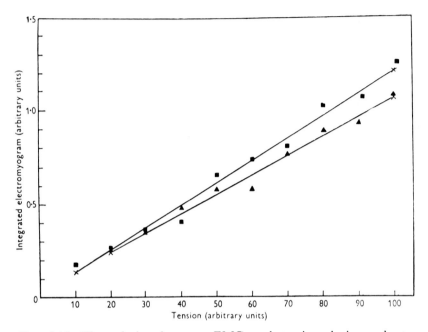

Fig. 8.13. The relation between EMG and tension during voluntary contraction in a single subject on two occasions. (From Lippold, 1952.)

tension, velocity and the EMG are interdependent, the integrated EMG providing a composite picture of the number and frequency of active muscle fibres.

Bergström (1959) of the University of Helsinki, Finland, has gone a definite step further than Lippold. After confirming the validity of the conclusion that the integrated potentials vary directly with the tension exerted, he showed that (in small muscles at least) these integrated potentials vary directly with the simple frequency of the spike potentials (fig. 8.15). He concluded that the counting of motor unit spikes (fig. 8.16) can be used to estimate the electical activity of, and thus the tension exerted by, the whole muscle. Close *et al.* (1960) have independently shown essentially the same thing using an electronic counting device.

In 1962, Bergström concluded from new experiments and calculations that there is a linear relationship between the number of impulses and the integrated *kinetic energy* of the muscular contraction. This work seems quite convincing.

Ahlgren (1966) showed that integrated EMGs in the muscles of mastication

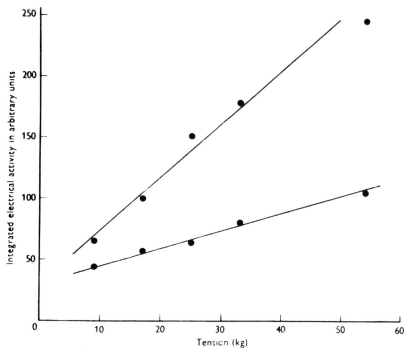

Fɪɢ. 8.14. The relation between integrated electrical activity (*via* surface electrodes) and tension in the human calf muscles. Shortening at constant velocity (*upper*) and lengthening at the same constant velocity. (From Bigland and Lippold, 1954a.)

rise linearly with the force of biting. For the muscle that opens the jaw (digastricus), Ahlgren and Lipke (1977) later found a direct relationship between force and integrated EMG from fine-wire electrodes. A similar relationship is apparent between the impulse frequency from human inter-costal muscles and the measured mechanical work of breathing (Viljanen, Poppius, Bergström and Hakumäki, 1964).

Acting as motors whose lines of pull change as to angle to the joint is being constantly changed, individual muscles cannot be expected to act with constant force through a whole movement. The EMG reflects this logic. Further, reflex phenomena can be expected to contribute to the recruitment of various muscles that act upon a joint during any movement. This is illustrated by the findings of Miwa and Matoba (1959). They found that during slow flexion biceps brachii is much more active at certain angles of

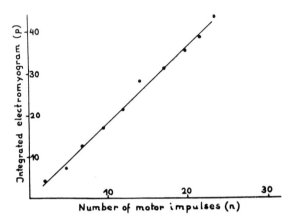

Fig. 8.15. The relation between the number of motor unit spikes (*n*) and the integrated area of the EMG (*p*). (From Bergström, 1959.)

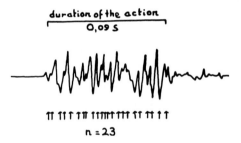

Fig. 8.16. Method of counting spikes (*arrows* showing the items counted). (From Bergström, 1959.)

the elbow: it reaches a peak of activity when the elbow is at 160° and falls rapidly to almost *nil* at 90°; it increases again at maximal flexion. Miwa, Tanaka and Matoba (1963) find similar changes occur in the activity of muscles in the thigh.

The maximum tension that can be produced voluntarily during sinusoidally produced brief isometric jerks at 5 Hz is the same as the maximum sustained tension (Soechting and Roberts, 1975). This emphasizes the importance of recruitment in the gradation of tension and synchronization of motor unit activity in the short sharp bursts. Milner-Brown, Stein and Lee (1975) of Edmonton, Canada hypothesize that training can lead to increasing synchronization and improved athletic power-performance.

With the angle of the joint unchanged, there is a linear relationship between the strength of a muscle and the amplitude of the integrated EMG, according to Liberson, Dondey and Asa (1962). They point out however—and this should be emphasized repeatedly—that the integrated output of *different* muscles cannot and must not be compared. Further, when the angle of the joint reaches certain limits, the results even within the same muscle may not accurately reflect the force exerted. The shape and orientation of certain muscles seem to complicate the picture further. Hof and van den Berg (1977) found in a study of triceps surae that the mean rectified EMG of any of the muscles involved (soleus, and the medial and lateral heads of gastrocnemius) is linearly proportional to the torque involved in slow isometric contractions acting on the ankle. The sum of individual muscle torques derived from the EMGs was equal to the mechanically measured total torque.

Bouisset, Denimal and Soula (1963) were able to show that within the limitation of their experimental set-up, the integrated EMG is linearly related to the speed of muscular shortening during contractions of biceps and triceps, both flexing and extending the elbow. With constant load, the activity rises linearly with acceleration; with constant maximal acceleration, the activity rises directly as the load is increased. More recently, Clarke (1967) in a paper that generally has been overlooked, demonstrated a correlation between the peak force of an isometric reflex contraction and its myopotential. Nevertheless, one cannot help but echo the cautious view of Ralston (1961): only under restricted conditions can force, speed and work output be judged from EMG data either recorded as spikes or integrated. During voluntary isometric contractions a comparison between integrated surface EMG and motor unit recruitment as a function of force reveals that the latter itself cannot account for the increased integrated global (surface) EMG, particularly for high values of force (Maton, 1976).

Considerable attention has been paid by a dynamic group of investigators in Paris and Lille to the quantitative relationship between surface-electrode-derived EMG and the force of contraction. Bouisset and Maton (1972) demonstrated that integrated EMGs from both unipolar and bipolar surface leads bear a direct relationship to those from fine-wire electrodes in the underlying muscles. Further, Bouisset and Maton (1973) showed that the activity of a muscle fibre near the surface of a muscle belly represents the activity of all the fibres involved in a voluntary movement. Cnockaert (1975) found that for the same amount of mechanical work the integrated EMG for positive work (concentric contraction) is 4 times that for negative work (eccentric contraction) for isolated movements against or with a yielding constant reactive force or resistance. With repetitive to-and-fro movements, it fell to 2.6 times higher. Earlier, the Finnish investigator Komi (1973) had found similar results in the branchioradialis muscle. Later, Komi and Rusko (1974) concluded that the fatigue effects on muscle tension in eccentric contraction may be of a mechanical origin, emphasizing the extreme loading.

In their successful effort to explain the linear relationship between the mean rectified surface EMG and the muscle force generated, Milner-Brown and Stein (1975) showed that neither amplitude nor frequency of myopotentials at the source vary directly with higher levels of force. But after the filtering by the tissues and the electronic processing, the relationship greatly improves. The peak-to-peak amplitude of the wave form contributed by each myopotential to the surface EMG increases approximately as the square root of the threshold force at which that unit recruited. The peak-to-peak duration of the wave form is independent of the threshold force. But firing rates increase at higher force levels.

Further, Grieve and Pheasant (1976) find that with zero to maximal voluntary effort the maximum electrical activity decreased when muscles are used in lengthened postures. Ralston, Todd and Inman (1976) in a study of rapid movements added a further caution: the time lag between the onset of electrical activity and the externally recorded tension or movement of the joint is about 30 to 40 msec. At relaxation, the discrepancies are even greater, the end of movement or tension lagging behind the cessation of potentials by 200 to 350 msec!

Chapter 9

Posture

E LECTROMYOGRAPHIC studies on postural muscles were begun by various investigators soon after the modern form of the technique was introduced near the end of World War II. Previously, it is true, rather crude attempts had been made to investigate the function of muscle groups by picking up and recording the electrical discharges that accompany the contraction of muscle. However, these attempts were thrown into the shade by the rise of modern electronics. Today, we must admit that, following the epoch-making stimulation studies of Duchenne in the nineteenth century, nothing very useful had been contributed to our knowledge of human posture by primarily electrical techniques until the past two decades.

It will be seen, therefore, that, narrowly defined, the subject to be discussed does not have a long history. Nonetheless, it cannot be divorced from an historical background because the object of our enquiry, i.e., posture, has been the concern of anatomists, biologists and orthopedic surgeons for many years. Therefore, this discussion will not be confined strictly to electromyography; neither will it be confined to studies on posture in its extremely limited sense.

The definition of posture can be altered for the sake of argument according to how broad or how narrow one wishes to make it. In the narrowest sense, posture may be considered to be the upright, well-balanced stance of the human subject in a "normal" position. In this sense, the electromyography of posture would deal with the maintenance of the erect subject's position against the force of gravity. The present account will, of necessity, emphasize this aspect of posture, but a broader, more generous and more palatable definition would not exclude the multiplicity of normal (and abnormal) standing, sitting and reclining positions that human beings assume in their constant battle against the force of gravity. In the final analysis, the intrinsic mechanisms of the body that counteract gravity make up the essence of the study of posture. One of these, the muscular mechanism, shall be our chief concern.

Posture of the entire body may be considered as a unit, and because such a consideration is rather artificial, it has inherent dangers. They often lead to a facile neglect of the posture in those parts of the body which do not intercept the main line of gravity for the trunk and lower limbs. As a result, posture in the upper limbs, both while hanging freely downwards and in various other positions, too often gets ignored. In still another direction, the posture of the mandible may also be ignored by the general anatomist, but it certainly is emphasized by the orthodontist for whom it assumes a considerable practical importance.

The problems of static posture, then, revolve around the truism that the balance or equilibrium of the human body or its articulated parts depends on a fine neutralization of the forces of gravity by counter-forces. These counter-forces may be supplied most simply both internally and externally by a supporting horizontal surface or series of horizontal surfaces that are inert. The "easiest" posture in which a human being can achieve equilibrium with gravity is the recumbent one. We should not lose sight of the fact that this is our normal posture for the first year or so of our lives and for about half of our lives thereafter. When we lie down, we bring the centre of gravity of the entire body as well as any or all of its parts closest to a supporting antigravitational surface.

Lundervold (1951) of Stockholm demonstrated by electromyography that healthy persons who do not tense their muscles can sit comfortably and relax in many positions, and can even work in many different manners without pronounced increase in muscular activity. Nervous subjects do not relax completely in more than a few positions and they cannot change their individual optimal working positions without a markedly increased exertion of muscle power.

Returning to the support of the erect body, we find that in the nineteenth century many laborious studies, some fruitful and others not, were performed to determine the line of gravity and the centre of gravity of the whole human subject. The simplest estimate and one that is most easily appreciated is that of von Meyer (1868) who found that the weight centre is situated at the level of the second sacral vertebra. (On the surface of the body this vertebra is at the level of the posterior superior iliac spines.) In the coronal plane the exact point lies 5 cm or less behind the line joining the hip joints, and, of course, it is in the midline.

It will be seen that to maintain an equilibrium in the standing posture with the least expenditure of internal energy, a vertical line dropped from the centre of gravity should fall downward through an inert supporting column of bones. This is the ideal and it is surprising how closely the human supporting mechanism approaches it, if only intermittently.

The idealized normal erect posture is one in which the line of gravity drops

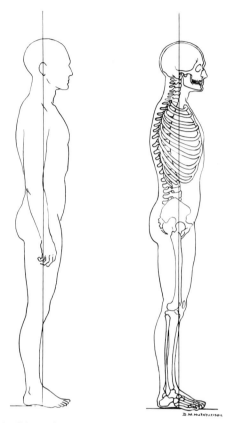

FIG. 9.1. Line of gravity in total erect man (see text)

in the midline between the following bilateral points: (1) the mastoid processes, (2) a point just in front of the shoulder joints, (3) the hip joints (or just behind), (4) a point just in front of the centre of the knee joints and (5) a point just in front of the ankle joints (fig. 9.1). Muscular activity is called upon to approximate this posture or, if the body is pulled out of the line of gravity, to bring it back into line. Using a force platform, Stribley *et al.* (1974) and Murray *et al.* (1975) found a large area of stability over which weight is or can be shifted and maintained. The pressure fluctuates incessantly around a mean point. It appears to become more stable in children between the ages of 4 and 8 years (Shambes, 1976).

Most people do not appreciate that, among mammals, man has the most economical of antigravity mechanisms once the upright posture is attained. His expenditure of muscular energy for what seems to the student of phylogeny to be a most awkward position is actually extremely economical. Most comparative anatomists certainly seem to be ignorant of this fact. A quadruped that is required to maintain the multiple joints of its limbs in a state of partial flexion by means of muscular activity demonstrates a much more wasteful antigravity machinery. An exception to this seems to be the elephant whose limbs serve as static columns to maintain an enormous weight. On the other hand, the specialization of the elephant's weight-bearing limbs is so great that it cannot produce a true jump for even short distances. Relative to its size, the muscles of its limbs are quite puny compared with man's. The reason for this disproportion is that, unlike the elephant, man is constantly challenging gravity by his continued wide range of postures, and great power is required only to achieve them. Thus we find that man's so-called antigravity muscles are not so much to maintain normal standing and sitting postures as they are to produce the powerful movements necessary for the major changes from lying, to sitting, to standing. Therefore it is wrong to equate the antigravity muscles of man with those of the common domestic animals which stand on flexed joints.

In man, the column of bones that carries the weight to the ground constitutes a series of links. Ideally, these links should be so stacked that the line of gravity passes directly through the centre of each joint between them. But even in man this ideal is only closely approached and never completely reached–and then only momentarily. As Steindler (1955) showed, a completely passive equilibrium is impossible because the centres of gravity of the links and the movement centres of the joints between them cannot be all brought to coincide perfectly with the common line of gravity. In spite of this, I believe that Steindler and many others have greatly exaggerated the amount of effort required to maintain the upright posture. The fatigue of standing is emphatically not due to muscular fatigue and, generally, the muscular activity in standing is slight or moderate. Sometimes it is only intermittent. On the other hand, the posture of quadrupeds, which is maintained by muscles acting on a series of flexed joints is highly dependent on continuous support by active muscular contraction. Of course the same is true for the human being in any but the fully erect standing posture.

Dudley Morton (1952) anticipated much of what has been recently proved by electromyography. Unfortunately, he incorrectly ascribed the fatigue of prolonged standing to a continuous activity of the muscles. This error is surprising because his calculations are otherwise quite valid. What he and others have ignored is that walking is usually less fatiguing than standing. Although extreme exertion can produce muscular fatigue, most fatigue in the lower limbs caused by standing is more intimately associated with the inadequacies of the venous and arterial circulation and with the direct pressures and tensions upon inert structures.

As Carlsöö (1961) has emphasized, certain muscle groups can be called "prime postural muscles." Among these are the neck muscles, sacrospinalis, hamstrings and soleus. Carlsöö also points out that such postural muscles are among the most powerful. However, one must note that this is not an absolute case.

Carlsöö found that during stooping most persons failed to use a well-balanced position, placing "too large a part of the load on the anterior part of the foot." They then powerfully engaged the soleus, gastrocnemius, flexor hallucis longus and peroneus. "Others placed too much of the load on the heels, so that the tibialis anterior and the peroneus muscles, which do not well tolerate continuous loading, were strongly activated."

Carlsöö considers the shifting from foot to foot in ordinary standing as a relief mechanism. "By assuming asymmetric working postures, and using the right and left leg alternately as the main support, the leg muscles are therefore periodically unloaded and relaxed." One should add that the relief to the inert structures is perhaps even more significant (p. 164).

Posture of Lower Limb

LEG. The function of the large muscles of the leg in relationship to posture has been studied by a number of investigators and quite early made the main subject of a book by Joseph (1960). Not infrequently, different conclusions have resulted from different techniques. For example, Joseph and Nightingale (1952, 1956) concluded from their study with surface electrodes that the soleus of all persons and the gastrocnemius of many show well-marked activity when the subject is standing at ease; and meanwhile, they claimed, the tibialis anterior is "silent." Their explanation, which agrees with the conclusion of Åkerblom (1948), is that the line of gravity is found to fall in front of the knee joint and ankle joint, necessitating activity in gastrocnemius. On the other hand, we showed with needle electrodes that there is actually a wide range of findings for each of these muscles, though, indeed, the posterior calf muscles are generally much more active than the tibialis anterior (fig. 9.2, *B*). Furthermore there is frequently a periodicity in the activity and this is apparently related to an almost imperceptible forward-and-backward swaying of the body (Basmajian and Bentzon, 1954). Periodicity was first noted in this regard by Floyd and Silver (1950) and it has been commented on by Portnoy and Morin (1956) and others. Granit's (1960)

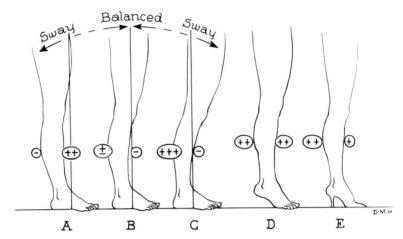

FIG. 9.2. Diagram of emg activity in anterior and posterior muscles of the leg under differing conditions.

statement that, in general, soleus is tonic while gastrocnemius is phasic may explain some of the discrepancies in the findings for the leg reported above. Carlsöö (1964) finds activity regularly in soleus during quiet, symmetric standing but never in tibialis anterior. With a heavy load held either in front of the thighs or carried on the back, activity becomes pronounced in soleus, apparently to counteract the forward leaning of the body. Tibialis anterior remains completely inactive.

As would be expected, any deliberate leaning forwards or backwards of a standing subject produces compensatory activity in the muscles to prevent the occurrence of a complete imbalance (fig. 9.2, *A-C*). A very finely regulated mechanism is in control and the slightest shift is reacted to through the nervous system by reflex postural adjustments; sometimes the motor responses are so fine that they can only be detected electromyographically.

Houtz and Fischer (1961) showed that these muscles respond to many influences such as postural changes, the shifting of body weight and the resisting of external forces applied to the upper part of the trunk. More recently, detailed studies have confirmed their findings (Okada, 1972, 1975).

In women who wear high heels, there appears to be a modification of the muscular response. Both our group (1954) and Joseph and Nightingale (1956) found that the wearing of high heels increases the activity of the calf muscles of individual subjects, apparently due to a shifting forwards of the centre of gravity (fig. 9.2, *E*). One would have expected that in women wearing high heels the well-known compensatory spinal lordosis would be a sufficient adjustment. Apparently it is not.

THIGH. The muscles of the thigh obey the same rules as those of the leg. By and large, the activity during normal, relaxed standing is usually slight. Indeed, it may be absent in most of the muscles for varying periods of time. The reports of Weddell, Feinstein and Pattle (1944), Åkerblom (1948), Arienti (1948), Wheatley and Jahnke (1951), Floyd and Silver (1951), Joseph and Nightingale (1954), Portnoy and Morin (1956), Oota (1956a), Joseph and Williams (1957) and Jonsson and Steen (1966), and the work in our laboratory seem to agree in principle. These overlapping and detailed studies include most of the large muscles of the gluteal region and thigh and no purpose would be served in recapitulating the details here. The main generalization to be extracted from all this is that the activity in these muscles is surprisingly slight during relaxed standing. The effect of swaying, mentioned above for leg muscles, appears in some hip muscles also (Jonsson and Synnerstad, 1967). Gluteus medius and tensor fasciae latae show such bursts but not the gluteus maximus which remains silent.

When subjects carry a load either held in front of the thighs or strapped to the back, Carlsöö (1964) found quadriceps remains completely inactive. Meanwhile, the ischiocrural muscles (hamstrings) show individual variations—from very active to completely inactive—apparently depending on the degree of flexion of the hip and on whether or not the line of gravity had been shifted anterior to the hip joint.

FOOT. The postural function of the muscles of the foot in relation to the normal support and the abnormal flattening of the arches has always posed a question of some fundamental interest. Using needle electrodes, we showed (Basmajian and Bentzon, 1954) that the intrinsic muscles are generally quiescent during normal standing but become extremely active when the subject rises on tip-toes and during the take-off stage of walking (fig. 9.3). This was confirmed in general by Sheffield, Gersten and Mastellone (1956) but it appears that others have not paid sufficient attention to this fundamental consideration in the posture of the foot.

In the past, the peroneal and tibial muscles have often been considered to play an important rôle in maintaining the longitudinal arches of the foot in standing. This theory seems to have been discredited by our findings and by the indirect contributory evidence of other investigators. During standing these muscles of the leg are generally quiescent. Furthermore, they remain inactive even when a subject suddenly lowers himself to a normal standing position from an elevated seated position. However, if in the standing posture the foot is obviously inverted by tibialis anterior, activity is quite intense.

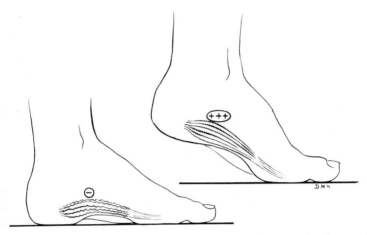

Fig. 9.3. Diagram of emg activity in intrinsic foot muscles (see text)

During locomotion, peroneal and tibial muscles show marked activity (fig. 9.2).

Apparently, the first line of defence against flat feet is a ligamentous one (Basmajian and Stecko, 1963). This is considered further on page 275. But the added stresses of walking require special mechanisms (Basmajian, 1955a, 1960). Independently, a similar view has been advanced by Dudley Morton (1952), who on the basis of various calculations predicted essentially the same thing. He showed that static strains upon the ligaments of the arch to sustain its elevated position are low in intensity and fall well within the capabilities of the ligaments. His calculations showed that only acute, heavy but transient forces (such as in the take-off phase of walking) required the dynamic action of muscle. Meanwhile, further confirmation was provided by Hicks (1951, 1954) who has demonstrated the importance of the plantar aponeurosis.

Hip and Knee. The hip and knee each has at least one muscle with a special postural function that can be demonstrated electromyographically. Experiments (Basmajian, 1958b) have shown that iliopsoas remains constantly active in the erect posture (in contrast to the large thigh muscles). It would appear that iliopsoas functions as a vital ligament to prevent hyperextension of the hip joint while standing (see p. 238). (The supraspinatus has a similar activity at the shoulder joint; see below.)

At the knee, Barnett and Richardson (1953) have shown a constant activity in the popliteus in the crouching or "knee-bent" posture. This apparently is related to a stabilizing postural function to help the posterior cruciate ligament prevent an anterior dislocation of the femur (fig. 9.4). There is no

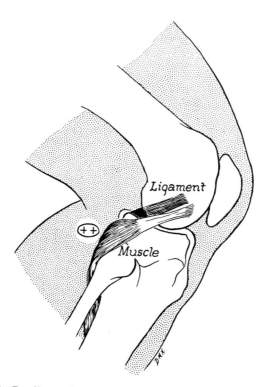

Fig. 9.4. Popliteus shows emg activity in the knee-bent stance

similar popliteal activity in the erect posture when dislocation is not threatening the joint.

Posture of Trunk, Neck and Head

SPINE. While standing erect, most human subjects require very slight activity and sometimes some intermittent reflex activity of the intrinsic muscles of the back according to Allen (1948), Floyd and Silver (1951, 1955), Portnoy and Morin (1956) and Joseph (1960). These authors showed that during forward flexion there is marked activity until flexion is extreme, at which time the ligamentous structures assume the load and the muscles become silent (fig. 9.5). Floyd and Silver (1955) proved (with both surface and needle electrodes) that in the extreme-flexed position of the back, the erector spinae remained relaxed in the initial stages of heavy weight-lifting. This observation appears to confirm strongly the dangers to the vertebral

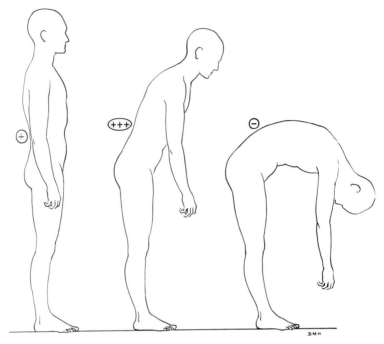

Fɪɢ. 9.5. Diagram of activity in erector spinae during forward bending (see text).

ligaments and joints of lifting "with the back" rather than with the muscles of the lower limb (see p. 282).

Asmussen (1960) first concluded from an emg study that continuous activity of the back muscles during standing is the rule because "the line of gravity passes in front of the spinal column." The last part of this conclusion is contrary to previous opinions and requires confirming. Later Asmussen and Klausen (1962) modified the earlier extreme view to conclude that "the force of gravity is counteracted by one set of muscles only, most often the back muscles, but in 20 to 25 per cent of the cases the abdominal muscles. The line of gravity passes very close to the axis of movement of vertebra L4 and does not intersect with the curves of the spine as often postulated." Carlsöö (1964) regularly found activity in sacrospinalis in the symmetric, rest position.

More recently, Klausen (1965) investigated that effect of changes in the curve of the spine, the line of gravity in relation to vertebra L4 and ankle joints and the activity of the muscles of the trunk. He concluded that the short, deep intrinsic muscles of the back must play an important rôle in stabilizing the individual intervertebral joints. The long intrinsic muscles and the abdominal muscles stabilize the spine as a whole. An increased pull of gravity always is counteracted by increased activity in one set of muscles only, i.e., either in the back or anterior abdominal wall.

Placing a load high on the back automatically causes the trunk to lean slightly forward. The increased pull of gravity is counteracted by an increased activity in the lower back muscles. A load placed low on the back reduces the activity of the back muscles. This last finding was duplicated by the independent work of Carlsöö (1964). However, Carlsöö found increased activity in sacrospinalis with a load held in front of the thighs. Thus the position of the load—either back or front—either aids the muscles or reflexly call upon their activity to prevent forward imbalance Nachemson (1966) finds the vertebral part of psoas major helps to maintain the posture of the lumbar vertebrae.

ABDOMEN. During relaxed standing, only slight activity has been recorded in the abdominal muscles by Floyd and Silver (1950), Campbell and Green (1955) and Ono (1958). The first investigators appear to have proved that the activity is greatest in the internal oblique to provide the protection that the muscles afford to the inguinal canal in the upright posture (see p. 325). Carlsöö (1964) reported that carrying a load on the back always increases activity in rectus abdominis, but carrying one in front of the thighs left rectus abdominis completely silent.

THORAX. Jones, Beargie and Pauly (1953) were the first to suggest that the intercostal muscles play a part in posture, or at least, in the maintenance of certain flexed postures and adjustments of position. Certainly worthy of further study is the interesting proposal advanced by Jones and Pauly (1957) that the intercostals have as their chief function the maintenance of a proper distance between the ribs while the rib cage is actively elevated by the neck muscles during inspiration (see p. 340). Credence in their theory is strengthened by the independent work of Campbell (1955) who has recorded activity in the scalenes and sternomastoid muscles during quiet respiration and that of Koepke *et al.* (1958). Surprisingly, these seemingly fundamental problems have not been attacked with any vigour by other investigators (see also p. 355).

MANDIBLE. Moyers (1950), Carlsöö (1952) and MacDougall and Andrew (1953) claimed that the muscles of the jaw which act against gravity in maintaining posture are all those that raise the mandible. Latif (1957), however, while working in our laboratory, demonstrated that temporalis alone is responsible for keeping the teeth in apposition and is constantly active in the upright posture. During the resting, mouth-closed position, there is strikingly greater activity in the posterior fibres which run almost horizontally backwards than there is in the anterior fibres which run vertically. Temporalis is discussed further in Chapter 18.

Posture of Upper Limb

Posture in the upper limb is chiefly a matter of maintaining the integrity of the series of joints in the hanging position. However, in the recumbent posture, if the upper limb is raised to a vertical position, many of the factors which normally govern the posture of the erect whole body come into play for the first time in the upper limb. In the standing position, on the other hand, the hanging limb poses different problems in posture because the force of gravity produces tensions rather than pressures. These tensions are easily carried by the bones which are rigid, but the logical question is often asked: what prevents dislocation of the series of joints? As we have seen already, the most frequent answer—that it is muscular action—is not correct.

SHOULDER. Not surprisingly, low grade postural activity occurs in the upper fibres of trapezius in supporting the shoulder girdle. I have made this observation incidental to other studies in normal subjects over a number of years. Minimal postural activity in the serratus anterior has also been described by Catton and Gray (1951) and confirmed by my own scattered observations during clinical electromyograms.

At the shoulder joint, we have found that the main muscular activity in resisting downward dislocation occurs in supraspinatus activity in resisting downward dislocation occurs in supraspinatus (and to a slight extent in the posterior, horizontal-running fibres of deltoid) (Basmajian and Bazant, 1959). The bulk of the deltoid, and the biceps and triceps show no activity in spite of their vertical direction. Surprisingly, this is true even when heavy weights are suspended from the arm. The function of supraspinatus is apparently associated with what I have described as a *locking mechanism* dependent upon the slope of the glenoid fossa. The horizontal pull of the muscle, along with an extreme tightening of the superior part of the capsule only when the arm hangs vertically, prevents downward subluxation of the humeral head (see page 164).

ELBOW. At the elbow joint, without an added load, there is no activity in the muscles, suggesting that the ligaments carry the weight (Basmajian and Latif, 1957; Basmajian and Travill, 1961). The addition of a small or moderate load does not produce any activity in biceps, triceps, brachioradialis or pronator teres. This is also true for the main elbow-crossing muscles when a human or gorilla subject is hanging from a trapeze (Elkus and Basmajian, 1973; Tuttle and Basmajian, 1973) (see page 167). Perhaps it is superfluous to note that in flexed positions of the elbow the maintenance of the flexed posture is shared by the brachialis and biceps muscles. However, brachioradialis shows little if any activity in maintaining flexed postures even against added loads because, as I have shown before, it is theoretically an example of a "shunt" muscle, as first postulated by MacConaill (1946)—see p. 152.

Wrist. At the wrist and hand, a minimum of activity is required to overcome the ordinary force of gravity. Repeated electromyograms have revealed a general silence of the forearm and hand while hanging at rest. The "postural" activity of the wrist flexors and extensors which accompany the making of a fist or the grasping of a handle is perhaps more properly a "synergistic" function and possibly outside the limits of our subject. In this regard, the only significant postural electromyography of this region was reported by Dempster and Finerty (1947). They loaded the hand of the horizontally held forearm and recorded the activity in various muscles that cross the wrist. When a muscle was in a superior position and working to support the load against gravity, its activity was three to four times as great as when it was below and maximally aided by gravity, i.e., when it was serving a stabilizing or synergistic function only (see also p. 97).

Recumbent Posture

In conclusion, let us return to the recumbent posture. Here, in this pleasantest of postures, the force of gravity is counteracted by mechanisms that are entirely passive. Repeated electromyograms by many investigators have demonstrated beyond the shadow of a doubt that resting muscles exhibit no neuromuscular activity (Basmajian, 1955a). It is time, then, that all anatomists and physiologists awoke to this fact and so altered their teaching. Contrary to widespread belief, there is no random activity of motor units in a resting muscle to provide what is often hazily called *muscular tone* (see section on Tone on p. 79).

Sitting Posture

In view of the confusing welter of theories regarding "ideal" sitting postures, the paucity of scientific emg experiments is surprising. Åkerblom (1948), who first published a monograph on the subject a quarter century ago in Sweden, remains almost alone in the field (see also Åkerblom, 1969). His investigations showed good relaxation of the spinal muscles when good support is given to the lumbar and thoracic regions, or when good lumbar support is given, and even when the subject is allowed to slump or sink into a ventriflexed position. Knutsson *et al.* (1966) duplicated Åkerblom's studies. Rosemayer (1971), working in Munich and in South Africa with Professor Lewer Allen, has extended this type of work to motorcar driving postures; related work is in progress in Naples under Carlo Serra (1968), with only preliminary results reported. Our work on back muscles of seated subjects (Donisch and Basmajian, 1972) is reported on p. 289.

Chapter 10

The upper limb

M ANY scattered emg studies of the upper limb followed upon the
original work in California of Inman, Saunders and Abbott
reported in 1944. This chapter will bring together most of the
available information in an organized form and deal with the actions of
groups of muscles topographically. In recent years we have seen a rapid
growth of studies in applied emg studies in gymnastics (e.g., Landa, 1974), in
athletics (e.g., Kemei *et al.*, 1971; Okamoto and Kumamoto, 1971; Toyoshima
et al., 1971; Yamashita *et al.*, 1971, 1972; Yoshi *et al.*, 1974; Matsushita *et al.*,
1974; Yoshizawa *et al.*, 1976; Okamoto, 1976; Okamoto *et al.*, 1976; Goto *et
al.*, 1976; Tokuyama *et al.*, 1976, 1977; Kazai *et al.*, 1976), and in ergonomics
(e.g., Jonsson and Jonsson, 1975, 1976; Jonsson and Hagberg, 1974; Carlsöö
and Mayr, 1974; Avon and Schmitt, 1975). The wrist, hand and fingers will
be dealt with in the next chapter.

Trapezius

Following the classical study of the above authors whose chief concern was
with the dynamics of the shoulder (see below), Yamshon and Bierman (1948)
and Wiedenbauer and Mortensen (1952) made emg studies of various parts
of the trapezius during voluntary movements in a series of normal adults.
The trapezius was found to be considerably active during elevation or
retraction of the shoulder and during flexion or abduction of the upper
extremity through a range of 180°. During scapular elevation the greatest
activity was recorded, as would be expected, from the upper parts of the
muscle; during retraction, from the middle and lower parts; and during
flexion, from the lower half. The greatest activity in trapezius appears during
abduction of the limb and chiefly in the lower two-thirds of the muscle. These
findings were confirmation of the California studies and have been confirmed
in detail by Thom (1965) of Heidelberg, and in general in our laboratory. I
cannot, however, find confirmation for Duchenne's belief that trapezius is a
respiratory muscle.

In his study of static loading, Bearn (1961b) discovered that the upper
fibres of trapezius, contrary to the universal teaching, "play no active part in

the support of the shoulder girdle in the relaxed upright posture." This was confirmed by Fernandez-Ballesteros *et al.* (1964). Some of Bearn's subjects initially showed a low level of activity in this part of the trapezius; but upon their being instructed to relax, the activity stopped entirely. As he notes, this observation is surprising. Indeed, the upper part of the muscle shows through the skin in thin people and appears to be under some tension even when no weight is borne by the limb.

When a load of 10 pounds (about 4.5 kg) is held in the hand, fully three-quarters of Bearn's subjects were able to relax the trapezius either immediately or within 2 minutes. The remainder showed very little activity compared with the result of slight shrugging movements. With a 25-pound (11.4 kg) load, a third of the subjects could support the weight without emg activity in trapezius. Bearn cautions against the interpretation that this is a desirable way to carry loads: moreover, he ascribes various abnormalities to the habitual depression of the clavicle.

Trapezius muscle is not just a postural or supporting muscle. Its rôle in the adjustments of the scapula during elevation of the upper limb is vital. In fact its activity is essential both in raising the arm (see Scapula Rotation, p. 191) and in preventing downward dislocation of the humerus (p. 196).

The only formal study of trapezius muscle is that of Ito *et al.* (1976) of Tokyo. With surface electrodes over six parts of the muscle in 10 healthy men they found the upper and lower parts have functional differences during varying movements. The upper part was markedly active during hyperextension; the lower, during lateral tilting of the pelvis.

Pectoral Muscles

Inman, Saunders and Abbott (1944) were the first to examine pectoralis major electromyographically. In abduction of the arm, no part of this muscle is active. In forward flexion, the clavicular head is the active part, reaching its maximum activity at 115° of flexion; the sternocostal head remains inactive. With the exception of the studies by Ravaglia (1958) in Italy, Scheving and Pauly (1959) in Chicago and Okamoto *et al.* (1967) of Kyoto, the pectoral muscles seem to have been otherwise ignored until de Sousa *et al.* (1969) and Jonsson *et al.* (1972) recently published their works. Ravaglia had been concerned with their alleged accessory functions in the respiration. He demonstrated the presence of moderate activity in them during forced inspiration but none with quiet breathing. Scheving and Pauly confirmed many of the findings of Inman *et al.*, and further confirmed the standard teaching regarding the important activity of the sternocostal head in adduction. However, they found that medial rotation must be against resistance for the pectoralis major to be called into action. With this de Sousa *et al.* (1969) disagreed emphatically, pointing out that the clavicular head is active almost always during either free or resisted medial rotations of the humerus. But they agreed that the sternocostal head remained relaxed except when adduction occurred; this is true regardless of the starting position of the limb (even behind the back) according to Jonsson *et al.* (1972). Shevlin *et al.* (1969) tested and confirmed standard textbook views of the functional differences between the two parts of the pectoralis major during isometric effort in different positions of the arm.

Following removal of the pectoral muscles during radical mastectomy, Flint *et al.* (1970) found a surprising lack of functional disturbance in ordinary activities. Surrounding muscles compensate except for stabilization of the shoulder joint in movements where the humerus is forced upwards and laterally.

Serratus Anterior

Though serratus anterior must be considered further under "Scapular Rotation," we should note the work of Catton and Gray (1951) who proved beyond a question of a doubt that this is not an accessory respiratory muscle. Their EMGs failed to demonstrate any activity in serratus anterior during voluntary deep breathing, during breathing that was obstructed by forcing the subject to breathe through a narrow tube, and even during coughing. The final blow to the concept of this being an accessory respiratory muscle was struck recently by Jefferson *et al.* (1960) who demonstrated that action potentials were generally absent during respiration in the nerve to serratus anterior.

Scapular Rotation

Just as scapular rotation is a distinct and important function, well-known since Duchenne, so the muscles which produce the movement are a distinct functional group. The emg studies of Inman *et al.* first drew special attention to them. As they showed, the upper part of the trapezius, the levator scapulae and the upper digits of serratus anterior constitute a unit whose main activities are in concert: they passively support the scapula (slight continuous activity), elevate it (increasing activity) and act as the upper component of a force couple that rotates the scapula (fig. 10.1).

The lower part of trapezius and lower half or more of the serratus anterior constitute the lower component of the scapular rotatory force couple; they were found to act with increasing vigour throughout elevation of the arm. The lower part of trapezius is the more active component of the lower force couple during abduction; but in flexion it is less active than serratus anterior apparently because the scapula must be pulled forward during flexion.

The middle fibres of trapezius are most active in abduction especially as the arm reaches the horizontal plane (90°). In forward flexion, the activity of the middle fibres of trapezius decreases during the early range but builds up toward the end. In general, then, the middle trapezius serves to fix the scapula but must relax to allow the scapula to slide forward during the early part of flexion.

The rhomboid muscles (major and minor) imitate the middle trapezius, being most active in abduction and least during early flexion.

Movements and Muscles of the Glenohumeral Joint

The chief muscles that act upon the shoulder (glenohumeral) joint are the deltoid, the pectorales (discussed above), the latissimus dorsi, teres major, and the four rotator cuff muscles—subscapularis, supraspinatus, infraspinatus and teres minor. The considerable interest these muscles have aroused amongst electromyographers is not surprising, for the movements and protection of the shoulder joint are of paramount importance. Both heads of

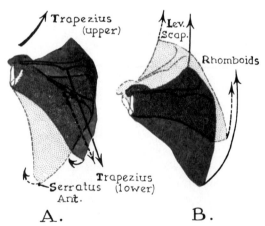

Fig. 10.1. Scapular rotation and force couples. *A*, glenoid "up"; *B*, glenoid "down." (From Basmajian, 1955a.)

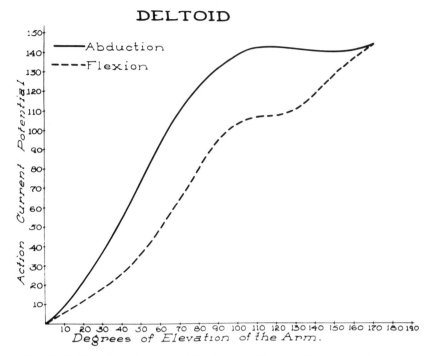

FIG. 10.2. Relation of emg activity in deltoid (in arbitrary units) to degree of elevation of arm. (From Inman, Saunders and Abott, 1944.)

biceps brachii play a modest rôle in shoulder function. Below are the composite results derived from the work of various authors.

ABDUCTION. The activity in the deltoid increases progressively and becomes greatest between 90° and 180° of elevation (fig. 10.2). The activity of supraspinatus increases progressively, too (Inman *et al.*). Thus, it is not simply an initiator of abduction as was formerly taught. Our own studies have conclusively substantiated these statements as have those of Wertheimer and Ferraz (1958) and Comtet and Auffray (1970). Quite surprising but rational is the discovery by Van Linge and Mulder (1963) that complete experimental paralysis of supraspinatus in man simply reduces the force of abduction and power of endurance. They concluded that in abduction supraspinatus plays only a quantitative and not a specialized role. No part of pectoralis major is active during abduction. The rôle of biceps brachii in abduction seems to be confined to a contribution in maintaining this position while the arm is laterally rotated and the forearm supine. When the arm is medially rotated and the forearm prone, biceps does not contribute to abduction (Basmajian and Latif, 1957).

FLEXION. The clavicular head of pectoralis major along with the anterior fibres of deltoid are the chief flexors (Shevlin *et al.*, 1969; Okamoto *et al.*, 1967). Okamoto (1968) has shown that there are substantial variations in local emg activity depending on slight changes in resistances applied in varying ways. Both heads of biceps brachii are active in flexion of the shoulder joint, the long head being the more active (Basmajian and Latif, 1957).

DEPRESSORS OF HUMERUS. Subscapularis, infraspinatus and teres minor were shown by Inman *et al.* to form a functional group which acts as the second or inferior group of the force couple during abduction of the humerus. They act continuously during both abduction and flexion. In abduction, activity in infraspinatus and teres minor rises linearly while activity in subscapularis reaches a peak or plateau beyond the 90° angle and then falls off.

ADDUCTION. Pectoralis major and latissimus dorsi produce adduction. The posterior fibres of deltoid are also very active, perhaps to resist the medial rotation that the main adductors would produce if unresisted (Scheving and Pauly, 1959). While we found the teres major acts only when there is resistance to the movement (Broome and Basmajian, 1971), Jonsson *et al.* (1972) report that when the arm is behind the coronal plane, adduction recruits teres major activity without added load.

TERES MAJOR AND LATISSIMUS DORSI. Teres major never exhibits activity during motion "but plays a peculiar rôle in that it only comes into action when it is necessary to maintain a static position, it reaches its maximum activity at about 90 degrees." These statements of Inman and his colleagues have been neither challenged nor confirmed until recently. Kamon (1966) finds that teres major is very active in movements of the free arm during gymnastics on the pommeled horse. On the other hand, even during the vigorous activity of shot-putting, the teres major remains relatively quiet (Hermann, 1962). De Sousa *et al.* (1969) reported that teres major was indeed active during active motion of the arm.

Our study (1971) was undertaken to clarify the disagreements because of the obvious importance of this large muscle. The closely related latissimus dorsi was studied simultaneously as a reference in 10 normal adult subjects.

Bipolar fine-wire (75 μ) Karma electrodes were inserted into the mass of the teres major 3 cm lateral and 3 cm superior to the inferior angle of the scapula, and into the latissimus dorsi muscle 4 cm inferior to the inferior angle of the scapula. Subjects were directed through a series of motions of the shoulder and arm: (1) standing at rest with the arms hanging at the side; (2)

medial rotation; (3) lateral rotation; (4) abduction; (5) adduction; (6) flexion; and (7) extension. These were first done without resistance and then against a resistance force that either allowed or prevented completion of the arc of the motion.

The teres major had no electrical activity in motions without resistance. But against active resistance, it consistently showed electrical activity during medial rotation, adduction and extension in both the static and the dynamic exercises. The latissimus dorsi had similar activity during both static tension and resisted motion; without resistance, in five of the seven acceptable subjects, latissimus dorsi had activity during medial rotation, adduction and extension.

The more precise and sensitive techniques used in this study appear to resolve the controversy over the functions of teres major. The key to the solution seems to be whether a movement or an attempted movement is always active. If added resistance is lacking, free movement of the shoulder joint in all its directions does not recruit the teres major although it usually recruits its close relation, latissimus dorsi; when the arm is behind the back, teres major activity does appear "without resistance" during (hyper)extension and in adduction (Jonsson *et al.*, 1972).

As far as latissimus dorsi alone is concerned, there is little doubt of the powerful extensor activity of this muscle and no one has found reason for doubting it during EMG. This is not the case for medial rotation. While an early study by Scheving and Pauly (1959) suggested the muscle was the essential medial rotator, more recent investigations by de Sousa *et al.* (1969) suggest that this view is fallacious and came from the concomitant adductor effect. Thus, de Sousa *et al.* clearly deny any rotatory function for latissimus dorsi, while our study confirmed earlier views.

Jonsson *et al.* (1972) have confirmed the common opinion that latissimus dorsi and the sternocostal part of pectoralis major depress the humerus.

DELTOID. Scheving and Pauly (1959) found that the three parts of deltoid are active in all movements of the arm, as did Yamshon and Bierman in an earlier and less sophisticated study (1949). In flexion and medial rotation, the anterior part is more active than the posterior; in extension and lateral rotation, the posterior is the more active; and in abduction the middle part is the most active. Scheving and Pauly suggest that, although one part of deltoid may act as the prime mover, the other parts contract to stabilize the joint in the glenoid cavity. They further recommend the inclusion of deltoid with the four rotator cuff muscles as stabilizers of the joint, but our work (reported below) does not endorse their recommendations.

Wertheimer and Ferraz (1958) and Shevlin *et al.* (1969) found that the anterior part of deltoid shows its principle action in forward flexion of the shoulder joint, but also participates in elevation and (slightly) in abduction of the arm. Deltoid does not participate in medial rotation. The intermediate portion acts strongly in abduction and elevation of the arm and also participates slightly in flexion and extension. The posterior part has its principal action in extension, but the action is inconstant and slight in abduction and elevation of the arm. Its participation in lateral rotation is minimal, being practically absent.

Hermann (1962), in a careful emg study of shot-putting, found that ideally the anterior deltoid is active during the entire manoeuvre. The greatest contracting occurs during the thrust phase of the shoulder and arm but before the shot is released from the hand. The middle fibres come into play—and into very strong action—after the thrust is initiated. The posterior fibres play no important rôle until the moment just as the shot leaves the hand.

Other applications of electromyography of the shoulder region in sports work situations were reported by Kamon (1966), Kamon and Gormley (1968) and Hinson (1969).

BICEPS BRACHII. Our finding of slight action of the biceps during flexion of the shoulder joint and *nil* activity during abduction with the arm medially rotated (Basmajian and Latif, 1957) confirmed the accepted teaching. More recently, Furlani (1976) agreed, but he added some useful details. During flexion *with resistance* (and with the elbow straight) both heads of biceps are always active. During abduction, resistance again recruits activity in both heads; the opposite movement, adduction against resistance, recruits activity in half of the short heads and nothing in the long head. The long head plays no rôle in rotatory movements of the shoulder, but the short head occasionally acts during medial rotation.

Prevention of Downward Dislocation of the Humerus

The part played by various muscles during movements of the shoulder joint has been the subject of investigation and argument for more than a century. Even though most of the important questions on movements in the shoulder area have been answered, little reliable information had been available regarding the rôle of such muscles in *maintaining joint stability*. In particular, the mechanism preventing downward dislocation or subluxation of the shoulder joint has not been adequately explained—indeed, it has been largely ignored. Cotton in 1921 and Fairbank in 1948, considering the matter in connection with fractures of the humeral neck, both assigned the greatest importance to the vertically running scapulohumeral muscles, for example, deltoid and biceps. During some incidental studies of the region, we were surprised to find the exact opposite. Therefore, to clarify the part played by the muscles and the capsule of the joint in preventing downward dislocation of the vertical or adducted humerus, the following two types of systematic investigation were performed (Basmajian and Bazant, 1959): (1) an electromyographic study of the deltoid, supraspinatus, infraspinatus, biceps and triceps of a series of young men, using multiple concentric-needle electrodes; and (2) a study of gross dissections of the shoulder joint.

Our findings do not support the hypothesis advanced by Cotton and endorsed by Fairbank. In fact, they completely disagree. It was apparent from the electromyographic results that the deltoid (the muscle one would expect to be especially active in preventing downward dislocation of the humerus) is inactive even with heavy pulls. Other muscles running vertically from the scapula to the humerus, particularly the biceps and the long head of triceps, are conspicuously inactive as well. Therefore, there now seems to be little, if any, reason to doubt that downward dislocation is prevented by the superior part of the capsule along with the supraspinatus (and to a lesser extent the posterior fibres of the deltoid). Strangely enough, these structures run in a horizontal and not in a vertical direction (fig. 10.3). Bearn (1961) confirmed the findings (in considerable detail) quite independently and using loads of 25 pounds.

The mechanism by which these horizontally placed structures succeed is dependent upon a well-known, but previously unexplained fact, namely, the

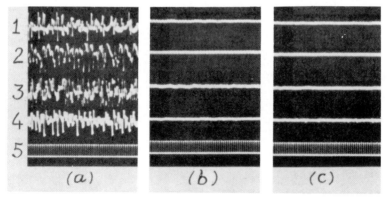

Fig. 10.3. EMGs, *a*, during abduction; *b*, unloaded arm hanging; and *c*, heavy downward pull applied to arm. Lines *1*, *2* and *3*—anterior, middle and posterior fibres of deltoid; line *4*—supraspinatus; line *5*—time marker: 10-msec intervals. (From Basmajian, 1961.)

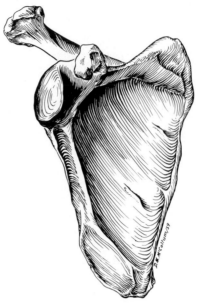

FIG. 10.4. Correct orientation of the right scapula viewed directly from in front at eye level. (From Basmajian and Bazant, 1959.)

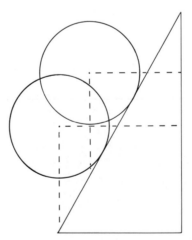

FIG. 10.5. Diagram to illustrate locking mechanism at shoulder. The farther down the slope the ball slides, the farther laterally it is displaced. If the lateral displacement can be prevented, the ball cannot move downward. (From Basmajian and Bazant, 1959.)

obliquity of the glenoid fossa. When the scapula is examined in its correct orientation, invariably it is found to face somewhat upward in addition to forward and laterally (fig. 10.4). It now appears that this slope of the glenoid fossa—particularly its lower part—plays an important rôle in preventing downward dislocation or subluxation. As the head of the humerus is pulled downward, it is of necessity forced laterally because of the slope of the glenoid fossa (fig. 10.5). If this lateral movement could be stopped, the result would be a stopping of the downward movement. The superior part of the capsule of the joint and the supraspinatus (as well as the posterior fibres of the deltoid) are so placed that they can—indeed they must—tighten to prevent the downward dislocation (fig. 10.6). Simple as this explanation may seem, it is dependent on our findings that (1) the vertically placed muscles definitely remain relaxed while (2) the supraspinatus (and posterior deltoid) become quite active and (3) the superior part of the capsule becomes taut. The very rareness of downward dislocation of the normal shoulder joint confirms the effectiveness of this locking mechanism.

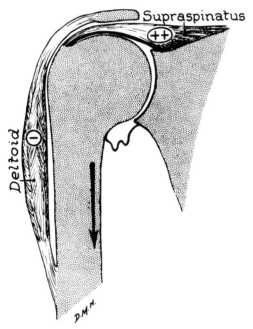

Fig. 10.6. Diagram contrasting presence of moderate activity in supra-spinatus with none in deltoid during loading in the direction of arrow (see text).

The ordinary effect of gravity on the unloaded arm is counteracted in many persons by the superior part of the capsule. In this area the coracohumeral ligament forms a real thickening and apparently this is an important function of the ligament. With moderate or heavy loads, the supraspinatus is called upon to reinforce the horizontal tension of the capsule; in some persons it is required even without a load. The posterior fibres of the deltoid, imitating in general the direction of the supraspinatus, must act in the same way but to a lesser extent.

The well-known subluxation that occurs in the humeral head of many patients after a stroke now can be explained and alleviated. The drooping shoulder region causes relative abduction of the humerus, unlocking the natural locking mechanism. Chaco and Wolf (1971) have gone further to demonstrate that in flaccid hemiplegia, actual supraspinatus weakness plays an important part in the subluxation. Recently Peat and Grahame (1977) have described an excellent method for emg evaluation of the shoulder region in hemiplegic patients.

The locking mechanism cannot operate when there is abduction of the humerus. As a result, the head can be easily subluxated when it is in the abducted position in the cadaver. Fairbank convincingly demonstrated that subluxation can be also produced in anesthetized normal men, but his roentgenograms of the subluxated shoulders show in each case a real degree of abduction of the humerus in relation to the position of the scapula. It would appear that in these unconscious subjects the subluxation was preceded by a drawing downward of the glenoid fossa (that is, relative abduction of the humerus and thus a neutralization of the locking mechanism). The muscles that prevent dislocation of the joint in its unstable position are the "rotator cuff" muscles—supraspinatus, infraspinatus, teres minor and subscapularis (and perhaps teres major and other muscles spanning the joint).

Elbow Flexors

Although they can be felt quite easily, the biceps brachii, the brachialis and the brachioradialis have not been fully understood as far as their integrated functions are concerned. Furthermore, both Duchenne and Beevor introduced and perpetuated errors that require correction. And again, although it is obvious that these muscles are primarily concerned with flexion, a variety of theories have obscured the rôle played by each during flexion and other movements of the elbow. It is probably fair to say that the few formal studies of these muscles (both with and without objective techniques) have been either too sharply circumscribed in approach or too highly generalized and deductive.

For these reasons, we made a detailed electromyographic study of both heads of the biceps, the brachialis and the brachioradialis in a long series of young adults (Basmajian and Latif, 1957).

A careful consideration of the time sequence of activity indicated there is a completely random selection in the sequence of appearance and disappearance of activity in these muscles. For example, during slow flexion with the forearm supine, all the muscles that showed any activity began this activity simultaneously in about half of the subjects. However, in only about one-quarter of them did the activity end simultaneously. In a small number in whom the activity did not begin simultaneously, it did, however, end simultaneously.

Any of the muscles that was to show activity during a movement functioned first or last in an unpredictable fashion, i.e., there was no set pattern. In the same way, the activity ceased in the muscles in an unpredictable order. Moreover, the muscle that was to show the greatest activity in individual subjects only occasionally began first and ended last.

Our results provided convincing evidence that in the movements produced by the biceps, the brachialis and the brachioradialis there is a fine interplay between them; this was to be expected. What is more striking, however, is the wide range of response from any one muscle in our series. Thus, although a general trend may be described, there is rarely any unanimity of action. For example, the brachialis is generally markedly active during quick flexion of the supine forearm, but in one of our subjects it was completely inactive.

These findings re-emphasize the general biological principle that there is a range of response in any phenomenon. It would seem that anatomists and clinicians have taken too little heed of this wide range of individual pattern of activity in something even so simple as elbow-flexion.

In our study, the long head of the biceps showed more activity than the short head in the majority of the subjects during slow flexion of the forearm, during supination of the forearm against resistance and during flexion of the shoulder joint (although there was little difference between the activity of the two heads during isometric contraction and during extension of the elbow). Sullivan, Mortensen, Miles and Greene (1950) reported similar findings in a more limited but fine experimental series with surface electrodes during flexion only.

The biceps is generally active during flexion of the supine forearm under all conditions and during flexion of the semiprone forearm when a load (of about 1 kg) is lifted (fig. 10.7). However, with the forearm prone, in the majority of instances the biceps plays little if any rôle in flexion, in maintenance of elbow flexion and in antagonistic action during extension, even with the load. Beevor (1903, 1904) stated that if the forearm is in supination the

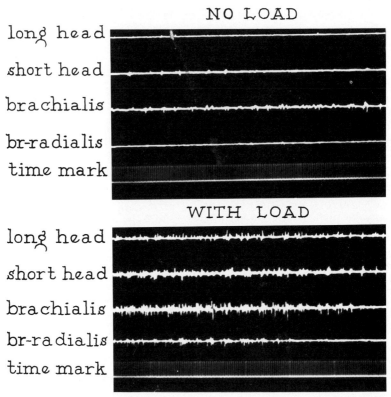

Fig. 10.7. EMGs of the two heads of biceps, the brachialis and the brachioradialis during slow flexion of semi-prone forearm.

biceps acts during flexion when there is a resistance of as little as 120 g, but that in a position of complete pronation it does not act until the resistance is at least 2 kg. The results of our emg study support Beevor's observations in regard to flexion.

Bankov and Jørgensen (1969) convincingly showed that both in isometric and in dynamic contractions the maximum strength in the elbow flexors is smaller with the pronated compared with the supinated forearm. The integrated EMG of biceps was considerably reduced in the pronated forearm. Confirmatory studies were done by Settineri and Rodriguez (1974).

The biceps is usually described as a supinator of the forearm. In our study, no activity in the muscle was demonstrated in the majority of the subjects during supination of the extended forearm through the whole range of movement except when resistance to supination was given. However, activity was observed in all the subjects when supination was strongly resisted. It follows that, generally, the biceps is not a supinator of the extended forearm unless supination is resisted.

It is necessary to explain why the biceps does not ordinarily supinate the extended forearm. It appears that because of the tendency of the biceps to flex the forearm, it is reflexly inhibited. Thus the extended position of the forearm is maintained while the supinator does the supinating. On the other hand, when supination is resisted, the biceps comes into strong action, and we have noted that usually the previously extended forearm is partly flexed as well during supination against resistance. For actions of biceps at the shoulder, see page 194.

The brachialis has been generally and erroneously considered by anatomists to be a muscle of speed rather than one of power because of its short leverage. We found it to be a flexor of the supine, semi-prone and prone forearm in slow or quick flexion, with or without an added weight. McGregor (1950) has correctly described it as a "flexor *par excellence* of the elbow joint." Apparently brachialis is called upon to flex the forearm in all positions because the line of its pull does not change with pronation or supination.

Maintenance of specific flexed postures of the elbow, i.e., isometric contraction, and the movement of slow extension when the flexors must act as antigravity muscles both generally bring the brachialis into activity in all positions of the forearm. This is not the case with the other two flexor muscles. Thus, the brachialis may also be designated the "workhorse" among the flexor muscles of the elbow.

A short burst of activity is generally seen in all the muscles during quick extension. This activity can hardly be considered antagonistic in the usual sense. Rather, it may provide a protective function for the joint. Biceps is particularly active during quick extension with an added load.

In the past, the brachioradialis has been described as a flexor, acting to its best advantage in the semi-prone position of the forearm. We found in most subjects that the brachioradialis does not play any appreciable rôle during maintenance of elbow flexion and during slow flexion and extension when the movement is carried out without a weight. When a weight is lifted during flexion, the brachioradialis is generally moderately active in the semi-prone or prone position of the forearm and is slightly active in the supine position. There is no comparable increase in activity with the addition of weight during maintenance of flexion and during slow extension. We also found that in most instances the brachioradialis is quite active in all three positions of the forearm during quick flexion and extension. It follows that the muscle is a reserve for occasions when speedy movement is required and when weight is to be lifted, especially in the semi-prone and the prone positions. In the latter position, the biceps usually does not come into prominent action. Furthermore, the activity of the brachioradialis in speedy movements is related to its function as a shunt muscle (see p. 152).

The brachioradialis has been described since Duchenne's day as a supinator of the prone forearm and a pronator of the supine forearm acting to the semi-prone position in both cases. Our study showed that it neither supinates nor pronates the extended forearm unless these movements are performed against resistance. Here, at most, brachioradialis acts only as an accessory muscle, coming into action when strength is required to supinate or to pronate the forearm. More probably, it acts only as a synergist.

Our observations strongly suggest that the biceps, the brachialis and the brachioradialis differ in their flexor activity in the three positions of the forearm (prone, semi-prone and supine). However, all three muscles act maximally when a weight is lifted during flexion of the semi-prone forearm. The semi-prone position of the forearm has been described as the natural position of the forearm, the position of rest and the position of greatest advantage for most functions of the upper limb (Basmajian, 1955a).

PRONATOR TERES AND ELBOW FLEXION. Our investigations have shown that pronator teres contributes to elbow flexion only when resistance is offered to the movement (Basmajian and Travill, 1961). It shows no activity during unresisted flexion whether the forearm is prone, semi-prone or supine (see fig. 10.10 on p. 208).

Before leaving the flexor muscles of the elbow, we might note that Wells and Morehouse (1950) believe that biceps and triceps (and latissimus dorsi) act as cocontractors in exerting a pull such as on an aircraft control stick. They found that the extent of the "contribution" each muscle makes is altered when the position of the arm is changed (see p. 97). When the arm is pulling in an extended position, the biceps dominates the action, but in the flexed or intermediate position, triceps is brought strongly into action. These findings can hardly be valid evidence of cocontraction because of many complicating factors in the set-up. However, Wells and Morehouse did show that, as far as muscular dynamics is concerned, "the best arm position of a pilot seated in a conventional upright position and operating a control stick is one which is intermediate between flexion and extension."

Triceps Brachii

My colleague, Anthony Travill (1962) found that the long head of triceps is surprisingly quiescent during active extension of the elbow regardless of the position of either the subject or his limb. The medial head, however, is always active and appears to be the prime extensor of the elbow; meanwhile, the lateral head shows some activity as well. Against resistance, the lateral and long heads are recruited. Therefore, we might compare the medial head of triceps to the brachialis which we noted above to be the workhorse of the elbow flexors; it is the workhorse of the extensors. The lateral and long heads are reserves for extension just as the two heads of biceps are reserves for flexion. Isometric contractions against immovable resistances at 60°, 90° and 120° produce a rising amount of myoelectric activity (Currier, 1972). There is also a considerable amount of decrease during learning acquired with repetition (Payton, 1974).

Travill confirmed Duchenne's view that, of the two superficial heads, the long head is the less powerful during extension. This is probably due to the lack of fixation of the scapular origin and the necessity of adducting the shoulder with the forearm either flexed or extended. Too strong a contribution from the long head would tend to give extension during adduction of the arm.

Pronation and Supination of Forearm

PRONATORS. Until recently no accurate authoritative information existed on the relative functional rôles of the two pronator muscles, although their gross anatomy is adequately described in the standard textbooks. Our investigation of the pronator teres and the pronator quadratus in a series of volunteers revealed that the few remarks on function that are presently available in books are largely misleading (Basmajian and Travill, 1961).

In each subject three groups of records were made from needle electrodes (fig. 10.8). In the first group, the elbow was kept in the extended, fully supported position on the table top (fig. 10.9, α). In the second group, the elbow was flexed to a right angle, with the forearm vertical and the arm and elbow supported (fig. 10.9, β). In the third group, the elbow was flexed to an acute angle while it was still fully supported (fig. 10.9, γ). In each of these three groups of tests, records from the two muscles were made during the following movements and positions: (1) slow pronation from the comfortable supine position to the fully prone position, (2) fast pronation through the above range, (3) "hold" in the fully prone position, (4) slow supination through the whole range to full (forced) supination and (5) fast supination through the above range.

Both pronator quadratus and pronator teres are active during pronation, the consistent prime pronating muscle being the pronator quadratus (fig.

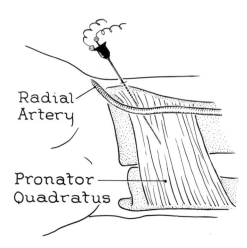

Radial Artery

Pronator Quadratus

Fɪɢ. 10.8. Diagram of electrode placement in pronator quadratus. (From Basmajian and Travill, 1961.)

10.10). This is true irrespective of the positions of the forearm in space or the angulation of the elbow joint. In general, the pronator teres is called in as a reinforcing pronator whenever the action of pronation is rapid. Similar reinforcement occurs during pronation against resistance.

Whether pronation is fast or slow, the activity in the pronator quadratus is markedly greater than that in the pronator teres. This observation conflicts with the opinions offered in a number of the internationally recognized textbooks (Steindler, 1955; Johnston, Davies and Davies, 1958; Hamilton and Appleton, 1956; Lockhart, Hamilton and Fyfe, 1959). Some writers sit squarely on the fence and suggest that both muscles pronate without preference. Only two books suggested before 1960 that the pronator quadratus is the main pronator (Hollinshead, 1958; Basmajian, 1960).

A number of authors have likewise expressed the view that the pronator teres displays its greatest activity either during midflexion of the elbow (Steindler, 1955; Lockhart, 1951) or during full extension (Hamilton and Appleton, 1956; Hollinshead, 1958). However, we were surprised to find that, regardless of whether the pronating action is carried out swiftly or slowly, the angle of the elbow joint has no bearing on the amount of activity of the pronator teres.

During slow supination there is no activity whatsoever in either of the pronators—though some have suggested that the deeper layer of the pronator

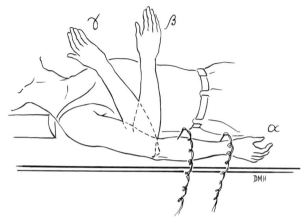

FIG. 10.9. Diagram of primary positions of limb during three groups of tests (see text). (From Basmajian and Travill, 1961.)

quadratus acts as a supinator. De Sousa *et al.* (1957, 1958), using a different approach, have independently arrived at the same conclusion.

During fast supination, there is negligible activity in the pronators. This is rather surprising in view of earlier work, mentioned above, on the electro-myographic activity of the biceps and triceps during flexion and extension of the elbow (Barnett and Harding, 1955; Basmajian and Latif, 1957). In those muscles, a sharp burst of antagonistic activity occurs during fast movements, this activity being thought to be the manifestation of a protective stretch reflex.

De Sousa and his colleagues (1957, 1958, 1961) in São Paulo, Brazil, showed that pronator quadratus is a pronator only. They agree with us that it participates in normal pronation, but there is some difficulty in reconciling with our own experience their finding that there is little early activity in pronator quadratus during the early stages of pronation. Their explanation of the difficulty, which indeed seems to be a valid one, is that at the beginning of unresisted pronation the natural elastic recoil from complete supination is quite enough. Our "pronations" were begun from the comfortably supine, not the fully supine, position. In any case, pronator teres acted no earlier than quadratus in their series. Moreover, the flexor carpi radialis, brachior-adialis and extensor carpi ulnaris were shown to have no pronating function.

SUPINATORS. Supination of the forearm in man is undeniably of fundamental importance and yet, all too often, the parts played by the chief supinator

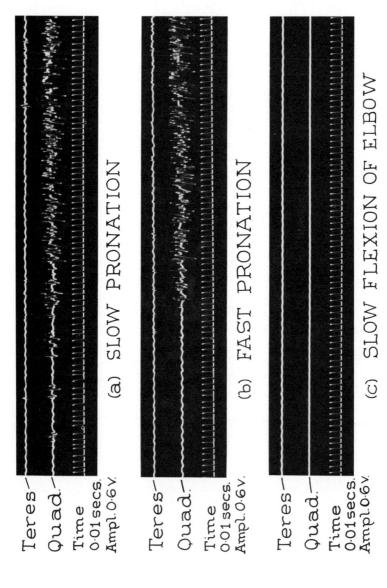

Fig. 10.10. Typical EMGs of *a*, slow pronation; *b*, fast pronation; and *c*, slow flexion of elbow. (From Basmajian and Travill, 1961.)

muscles are either ignored or taken for granted. Throughout the textbooks, there is no thread of consistency and the truth appears to be so tangled with hopeful guesses it cannot be recognized. For example, only in recent years did we confirm Beevor's strong insistence in 1903 that brachioradialis (known for years as "supinator longus") is not a true supinator (see p. 204).

Therefore, we carried out an electromyographic study of the supinator and the biceps brachii in a series of young volunteers (Travill and Basmajian, 1961). This study was complementary to and, in part, overlapped by that on the pronator muscles outlined above. A needle electrode was inserted into the middle of each of the following four muscles: the supinator, the biceps brachii, the pronator quadratus and the pronator teres, the pronators being tested simultaneously as controls only.

Two series of recordings were made from each subject: the first series with the elbow extended, and the second with the elbow flexed to 90°. With each of these two positions recordings were made during: (1) the movement of supination from full pronation, (2) the "hold" position of maximum supination and, finally, (3) the return movement of pronation to the original comfortably supine position. Recordings were made during slow movements, fast movements and forceful movements against the resistance offered by the grip of an observer.

Slow unresisted supination, whatever the position of the forearm, is brought about by the independent action of the supinator (fig. 10.11). Similarly, fast supination in the extended position requires only the supinator; but fast unresisted supination with the elbow flexed is assisted by the action of the biceps. All movements of forceful supination against resistance require the cooperation of the biceps in varying degrees.

This last-mentioned cooperative activity of the supinator and the biceps during resisted supination, especially when the elbow is flexed, has never been seriously questioned in the past. Primacy of the supinator during the unresisted movement has not, however, received such universal acceptance, even though it was first suggested by Duchenne and broadly hinted at by Bierman and Yamshon (1948). For example, Steindler (1955) and Gardner, Gray and O'Rahilly (1960) emphasize only the power of the biceps against resistance; this power is undeniable.

During supination, the action of supinator is augmented by that of the biceps. This was similar to our earlier findings for pronation, where pronator quadratus is augmented by the pronator teres when required.

Both the supinator muscles are completely relaxed during pronation (slow, fast or resisted). This is again similar to our findings for pronation, where

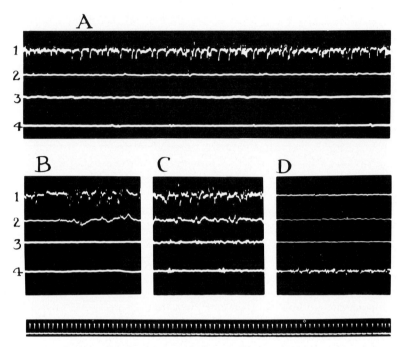

Fig. 10.11. EMGs during *A*, slow supination; *B*, fast supination; *C*, forceful supination against resistance; and *D*, pronation of the forearm. Channel *1*, supinator; *2*, biceps brachii; *3*, pronator teres; and *4*, pronator quadratus. Time marker: 10-msec intervals. (From Travill and Basmajian, 1961.)

complete relaxation of the pronator quadratus and the pronator teres during supination is the rule.

The "hold" or static position of supination depends on activity in supinator for the maintenance of the supine posture. Against added resistance, however, the biceps always becomes active. The movement of supination is initiated, and mostly maintained, by the supinator; it is only assisted by the biceps as needed to overcome added resistance.

ANCONEUS. This small muscle has been the subject of controversy since Duchenne suggested a special rôle for it, that of abduction of the ulna during pronation of the forearm. On the other hand, one never sees cases of localized paralysis of anconeus from which its rôle as an abductor could be confirmed. Furthermore, pronation is equally efficient with or without abduction. However, some simple observations reveal that the usual way in which most persons carry out pronation includes the slight abduction of the ulna. Thus the hand can be "turned over" without its shifting away from its original postion. The anconeus is in the ideal position to perform this secondary movement.

Under the direction of Prof. de Sousa of São Paulo, Brazil, Da Hora (1959) of the University of Recife re-investigated anconeus morphologically and electromyographically. From the latter study he concluded that anconeus was always active during extension of the elbow. What is surprising, however, is his finding that it is active in both pronation and supination whether these movements are resisted or unresisted. He also reported activity during flexion of the elbow, especially against resistance.

Ray, Johnson and Jameson (1951) reported that anconeus was very active during the whole of pronation. However, Travill (1962) concluded from his emg study of this muscle that it is only active when resistance is offered to the movement—but it is quite as active during resisted supination. All authors confirm the classical view that anconeus is most active during extension of the elbow.

Because bipolar fine-wire electrodes and other refinements in electromyography might solve the enigma and because orthopedic surgeons continue to embarrass us with questions on the anconeus, we undertook a formal electromyographic study of the anconeus in 10 subjects as well as their related muscles for comparison (Basmajian and Griffin, 1972). Our special bipolar fine-wire electrodes were inserted into the middle of the bellies of the anconeus, supinator, triceps (medial head) and pronator teres. An electrogo-

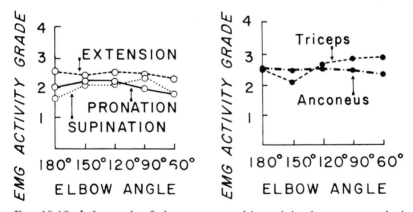

FIG. 10.12. *Left*, graph of electromyographic activity in anconeus during supination, pronation and extension of the elbow. *Right*, graph comparing electromyographic activity in anconeus and triceps (medial head) during extension of elbow. (From Basmajian and Griffin, 1972.)

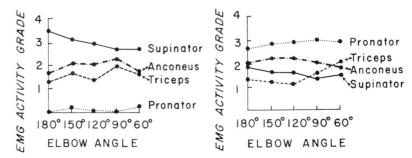

FIG. 10.13. *Left*, graph comparing activity in four muscles during supination of the elbow at various set angles of the elbow. *Right*, graph comparing activity in four muscles during pronation of the elbow at various set angles of the elbow. (From Basmajian and Griffin, 1972.)

niometer monitored and recorded elbow angle, which was held constant at a series of angles (fig. 10.12) by a special holding device.

The electromyographic results for supination, pronation and extension in anconeus alone are summarized in figure 10.12, *left*. It shows that extension recruits somewhat more than moderate activity while pronation and supination elicit moderate activity in the anconeus through most of the range of elbow motion.

Comparing the anconeus with the medial head of triceps shows similarity in activity levels with some mild variations in detail (fig. 10.12, *right*).

During supination the (expected) marked activity in supinator and relative quiet of the pronator teres are accompanied by moderate activity in the anconeus and slight to moderate activity in the triceps. During pronation, moderate activity in the anconeus is accompanied by slight to moderate activity in the supinator and triceps (see fig. 10.13).

A definite function for the anconeus is still not proved, but its moderate activity during both pronation and supination (especially at their extremes), also described by Da Hora and Travill, gives little or no support to Duchenne's proposed function of ulnar deviation during pronation. Rather, it suggests some form of joint stabilization. This synergy is apparently shared by the closely related medial head of triceps and perhaps the supinator (Basmajian and Griffin, 1972).

Forearm-Hand Synergies

All forearm muscles contribute to hand function, including the supinators and the pronators; they contract to various degrees during finger movements (see page 219).

Chapter *11*

Wrist, hand and fingers

T HE two radial extensors of the wrist have been the special subject of electromyographic examination by Tournay and Paillard (1953). They showed that during pure extension of the wrist the extensor carpi radialis brevis is much more active than the longus whether the movement is slow or fast. Actually, except with fast extension, the longus was essentially inactive. However, the rôles of the two muscles are completely reversed during prehension or fist-making; now the longus is very active as a synergist. This appears to be an extremely important observation. The two muscles are both quite active during abduction of the wrist, as one would guess from their positions.

The work of Gellhorn (1947) and of Dempster and Finerty has already been described in Chapter 4 (p. 97).

Bačkdahl and Carlsöö (1961) found that during extension of the wrist there is a reciprocal innervation between extensors and flexors. Extensores carpi radiales (longus et brevis) and extensor carpi ulnaris, as well as the extensor digitorum, work synchronously; none seems to be the prime mover. This was confirmed by McFarland *et al.* (1962). During forced extreme flexion of the wrist, there is a reactive cocontraction of the extensor carpi ulnaris, apparently to stabilize the wrist joint; this does not occur with the extensor digitorum and extensores carpi radiales.

During flexion of the wrist, the flexores carpi radialis, ulnaris and digitorum superficialis act synchronously, according to Bäckdahl and Carlsöö—none is the prime mover. Flexor digitorum profundus plays no rôle. Two possible muscles in the antagonist position (the radial extensors of the wrist and the extensors of the fingers) are passive, even in extreme flexion of the wrist; but the extensor carpi ulnaris shows marked activity as an antagonist.

In abduction and adduction, the appropriate flexors and extensors act reciprocally as one might expect, the antagonist muscles relaxing. Extensor digitorum contracts during abduction (radial abduction), but Bäckdahl and Carlsöö found that this contraction is not limited to the radial part of the muscle. Apparently this last activity has a synergistic function. McFarland *et al.* also found activity in extensor digitorum during extremes of abduction and adduction of the wrist; moreover, the flexor digitorum superficialis was active too. This group of investigators also emphasize the uniform occurrence of antagonist activity in the flexors when the wrist is extended and the metacarpophalangeal joints hyperextended.

Fingers

Increasing scientific attention to the movements of the fingers has appeared. Person and Roshtchina (1958) of Moscow were concerned with the nervous mechanisms that enable a person to perform isolated movements of a single finger. They studied the common flexors and extensors of the fingers during rhythmic flexions and extensions. When all the fingers are moving simultaneously the activity of the "antagonist" muscles conforms to the principle of

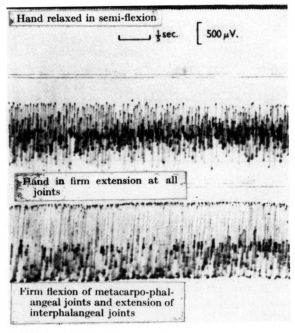

FIG. 11.1. EMGs of a lumbrical muscle in several positions of the hand. (From Backhouse and Catton, 1954.)

reciprocal inhibition (p. 94). When only the little finger or ring finger is moving while the others are extended, the extensor is active during both extension and flexion. When one finger is moved while the others are kept bent, the flexor is active during both movements.

If a single finger moves, the "antagonist" must remain active to immobilize the other fingers. However, if the other fingers are held immobile by an observer, there is no activity in the antagonist muscle.

Backhouse and Catton (1954) studied the lumbricals of the hand and proved conclusively that they are only important in extension of the inter-phalangeal joints reinforcing the action of the extensor digitorum and inter-ossei. They agree in large measure with the Australian anatomist, Sunderland (1945), who suggested that the importance of lumbrical-interosseus extension at the interphalangeal joints is in the prevention of hyperextension of the proximal phalanx by the extensor digitorum. This preventive action allows a more efficient pull on the dorsal expansion which extends the interphalan-geal joints (figs. 11.1 to 11.3).

Metacarpophalangeal flexion is performed by a lumbrical only when the interphalangeal joints are extended. Backhouse and Catton concluded that a lumbrical has no effect on rotation or radial deviation of its finger during opposition with the thumb (as first suggested by Braithwaite, Channell, Moore and Whillis, 1948, in their classic morphological study).

$\left[\right.$ 250 μV.

Forced opposition against thumb in extension
compared with

Muscle relaxed   Muscle fully active

FIG. 11.2. EMGs of second lumbrical muscle while middle finger is in different positions. (From Backhouse and Catton, 1954.)

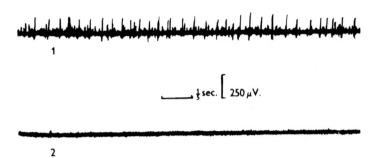

FIG. 11.3. EMGs of lumbrical. *1*, relaxed in extension with ulnar deviation, and *2*, radial deviation against resistance. (From Backhouse and Catton, 1954.)

At Washington University in St. Louis, Lake (1954, 1957) made a simultaneous study of the extensor digitorum (communis), flexor digitorum superficialis (or sublimis) and the second and third dorsal interossei. She found that the extensor digitorum begins or increases its activity with the inception of interphalangeal (IP) joint extension regardless of the position of the metacarpophalangeal (MP) joints. During extension or hyperextension of the MP joint, extensor digitorum alone was active.

Flexor digitorum superficialis is active during flexion of the middle phalanx (proximal IP joint), and it is active in flexion of the MP joint providing the next distal joint is stabilized. Surprisingly, the superficialis is active during rapid, forceful IP extension regardless of the position of the MP joint.

The interossei in Lake's research were found to be markedly active from the very onset of flexion of the MP joint even with moderate effort. IP joint position was of no consequence in this. The interossei also showed activity before the onset of visible extension in either the proximal or distal IP joint. In the case of extension of the proximal joint, the distal joint had to be extended simultaneously, but the position of the MP joint was not important.

Meanwhile, Brown, Long and Weiss (1960) and Long *et al.* (1960, 1961, 1970) at Case Western Reserve University in Cleveland have completed a comprehensive and ingenious study of the hand musculature, the results of which add to our understanding of its kinesiology. Using multiple, indwelling, pliable wire electrodes, this group has shown that the interossei of the hand act as MP flexors only when their other action of IP extension does not conflict. Therefore they act best and strongest when combined MP-flexion—IP-extension is performed. During all IP extension, the intrinsic muscles of the hand contract regardless of MP posture.

They concluded that the long tendons of the fingers provide the gross motion of opening and closing of the fist at all the joints simultaneously.

However, the intrinsic muscles perform their major function during any departure from this simple total opening or closing movement. Thus, they are the primary IP extensors while the MP joints are flexing.

Long and Brown (1962, 1964) confirmed and expanded on their preliminary findings for the lumbricals. In general, these findings confirm those of Backhouse and Catton (see above). The lumbricals are silent during total flexion of the entire finger, but are very active whenever the proximal or distal IP joints are extended actively or are held extended while the MP joint is flexed actively. The lumbricals can be kept very quiet during MP movements in any direction by keeping the IP joints fully flexed.

To summarize, Long and Brown conclude that the interossei and the lumbrical of one finger do not form a functional unit—they act discretely. The interossei participate in IP extension only when the MP joint is either flexing or held flexed. The lumbrical always takes part in interphalangeal extension. It shows a crescendo of activity throughout the movement, reaching a peak at full extension. This suggests that its function may include prevention of hyperextension at the MP joint. Neither the interossei nor lumbrical of the middle finger acts during closing of the full hand, suggesting that in this total movement they are not synergists.

The activity of the long extensors and flexors occurs in special sequences. Extensor digitorum acts during MP extension—in both the movement and the "hold" position. But it is also active in many flexion movements of that joint, apparently acting as a brake. The flexor profundus is the most consistently active flexor of the finger. Joined by the flexor superficialis, the profundus may act as a flexor of the wrist joint also. The superficialis has its maximal action when the hand is being closed or held closed without flexion of the distal IP joint.

Power Grip and Precision Handling

The important paper by Long *et al.* (1970) is the bed-rock of our recent emg confirmation and clarification of the fine functioning of the human hand. One can do no better than to quote or paraphrase their main conclusions and to recommend consulting the original to readers who want further details.

Power Grip

In power grip the *extrinsics* provide the major gripping force. All of the extrinsics are involved in power gripping and are used in proportion to the desired force to be used against the external force.

The major *intrinsic muscles* of power grip are the interossei, used as phalangeal rotators and metacarpophalangeal flexors. The lumbricales, with exception of the fourth, are not significantly used in power grip. The thenar muscles are used in all forms of power grip except hook grip.

Precision Handling

In precision handling, specific extrinsic muscles provide gross motion and compressive forces.

In rotation motions the interossei are important in imposing the necessary rotational forces on the object to be rotated; the motion of the metacarpophalangeal joint which provides this rotation is abduction or adduction, not rotation of the first phalanx. The lumbricales are interphalangeal joint extensors as in the unloaded hand, and additionally are first phalangeal abductor-adductors and rotators.

In translation motions towards the palm, the interossei provide intrinsic compression and rotation forces for most efficient finger positioning; the lumbricalis is not active. Moving away from the palm the handled object is driven by interossei and lumbricales to provide intrinsic compression and metacarpophalangeal-joint flexion and interphalangeal-joint extension.

The thenar muscles in precision handling act as a triad of flexor pollicis brevis, opponens pollicis and abductor pollicis brevis to provide adduction across the palm, internal rotation of the first metacarpal and maintenance of web space depth.

The adductor pollicis is used in specific situations when force is required to adduct the first metacarpal towards the second.

Pinch

In pinch, compression is provided primarily by the extrinsic muscles. Phalangeal rotational position is adjusted by the interossei and perhaps also by the lumbricales. Compression is assisted by the metacarpophalangeal-joint flexion force of the interossei and flexor pollicis brevis and by the adducting force of the adductor pollicis. The opponens assists through rotational positioning of the first metacarpal (Long *et al.*, 1970).

Brandell (1970) used the manipulation of a ruler, pencil and paper to determine the fine function of the index musculature. He found that the first dorsal interosseus muscle was the most consistently active when the ruler was carried with the three joints of the index finger flexed, but the flexor digitorum superficialis and extensor digitorum muscles appeared to play an important rôle in the maintenance of the grip when the ruler was carried with the interphalangeal joints of the index finger straightened.

Of the five subjects Brandell tested, one had practised the procedure over a period of several weeks before the combined records were made, while the performance of the remaining four was entirely unrehearsed. The methods used by the trained and untrained subjects to grasp and release the ruler involved contrasting uses of the index finger and its musculature. The straight, or straightening, index finger, which was correlated with flexor digitorum superficialis activity, performed an unexpectedly important rôle in the basic phases of ruler manipulation.

Forearm-Hand Synergies

In a comprehensive multi-channel study with fine-wire electrodes, Sano *et al.* (1977) of Tokyo revealed the integrated rôle of forearm and finger muscles during movements of fingers in different postures. Some forearm muscles participate as synergists early in a voluntary finger motions, others only late in the movement. During movements of the index finger with various forearm positions the supinator muscle appeared to take part in stabilization of the proximal radio-ulnar joint and the pronator quadratus in that of the distal radio-ulnar joint (especially when the forearm is pronated).

Sano *et al.* found anconeus muscle active in all movements of the index, apparently for stabilization of the humero-ulnar joint. Flexor carpi radialis, known to be a synergist during finger extension, also is active during extension. Both in wrist extension and in index-finger movements, extensor carpi radialis brevis is more active than the longus. These muscles are apparently as active as the extensor digitorum itself, according to Sano *et al.* They emphasized the complex activity of all the forearm synergists during unrehearsed normal movements of the digits.

Deformities of Fingers

In the rheumatoid hand with its developing deformities, Backhouse (1968) and Swezey and Fiegenberg (1971) have shown how electromyography of intrinsic hand muscles reveal the rôle of inappropriate muscle actions.

Thenar and Hypothenar Muscles

The electromyography of the thenar and hypothenar muscles has been neglected until very recently, possibly because of the close packing of these small muscles. The French pioneers, Tournay and Fessard (1948), and Weathersby (1957) made useful, but brief, preliminary reports; this was followed by a longer report by Weathersby *et al.* Otherwise, no substantial work appeared until 1965 when Forrest and I published a systematic study with fine-wire electrodes; this paper will be heavily drawn upon in the subsequent paragraphs.

The report of Tournay and Fessard was concerned with general phenomena rather than with particular actions of the thumb muscles. Weathersby found with surface and needle electrodes that each of the thenar muscles was involved to some extent in most of the movements of the thumb. The abductor pollicis brevis contracts strongly during opposition and flexion of the thumb as well as in abduction. The opponens shows strong activity in abduction and flexion of the metacarpal as it does in opposition. The flexor pollicis brevis shows considerable activity in opposition as well as flexion, and in adduction. The adductor pollicis (which is, of course, not properly a thenar muscle) is active in adduction and opposition, and, to a slight extent in flexion of the thumb.

It is interesting to note that Sala (1959) of Pavia, Italy, used electromyography combined with nerve stimulation to determine the innervation of the muscles of the thenar eminence. He found that one out of every four flexor brevis muscles was exclusively supplied by the ulnar nerve, while in almost all the remainder both ulnar and median nerves shared in the supply. These findings are quite contrary to classical teaching. Opponens pollicis was supplied exclusively by the median nerve in two-thirds of cases only, most of the remaining one-third having a double supply. He reported frequent bilateral asymmetry; this is disturbing, to say the least, when one considers the practical difficulties introduced into clinical examinations.

Our study of the thenar and hypothenar muscles (Forrest and Basmajian, 1965) included 25 young adult subjects. A preliminary detailed study in the dissecting room gave a thorough knowledge of the anatomy, relationship and landmarks of the muscles concerned and provided an opportunity to practise and perfect the placement of electrodes. Their location in the middle of a muscle belly gives the truest picture of the general activity of that muscle. In the case of the flexor pollicis brevis, the electrode was inserted into superficial fibres corresponding to the portion of the muscle described by Jones (1942) as the superficial or external head; this arises from the flexor retinaculum and trapezium, passes along the radial side of the tendon of the flexor pollicis longus, and inserts into the radial sesamoid bone of the metacarpophalangeal joint and the base of the first phalanx of the thumb. In a later study, my colleagues Forrest and Khan (1968) showed clear differences between this superficial head and the deep head insofar as specific fine movements are concerned.

All movements began with the subject's hand in the rest position, in which, of course, there were no action potentials. The subject then moved his hand into a series of prescribed positions. Movement was performed slowly, each position was held for several seconds, and then the hand was returned slowly to the rest position. Positions of opposition were held either softly (with thumb and finger just touching), or firmly (with just enough pressure to resist the withdrawal of a sheet of paper from between the thumb and finger). Objects (a cup, glass or dowel) were held firmly, that is, securely, but with much less than maximum strength and effort. Although recordings were made during the entire movement, only those from the active positions are considered here, *viz*:

1. *Of the thumb:*

A. Extension (movement away from the radial side of the palm and index finger in the plane of the palm);

B. Abduction (movement away from the radial side of the palm and index finger in a plane 90° to that of the palm);

C. Flexion (flexion of the interphalangeal, metacarpophalangeal and carpometacarpal joints of the thumb in a plane parallel to that of the palm so as to scrape the ulnar side of the thumb lightly across the palm);

2. *Of the little finger:*

A. Extension (full extension of all the joints of the little finger);

B. Abduction (movement away from the ring finger in the plane of the palm);

C. Flexion (90° flexion of the little finger at the metacarpophalangeal joint with both interphalangeal joints almost fully extended);

3. *Eight positions of opposition* in which the thumb was held *softly* opposed to each finger in two ways—with the pad of the thumb to the lateral side of the bent finger near its tip, and with the thumb and finger tip-to-tip, roughly forming the shape of the letter O (this series began with position one, opposition to the side of the index finger—as in figure 11.4, *a*—and position two, opposition to the tip of the same finger, and then proceeded in a similar fashion to the long, ring and little fingers, ending with position eight, tip-to-tip opposition to the little finger—as in figure 11.5, *a*);

4. *The same eight positions* with *firm* opposition;

5. *Clasping* firmly a wooden dowel one inch in diameter;

6. *Holding*, in turn, 2 inches above the table, first a glass of water and then a

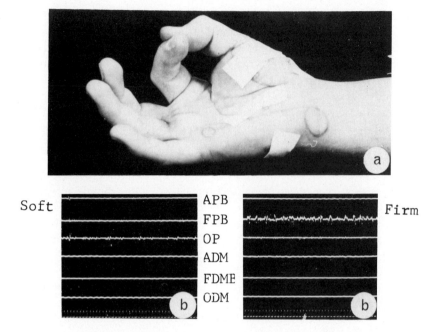

FIG. 11.4. *a*, opposition of the thumb to the side of the index finger (position one); *b*, electromyographic recordings during soft and firm opposition, respectively. (From Forrest and Basmajian, 1965.)

cup of water by the handle while the subject sat with elbow unsupported and flexed to 90°.

Postures of Thumb

During extension, only the opponens pollicis and abductor pollicis brevis showed appreciable activity, which was moderate on the average. During abduction, the same two muscles showed marked activity on the average whereas the activity of the flexor pollicis brevis was slight. During flexion, the mean activity of the flexor pollicis brevis was moderate to marked, but the opponens pollicis was only slightly active and the abductor pollicis brevis was essentially inactive.

The occurrence of equal levels of activity in both the abductor pollicis brevis and the opponens pollicis during extension and abduction of the thumb cannot be rationalized on the basis of their insertions. These are such

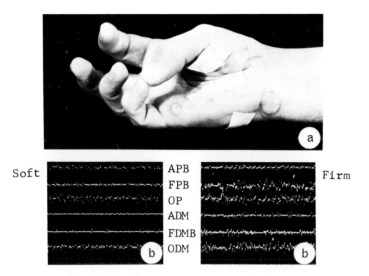

FIG. 11.5. *a*, opposition of the thumb to the little finger, tip-to-tip (position eight); *b*, electromyographic recordings during soft and firm opposition, respectively. (From Forrest and Basmajian, 1965.)

that these muscles would be expected to move the thumb in opposite directions, especially during extension and to a lesser extent during abduction. Weathersby, Sutton and Krusen (1963) suggested that stabilization of the part in order to produce a smooth, even motion was a possible explanation for the significant activities of muscles in situations such as this. This would seem to be a valid explanation.

Not all thenar muscles were active during extension and flexion of the thumb. Only three subjects showed more than slight activity in the flexor pollicis brevis during extension, the mean activity being *nil*-to-negligible. During flexion, the abductor pollicis brevis exhibited negligible activity; the opponens pollicis, slight activity on the average; and the flexor pollicis brevis, moderate-to-marked activity. Indeed, in the position of flexion, 10 of the 25 subjects had *nil* or negligible activity in both the opponens and abductor while the flexor was significantly active.

In one other position there was coincident activity and inactivity in the thenar muscles. During firm pinch between the thumb and side of the flexed index finger (position one), only negligible activity was recorded from the abductor pollicis brevis. Yet the opponens pollicis, and, in particular, the flexor pollicis brevis were significantly active.

Postures of Little Finger

During extension of the little finger, all three hypothenar muscles were rather inactive on the average; but in many subjects the activity in one or more of the three muscles was negligible or *nil*. During abduction, although the abductor digiti minimi fulfilled the function indicated by its name and was the dominant muscle (with a mean of moderate-to-marked activity), the two other hypothenar muscles were also significantly active. During flexion, moderate-to-marked activity occurred in all three hypothenar muscles.

The abductor digiti minimi was very active during flexion of the little finger at the metacarpophalangeal joint. (The participation of this muscle in this position of the finger is obvious also by palpation.) Part of the explanation for this activity depends on the muscle's insertion into the ulnar side of the base of the proximal phalanx. The abductor digiti minimi was also significantly active when the thumb was held opposed to either the ring or little finger. Some of this activity is possibly associated with the small degree of flexion at the fifth metacarpophalangeal joint that is required when the thumb and little finger are opposed. Yet, such flexion is obviously not required during opposition of the thumb and ring finger. Some of the activity of the abductor digiti minimi, then, may be to provide stability; and simple abduction of the little finger may be the least important function of the abductor of this finger.

Positions of Opposition

During soft opposition of the thumb to the side and tip of each finger (positions one through eight), gradual increases in activity were recorded from all six muscles, starting at position one in the case of the thenar muscles and beginning at position five in the case of the hypothenar muscles (figs. 11.5, *b* and 11.6). The opponens was the most active of the thenar muscles; the flexor was the least active (fig. 11.4, *b*). The opponens digiti minimi was the most active hypothenar muscle. All the thenar muscles were more active than the hypothenar muscles.

When opposition was firm (fig. 11.7) the flexor pollicis brevis replaced the opponens pollicis as the dominant muscle, particularly in positions one to four (index and long fingers) (fig. 11.4, *b*). In positions five to eight, the activity of the opponens pollicis approached and then equalled that of the flexor pollicis brevis. The abductor pollicis brevis was the least active of the thenar muscles (fig. 11.5, *b*).

The steady increase in thenar-muscle activity from position one through eight seen during soft opposition was not observed during firm opposition.

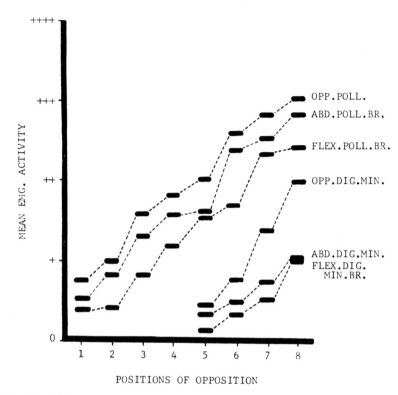

FIG. 11.6. Mean electromyographic activities during soft opposition of the thumb to side and tip of each finger, beginning with the side of the index finger (position one) and ending with the tip of the little finger (position eight). (From Forrest and Basmajian, 1965.)

Instead, higher levels of activity were usually recorded during firm opposition of the thumb to the side of the finger as compared with the activity during tip-to-tip opposition with the same finger.

The hypothenar muscles showed a steady increase in activity during firm opposition beginning at position four of the thumb. The opponens digiti minimi was again the most active muscle, and its mean activity was even slightly greater than that of the abductor pollicis brevis in position eight (fig. 11.5).

When the thumb was opposed firmly to the index and long fingers, the flexor pollicis brevis replaced the opponens pollicis as the most active of the six muscles (fig. 11.4, *b*). The opponens, however, approached and then equalled the flexor in its activity during firm opposition to the ring and little

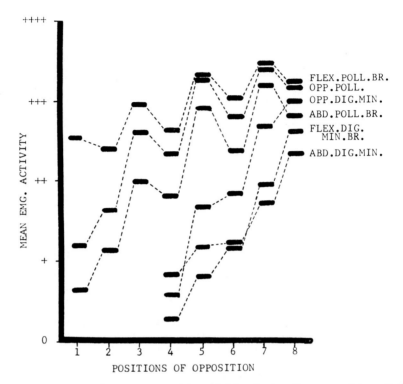

FIG. 11.7. Mean electromyographic activities during firm opposition of the thumb to the side and tip of each finger, beginning with the side of the index finger (position one) and ending with the tip of the little finger (position eight). (From Forrest and Basmajian, 1965.)

fingers (positions five to eight). Firm pinch between the thumb and the index and long fingers is a grip position of day-to-day importance.

Little (1960) attributes to the flexor pollicis brevis and to the mechanically advantageous position of the adductor pollicis the great power of the thumb, which enables this digit to balance the combined power of the fingers. Ignoring the contribution of non-thenar muscles such as the adductor (which is expanded on below) our findings emphasize the importance of the short flexor in thumb power. Weathersby, Sutton and Krusen (1963) also noted an increase in flexor activity as the subject pressed lightly with the thumb and index finger in a position corresponding to our position two.

One may observe that the greater the medial rotation of the first metacarpal, the greater is the tendency of the head of the fifth metacarpal to be drawn in an anterolateral (volar-radial) direction. The opponens digiti minimi is mainly responsible for this movement of the fifth metacarpal, and its action is almost reflexive in nature. The more active the opponens pollicis is in medially rotating the first metacarpal, the more active the opponens digiti minimi becomes. But the opponens pollicis is always the more active muscle. It is possible that beyond a certain degree of medial rotation of the first metacarpal, the two opponens muscles begin to act in unison to form the transverse metacarpal arch mentioned by Littler (1960) and by others. Indeed, this might be expected when one views the two opponens muscles, with the flexor retinaculum between, linking up the first and fifth metacarpal bones.

Positions of Grip

The important rôle of the flexor pollicis brevis in firm grasp is illustrated in the positions of firmly clasping a dowel and of holding a cup of water (fig. 11.8). Although the flexor pollicis brevis was the most active muscle while

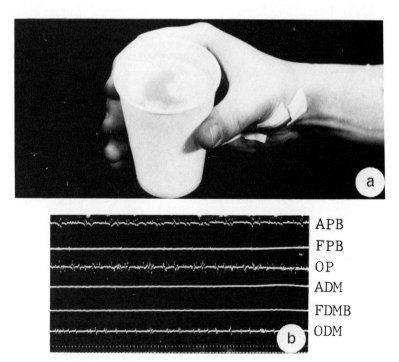

APB
FPB
OP
ADM
FDMB
ODM

Fig. 11.8. *a*, holding a glass of water; *b*, electromyographic recording. (From Forrest and Basmajian, 1965.)

the dowel was grasped firmly, this was not the case when the glass of water was also held firmly. Both the opponens pollicis and the abductor pollicis brevis were then more active (fig. 11.8). This finding, and other preliminary work that we have done, has led to a tentative conclusion that the more the thumb is abducted (as in holding the glass), the less the flexor brevis contributes to a firm grip. The activity of this muscle, which provides firmness of grip when only a small degree of abduction exists (as in holding the cup), is replaced by that of the opponens when a large amount of abduction is present (figs. 11.9, 11.10). In the absence of significant flexor activity, this activity of the opponens, coupled with that of the abductor, provides the power of a firm grip.

In summary, not all thenar muscles are active in all thumb positions; but all hypothenar muscles are active in three basic postures of the little finger. Two somewhat different patterns of activity occur when the thumb is first softly and then firmly opposed to each of the fingers in a sequence that begins at

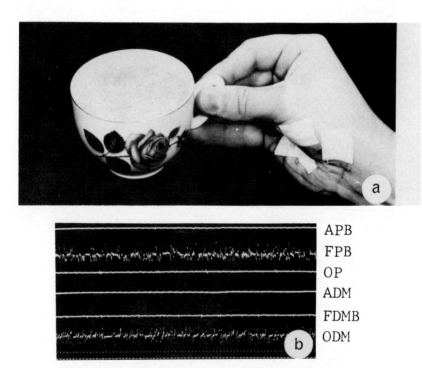

APB
FPB
OP
ADM
FDMB
ODM

Fig. 11.9. *a*, holding a cup of water; *b*, electromyographic recording. (From Forrest and Basmajian, 1965.)

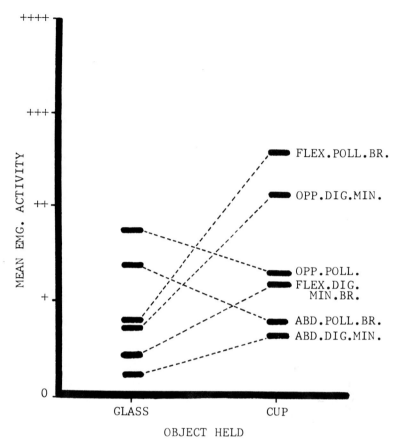

Fig. 11.10. Holding a glass and cup of water: mean emg activities compared. (From Forrest and Basmajian, 1965.)

the index and ends at the little finger. The flexor pollicis brevis is dominant in firm grip, particularly in grip between the thumb and two radial fingers; but a large degree of abduction of the thumb might possibly be a limiting factor in the activity of this muscle. The two opponens muscles seem to act as a unit in opposition of the thumb to both the ring and little fingers. Certain activity in some of the six muscles and inexplicable on a morphological basis probably serves to provide stability.

Forrest and Khan (1968) and Khan (1969) of my former department at Queen's University performed many elegant experiments to carefully discriminate between the various fine functions of the superficial and deep heads of the flexor pollicis brevis and of the transverse and oblique heads of the adductor pollicis. These "four" muscles radiate from their packed insertions like a fan. Then Johnson and Forrest (1970) elaborated the interplay of functions among the abductors and flexors of the thumb, including both intrinsic and extrinsic muscles.

Flexor Pollicis Brevis vs Adductor Pollicis

Patterns of activity emerged for each part of the flexor brevis and adductor—which is not a thenar muscle (Forrest and Khan, 1968). As might be expected, there were marked differences between the rôle of the transverse head of the adductor and the superficial head of the flexor during most functions. Most of the time, but not always, the two heads of each muscle act in concert. Surprising to those who hold that the deep head of the flexor belongs either with or to the oblique adductor was the finding that they often acted apart; this justifies the view that they are clearly different muscles belonging to flexion or to adduction, respectively. Forrest and Khan showed that in very fine functions each part of the two muscles is a functional entity with precise influence on the stabilization of the thumb.

Khan's (1969) detailed findings include the following.

The deep flexor is more flexor than adductor in its direction of fibres and observed activity, making its present name appropriate. Similarly, the oblique adductor is more adductor than flexor in its direction of fibres and observed activity; its present name is also appropriate.

Adductors are more active than flexors during combined adduction-lateral axial rotation of the thumb. Conversely, flexors are more active than adductors during combined abduction-flexion-medial axial rotation of the thumb.

The introduction of moderate resistance to a series of positions of simple pinch between the thumb and first two fingers resulted in no appreciable increase of activity in all muscles except the superficial flexor. This may be evidence, though inconclusive in this study, that the prime function of some of the intrinsic muscles of the thumb is to move and position it, not to provide it with strength.

Flexors are quite active during opposition of the thumb to the little finger and medial side of the hand, but significant adductor activity in this situation does not occur unless pressure is exerted by the thumb.

A small decrease in the activity of each of the four muscles apparently occurs as the thumb exerts moderately firm pressure against a series of objects requiring gradually increasing abduction of the thumb. A future study should deal with the rôle of the thumb and its muscles during grips of the hand.

Intrinsics vs Extrinsics

Johnson and Forrest's (1970) study of the intrinsic-extrinsic muscle interplay of paired muscles involved 20 subjects and employed sophisticated emg and electrogoniometric devices. A comparison of the first pair, the abductor pollicis brevis and longus, revealed that the brevis was usually more active than the longus; this was true in simple abduction of the thumb and in movements of the thumb having abduction as one component, with or without a load applied. In thumb extension, however, the longus was more active than the brevis when no load was applied; with a load, activities were equal. In comparing the second pair, the flexor pollicis brevis (superficial head) and longus, the brevis was usually more active than the longus in various postures of the hand requiring flexion-medial rotation of the thumb. In full flexion of the thumb (where medial rotation is minimal) the longus exhibited greater activity than the brevis. An interesting finding with respect to the two flexors was that they flexed the metacarpophalangeal joint of the thumb either independently or together, depending upon the position of the interphalangeal joint or the load applied.

The detailed findings reported by Johnson (1970) in his thesis include the following most significant points.

Flexor Pollicis Longus vs Brevis

(1) The position of the distal phalanx in movements of the thumb apparently determines, to some extent, the interrelationships of these two muscles with respect to their action on the metacarpophalangeal joint of the thumb. As the interphalangeal joint is increasingly flexed, the flexor pollicis longus becomes the prime mover of the metacarpophalangeal joint. This is particularly evident when the thumb is flexed, and when its tip is placed near the bases of the two radial fingers. When the thumb is positioned near the distal ends of fingers, the short flexor appears to act as the prime mover of the proximal phalanx at the metacarpophalangeal joint.

(2) Most daily activities involving the thumb require that it be positioned near the tips of the fingers. The relatively high levels of activity seen in the short flexor as the thumb assumes these positions (no load applied) indicates that its primary rôle may be to position the thumb with the assistance of other thenar muscles and adductor pollicis. Throughout these movements, the long flexor remains relatively quiet and is probably involved only minimally in these positioning activities.

(3) The long flexor appears to provide much of the force necessary in overcoming moderate loads applied to the thumb while its tip is positioned

near the tips of fingers. This is true regardless of the position of the distal phalanx. With a load applied to the thumb during movements such as opposition to the tips of fingers, activity in the long flexor increases sharply—more so than in the short flexor. This conclusion is further reinforced by a morphological characteristic of this muscle. Some of its tendinous fibres insert onto the midportion (palmar surface) of the distal phalanx. As the distal phalanx is flexed, this factor apparently brings about an increase in the muscle's line of application of force to the centre of rotation of the joint.

Abductor Pollicis Longus vs Brevis

The brevis is more active than the longus in all movements of abduction or movements with abduction as a component without or with a load applied. This includes those postures of the thumb described as opposition and lateral pinch.

Flexor Pollicis Longus vs Extensor Longus

During adductory movements of the thumb, the intrinsic muscles seem to require the assistance of the flexor policis longus and the extensor pollicis longus when a load is applied. In process of balancing the other muscle's tendency to flex or extend the thumb, the resultant action, apparently, of the two muscles is to adduct the thumb.

Functions of Individual Muscles

According to Johnson (1970) the following clear functions are apparent.

ABDUCTOR POLLICIS LONGUS. In addition to its rôle as a primary mover in abduction of the carpometacarpal joint of the thumb, this muscle appears also to bring about some flexion of the joint. This conclusion was previously based on non-electromyographic evidence only, and is now supported by his study.

EXTENSOR POLLICIS LONGUS ET BREVIS. These muscles are the prime movers in extension of the thumb. Perhaps their most valuable contribution to the function of the thumb lies in their ability to assist the abductor muscles in repositioning the thumb from positions of opposition, or by continued action, to spread the thumb out widely in order that the hand may grasp large objects. These muscles evidently act also to stabilize the interphalangeal and metacarpophalangeal joints of the thumb during some movements of opposition, perhaps enabling the thumb to perform fine, smooth and precise movements.

ABDUCTOR POLLICIS BREVIS. Johnson's (1970) conclusions confirm findings made by previous authors using electromyography in their investigations. (1) The abductor pollicis brevis through its insertion into the extensor expansion of the thumb assists the extensor pollicis longus to extend the distal phalanx—particularly when Johnson applied a load. (2) Through its insertion into the proximal phalanx, it assists the flexor pollicis brevis and opponens pollicis in flexion—medial-rotation of the thumb.

Conclusion

This section on the muscles of the hand must be ended by admitting the need for much systematic work on this most important functional region. Finally, a careful review of Duchenne's classical experiments is time well spent for those who wish to enlarge further upon the now available electromyographic findings reported above. There they will discover that much of our current understanding of the interossei and long muscles of the fingers actually springs from ancient times. Indeed, they will learn that it was Galen, in the second century A.D., who first described the actions of the lumbricals and interossei.

Chapter 12

Lower limb

B ECAUSE of its importance in posture and locomotion and because of its accessibility and large size, the lower limb has been the subject of electromyography from the earliest days of this science. The quality of the research done in this region of the body has been spotty and many unwarranted conclusions have been made and—quite fortunately—largely ignored by textbook writers. Part of the trouble stems from poor technique that in turn arises from inexperience. Novices seem to be especially prone to doing their earliest studies on the lower limb. Furthermore, some of them seem to have become completely overcome by their initial effort and stopped publishing completely. One can only hope that this was occasioned by remorse.

The muscles of the limb will be discussed from above downwards. Reference has already been made to the postural functions of these muscles in Chapter 9. Therefore, some repetition is unavoidable and, indeed, desirable. Locomotion is treated as a special chapter (p. 295) but will be referred to wherever necessary in this chapter also.

Hip Region

The muscles of this region which have been studied by various investigators and which will now be considered are: iliopsoas, the gluteal muscles and tensor fasciae latae. Other muscles that cross the hip joint (adductors, hamstrings, rectus femoris, and sartorius and gracilis) will be considered with the muscles of the thigh, p. 244.

Iliopsoas

Recent advances in the surgery of the hip joint have focussed attention on the muscles of the region. Interest in the functions of iliopsoas in particular was first renewed by the novel surgical procedure introduced by Mustard of Toronto (1952) and modified by Sharrard (1964) in which the insertion of iliopsoas is transplanted to the greater trochanter to substitute for paralyzed abductor muscles. Surgeons found that the resulting restoration of stability to the pelvis greatly outweighs the reduction of flexor power. The muscle remains alive in most cases (Broome and Basmajian, 1971) and the remaining flexor muscles are quite capable of providing any needed flexion for ordinary functions (Mustard, 1958).

When one begins to search the usual source-books for precise information about the actions and functions of iliopsoas, the only point that is agreed upon by all is that the muscle is obviously a flexor of the hip and probably has some influence on the lumbar vertebrae. There is a confusing disagreement about the other influences produced by the muscle. Last (1954) ascribes medial rotation of the hip mainly to the "powerful pull of ilio-psoas." The American *Gray's Anatomy* edited by Goss (1959) states that it rotates the hip medially while the British *Gray's* edited by Johnston and Whillis (1954) is more cautious with "it produces a slight degree of medial rotation . . . . " Lockhart (1951) in *Cunningham's Textbook* agrees with this. Woodburne (1957) agrees with Steendijk (1948) that iliopsoas rotates medially when the limb is extended and laterally when it is flexed. To complete the spectrum of opinion, at least a dozen major reference works (see Steendijk, 1948) state that iliopsoas is a lateral rotator.

In this running controversy that is now more than a century old, almost everyone has lost sight of the principle that a muscle so close to a joint must have an important postural or stabilizing function.

There has been some disagreement in the electromyography of this muscle partly resulting from different techniques. Using surface electrodes over this deep and almost inaccessible muscle, Joseph and Williams (1957) concluded erroneously that iliopsoas is inactive in standing subjects. This was confirmed with fine-wire electrodes for iliacus by LaBan, Raptou and Johnson (1966) although my own investigation of iliacus with long needle electrodes (1958b) indicated a continuous slight to moderate activity during relaxed standing in four persons (fig. 12.1, *A*). Using bipolar fine-wire electrodes, we found the psoas major also shows some slight to moderate activity during relaxed standing in many normal young adults. The bipolar, fine-wire electrodes (p. 32) were introduced from behind directly into the psoas in the mid-lumbar region (Basmajian and Greenlaw, 1968). Similar findings have been made by Keagy *et al.* (1966) of Northwestern University in five subjects and in several normal subjects by Close (1964). Although there seems to be a wide division of opinion here, the truth may be that there is very little real difference. What disagreement there is probably arises from differences both in technique and in the stance of subjects. In any case, the activity is not very marked.

ILIACUS. As one would expect, action potentials are recorded during flexion

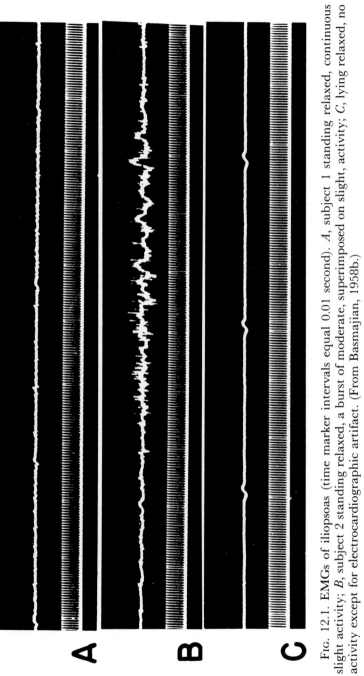

Fig. 12.1. EMGs of iliopsoas (time marker intervals equal 0.01 second). *A*, subject 1 standing relaxed, continuous slight activity; *B*, subject 2 standing relaxed, a burst of moderate, superimposed on slight, activity; *C*, lying relaxed, no activity except for electrocardiographic artifact. (From Basmajian, 1958b.)

of the hip in almost any posture of the whole subject and in almost the whole range. The amount of activity varies directly with the effort or resistance. We find marked activity in iliacus throughout flexion of the hip during "sit-up in the supine position" (Greenlaw and Basmajian, 1968, unpublished). However, LaBan *et al.* found there was little or no activity in iliacus during the first 30° of hip flexion. But, during "sit-up" from the "hook-lying" position, considerable activity occurred during the entire movement. Flint (1965) reported considerable variation in the styles of doing sit-ups but generally got little activity through surface electrodes in the first 45° of flexion.

Sometimes the iliacus muscles show intermittent short bursts of marked activity at irregular, short intervals during quiet standing; these apparently occur with invisible changes of position of either the limb or the trunk (fig. 12.1,*B*).

When our subjects lie down or sit at ease, and during extension, and adduction of the joint, there is no activity (fig. 12.1,*C*). Both medial and lateral rotation of the hip joint may produce some slight activity, whether the joint is passively or actively held in any of the extended, semi-flexed or flexed positions (fig. 12.2), but usually there is clearly more on lateral rotation. LaBan *et al.* confirm these findings in general.

Psoas Major. Our direct recordings from psoas are strikingly similar to those from iliacus. Thus we get a slight activity during relaxed standing, strong activity during flexion in all of many trial postures, slight-to-moderate activity in abduction and in lateral rotation (depending on the degree of accompanying flexion), none during most medial rotations, and little during most other conditions involving the thigh. Tönnis (1966a, b) agrees with our rotation findings, but, although he got similar results for abduction, he discounts his results, feeling the iliopsoas "cannot" be an abductor but rather must be an "antagonist to the glutei"—a tortuous and unnecessary conclusion. We found that the only lumbar movement which consistently recruits psoas is a deliberate increase in lumbar lordosis while standing erect. (The details of this study may be enlarged upon in a future paper by Greenlaw and myself.) Robert Close (1964) emphasizes the abductor function of iliopsoas, which he found to be quite active during extreme abduction.

Nachemson's (1966) study of the vertebral part of psoas with coaxial needle electrodes dealt with vertebral effects. He concluded that psoas has a significant rôle in maintaining upright postures.

Iliopsoas—both its parts—appears to be an active postural or stabilizing muscle of the hip joint as well as a flexor (fig. 12.3). The controversy as to whether it is a medial or a lateral rotator should be abandoned because, in fact, it is only a weak lateral rotator. Indeed, reviewing the work of Duchenne,

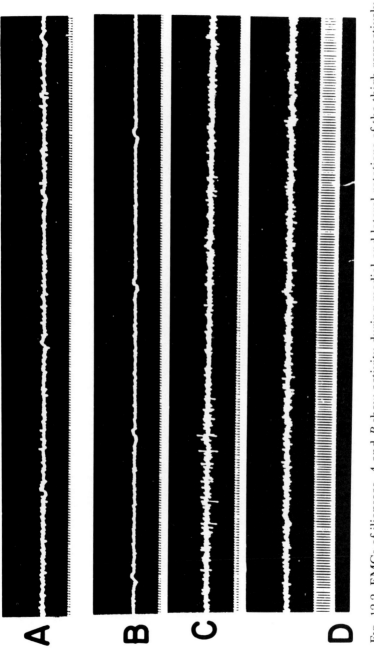

FIG. 12.2. EMGs of iliopsoas. *A* and *B* show activity during medial and lateral rotations of the thigh respectively in one subject. *C* and *D*, medial and lateral rotation in another subject. (From Basmajian, 1958b.)

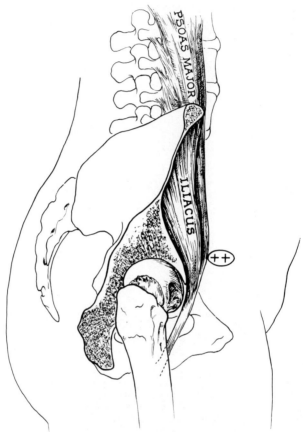

Fig. 12.3. Diagram of postural function of iliopsoas. It is slight-to-moderately active continuously during standing.

one would find that he did not disagree with this conclusion although his view has often been misrepresented.

A number of clinical studies have appeared which reflect on normal functioning and may be of special interest to some readers (Sutherland *et al.*, 1969; Baumann, 1969, 1971, 1972; Baumann and Behr, 1969; Stotz and Heimstädt, 1970).

The Glutei and Tensor Fasciae Latae

The gluteus maximus has usually been considered separately even where many muscles have been studied simultaneously, while the gluteus medius and gluteus minimus have usually been considered together because of their close association. Tensor fasciae latae is, of course, closely associated with the glutei.

GLUTEUS MAXIMUS. Wheatley and Jahnke (1951), Karlsson and Jonsson (1965) and Greenlaw and I concluded that the gluteus maximus was active only when heavy or moderate efforts were made in the movements classically ascribed to this muscle (Greenlaw, 1973). It was active during extension of the thigh at the hip joint, lateral rotation, abduction against heavy resistance with the thigh flexed to 90° and adduction against resistance that holds the thigh abducted. Lateral rotation (but not the opposite) also produced activity in gluteus maximus. While the whole muscle is engaged in extension and lateral rotation, only its upper part is abducent. As an abductor, gluteus maximus is a reserve source of power. Furlani, Berzin and Vitti (1974) are in general agreement with these findings.

The studies of Joseph and Williams (1957) show that the gluteus maximus is not an important postural muscle even during forward swaying. In bending forwards it exhibited moderate activity. When straightening up from the toe-touching position it showed considerable activity throughout the movement. Our own tests of this muscle with fine-wire electrodes tend to confirm all the above findings which were made with surface electrodes. They also confirm the report of Inman (1947) that the gluteus maximus is not a postural abductor muscle when the subject is standing on one foot (as are the medius and minimus). However, Karlsson and Jonsson found that when the centre of gravity of the whole body is grossly shifted, activity of gluteus maximus occurs. In positions where one leg sustains most of the weight, the ipsilateral muscle is active in its upper or "abducent" part; apparently this is to prevent a drooping of the opposite side. They also found that, during standing, rotation of the trunk activates the muscle that is contralateral to the direction of rotation (i.e., corresponding to lateral rotation of the thigh). Forward bending at the hip joint and trunk recruits gluteus maximus apparently to fix the pelvis. One of the chief values of the work of Karlsson and Jonsson is their showing the *range* of responses from their subjects—who showed considerable normal variation.

Duchenne's observation that complete paralysis of gluteus maximus in no way disturbs relaxed walking has often been noted but we should emphasize that this does *not* mean that normal walking does not recruit activity in the muscle. In fact, with fine-wire electrodes in the upper, middle and lower parts of gluteus maximus in a long series of subjects we found two considerable bursts in specific sections of the walking cycle (Greenlaw, 1973). This is enlarged upon under the section on gait on page 305. Finally, Houtz and Fischer (1959) have found it to be unimportant in bicycle pedalling.

Hominid Evolution. From our emg study of gorillas, we concluded that no major factual impediment exists in its functions to theorizing that the human gluteus maximus evolved from an ape-like condition (Tuttle, Basmajian and Ishida, 1975). In the gorilla, the upper and middle parts act prominently as a hip extensor and lateral rotator. The lower part of the muscle is active as a hip extensor.

GLUTEUS MEDIUS AND MINIMUS. Though Inman's demonstration (1947) of marked emg activity in the abductors when the subject stands on one foot is hardly surprising, the reader is referred to his valuable 1949 paper in which he described various other factors as well (fig. 12.4). Similarly, the finding of Joseph and Williams (1957) that gluteus medius and minimus are quiescent during relaxed standing is to be expected. Our detailed studies with fine-wire electrodes have also confirmed the usual teaching about these abductors—emphasizing their importance in preventing the Trendelenburg sign, during abduction of the thigh, and in medial rotation (anterior fibres). This last action is the only controversial one and seems now to be confirmed

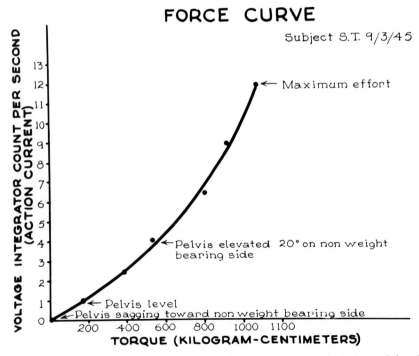

FIG. 12.4. A typical force curve, relating torque to action potentials of abductor muscles of the hip. (From Inman, 1947.)

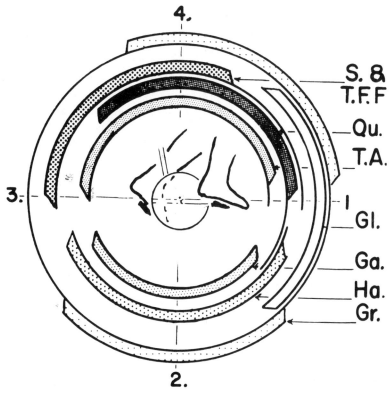

FIG. 12.5. Diagrammatic summary of emg activity in lower limb during one cycle of bicycling. (From Houtz and Fischer, 1959.) Greatest activity is indicated by *shaded areas*, but where cycle is completed with a single line, this means that slight activity continues. *Gr.*, gracilis; *S. & T. F. F.*, sartorius and tensor fasciae latae (femoris); *Qu.*, quadriceps; *T. A.*, tibialis anterior; *Gl.*, gluteus maximus and medius; *Ga.*, gastrocnemius; *Ha.*, hamstrings.

by our work. Our locomotion studies with wire electrodes (see p. 308) show triphasic activity for gluteus medius and biphasic for gluteus minimus during each cycle of walking (Greenlaw, 1973). Houtz and Fischer (1959), from their electromyographic studies, concluded that the activity in all the glutei was minimal in bicycle pedalling (fig. 12.5).

During elevation (flexion) of the thigh in the erect posture, Goto *et al.* (1974) found the anterior part of the gluteus medius was also active in the initial stage only. With resistance it increased.

TENSOR FASCIAE LATAE. Wheatley and Jahnke (1951), Carlsöö and Fohlin (1969), Goto *et al.* (1974) and Carvalho *et al.* (1972) found action potentials in this muscle during flexion, medial rotation and abduction of the hip joint. It was a medial hip rotator in all positions. Duchenne clearly stated that the power of tensor fasciae latae as a rotator (in response to faradic stimulation) is weak, and with this I agree. Carlsöö and Fohlin discount the rotary influence of tensor fasciae latae at the knee, finding no activity. During walking, Greenlaw and I found the muscle active biphasically during each cycle (see p. 309). Unlike the glutei, tensor fasciae latae is active during bicycling, showing its greatest activity as the hip is being flexed (Houtz and Fischer, 1959).

Thigh Muscles

The groups and single muscles to be considered now are the adductors (longus, brevis, magnus, gracilis and pectineus), the hamstrings (semimembranosus, semitendinosus and biceps femoris), sartorius, rectus femoris, the vasti (medialis, lateralis and intermedius) and the popliteus. Some of these cross the hip joint only (adductors), others cross both hip and knee (hamstrings, rectus, gracilis and sartorius), and still others cross the knee only (vasti and popliteus).

Adductors of the Hip Joint

In the first edition of this book, it was necessary to admit that "a surprising hiatus appears in our knowledge of the adductors. Forming an enormous mass on the medial side of the upper thigh, they must have considerable importance. In spite of this, their exact function is usually a matter of guesswork."

Since these words were written, in addition to the extensive studies of Greenlaw and myself (1968, unpublished), Janda and Véle (1963) and Janda and Stará (1965) have helped to correct the situation. They studied the rôle of the adductors in children and adults during flexion and extension of both the hip and the knee, with and without resistance. (Care was taken to avoid rotation.)

In almost every child the adductors were activated during flexion or extension of the knee and they were very active against resistance. Most adults showed activity during flexion of the knee, but only a minority were active during extension. With resistance almost all adult subjects showed great activity.

During movements of the hip the rôle of the adductors was localized to their upper parts. During flexion against resistance, all the children and half the adults showed activity. During resisted extension, all were active.

Janda and Stará suggest that this labile response of the adductors is related to a postural response. They believe that these muscles are facilitated through reflexes of the gait pattern rather than being called upon as prime movers. With this view one can readily agree. Spruit's (1965) theoretical analysis of the adductors adds conviction to the opinion.

Recently, Machado de Sousa and Vitti (1966) studied the adductores longus and magnus (upper and lower parts) during movements of the hip joint. During free adduction the longus is always active while magnus is almost always silent unless resistance is offered. Both muscles are active during medial rotation but not during lateral rotation of the hip, settling a classic argument that usually leaned in the other direction. The upper fibres of the adductor magnus showed the greatest activity.

During flexion of the thigh, de Sousa and Vitti found the main activity occurring in the adductor longus while the magnus is often completely silent. The results of Goto *et al.* (1974) tend to agree. While standing in a relaxed natural posture, both muscles are inactive. However, weak activity sometimes appears when standing on one foot. Okamoto *et al.* (1966) found similar results.

Using fine-wire electrodes, Greenlaw and I examined a long series of normal adult subjects during both free test movements and various postures and locomotions. Electrodes were in adductor longus, adductor brevis and the upper and lower parts of adductor magnus. Even when standing on one foot, the adductors on that side remain silent. Medial rotation (but not lateral) recruits them all (except perhaps for the vertical part of the magnus). During walking, the adductors showed different types of phasic activity. There is marked difference between the two parts of the adductor magnus; the upper (truly adductor) part is active almost through the whole cycle. Adductor brevis and longus show triphasic activity with the main peak at toe-off (Greenlaw, 1973).

Gracilis which belongs "officially" to the adductor group will be considered separately below.

Hamstrings

The three hamstrings, biceps femoris, semimembranosus and semitendinosus, act on both hip joint and knee joint. Various studies including my own have shown that the first of these is active in ordinary extension of the hip joint (in contrast to gluteus maximus which acts only against resistance), and in flexion and lateral rotation of the tibia at the knee. Wheatley and Jahnke (1951) have shown biceps is active also in lateral rotation of the extended hip and in adduction against resistance of the abducted hip. Furlani *et al.* (1973, 1977) found with a needle electrode in each head of the biceps femoris of normal men that both heads were active in less than half of cases during lateral rotation of the knee. The muscle is an obvious flexor of the knee joint. During the standing-at-rest position and standing on one foot, both heads fell silent, as they did with adduction of the thigh (Furlani *et al.*, 1973, 1977).

The semimembranosus and semitendinosus are quite active in extension and adduction against resistance of the abducted hip, and flexion and medial rotation of the tibia at the knee joint. With medial hip rotation, recruitment is slight (Greenlaw and Basmajian, 1968, unpublished; Greenlaw, 1973).

Joseph and his colleagues at Guy's Hospital Medical School (1954 *et seq.*) demonstrated the much greater stabilizing function of the hamstrings as compared with gluteus maximus, but emphasized their quiescence in ordinary standing. Portnoy and Morin (1956) agreed with them, as do Greenlaw and I; even standing on one foot does not recruit much activity in hamstrings.

In flexion at the hip and in leaning forwards, the hamstrings are much more active as supports against gravity. Arienti (1948a, 1948b) of Milan, Italy, showed by emg studies performed while the subjects were walking on a treadmill that the hamstrings come into action at different stages of walking. It is not possible to forecast the exact phase of activity in a muscle during walking by only examining it while the limb is put through artificial tests of prime movers. For example, if semitendinosus and semimembranosus are examined while the two are producing a deliberate test movement, such as flexion of the knee, they are found to act synchronously. On the other hand, the semitendinosus has a triphasic pattern during each walking cycle while semimembranosus is biphasic (see p. 309).

Arienti believed that, although both heads of biceps femoris act synchronously during a free-moving test of flexion, the short head acts during the swing phase of walking while the long head acts as a stabilizer when the foot is on the ground. Greenlaw and I found that an increase of walking speed changes a biphasic pattern to a triphasic one in the long head (p. 309).

Hirschberg and Nathanson of New York University made similar studies, reported in 1952. The hip muscles, quadriceps and hamstrings showed specific individual patterns being (in general) active during the transition from the swing phase to the stance phase (fig. 12.6). Only gluteus medius continued to contract beyond the middle of the stance phase; the others stop contracting within the first third of the stance phase. At the transition from the stance to swing, the adductor muscles (and sometimes the hamstrings) contract, according to Hirschberg and Nathanson.

There is now no doubt that the hamstrings do not by regional contraction act only on one joint. Our studies of these muscles described in Chapter 8 (p.

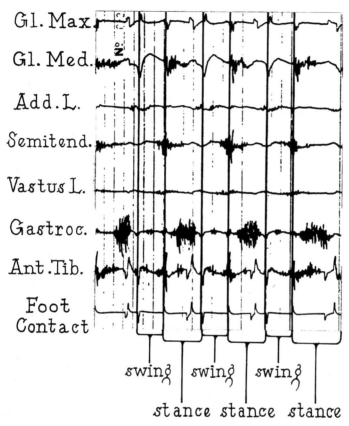

Gl. Max.
Gl. Med.
Add. L.
Semitend.
Vastus L.
Gastroc.
Ant. Tib.
Foot Contact

swing swing swing

stance stance stance

FIG. 12.6. EMG of normal gait. (From Hirschberg and Nathanson, 1952.)

156) showed that the entire muscle contracts regardless of whether the upper or lower joint was moved. Which joint is to move as a consequence depends on the immobilization of the other joint by other agencies.

Rectus Femoris

Undoubtedly, the rectus femoris is a flexor of the hip and extensor of the knee joint, and electromyography can only contribute information of secondary importance such as the timing of its activity in various movements. Because flexion of the hip is closely associated with extension of the knee, one is not surprised to fine a single muscle performing these two movements. As with the hamstrings, my studies on two-joint muscles demonstrated that normally the whole muscle contracts even with isolated movements of only one of the two joints. With truly isolated movements of single joints, Carlsöö and Fohlin (1969) revealed that *no activity* occurs in rectus femoris during knee extension. Okamoto (1968) found that the hip joint must be stabilized for full function to occur in rectus during knee extension.

We found rectus femoris negligibly active when standing on one foot; it is a lateral rotator but not a medial rotator. During moderate walking it shows slight biphasic activity (Greenlaw, 1973).

Rectus femoris also aids in abduction of the thigh (Wheatley and Jahnke, 1951), and apparently is relaxed in ordinary standing. The studies of Joseph and Nightingale (1954), Joseph and Williams (1957), Portnoy and Morin (1956), Floyd and Silver (1950) and my own observations all agree on this. Houtz and Fischer (1959) showed a marked activity in rectus together with the vasti during the thrusting motion of bicycle-pedalling but not during the flexion of the hip joint.

Gracilis and Sartorius

Though belonging to the adductor mass, the gracilis crosses and therefore acts on both hip and knee. It is active in flexion of the hip with the knee extended but is inactive if the knee is allowed to flex simultaneously (Wheatley and Jahnke, 1951). It adducts the hip joint (and therefore it rightfully belongs to its parent group, the adductors) and it rotates the femur medially (Jonsson and Steen, 1966; Greenlaw and Basmajian, 1968, unpublished; Greenlaw, 1973). Jonsson and Steen find that during flexion of the hip joint gracilis is most active during the first part of flexion both in free "basic movements" and during walking and cycling. In walking on a horizontal level and on a staircase, its activity occurs during the swing phase. At the knee it is a flexor and medial rotator of the tibia, although in medial rotation its activity appears to be slight, according to Jonsson and Steen. They and we also find it is insignificant in maintaining the standing posture. In bicycle pedalling it is not very active (Houtz and Fischer, 1959), but during walking, gracilis, like adductor longus, shows prolonged activity through most of each walking cycle (Greenlaw, 1973).

Sartorius is active during flexion of the thigh regardless of whether the knee is straight or bent and during flexion of the knee joint or medial rotation of the tibia (Wheatley and Jahnke, 1951). Both sartorius and gracilis may play a rôle in the fine postural adjustments of the hip and knee although Joseph and Greenlaw and I found no activity in sartorius during relaxed standing. In the bicycling experiments of Houtz and Fischer (1959), sartorius showed its maximal activity during the thrusting phase of pedalling, as would be expected. We found only one real peak of activity during walking—it came a short time after toe-off (Johnson *et al.*, 1972; Greenlaw, 1973). Using fine-wire electrodes, we found that during level walking, there was some activity throughout the swing phase rising to about the middle of the swing phase. During the remainder of the gait, there was only *nil*-to-slight activity. The same pattern occurred in descending steps. However, in ascending steps, the sartorius appeared to be more active immediately before heel strike (Johnson, Basmajian and Dasher, 1972; Carvalho *et al.*, 1972).

We found that flexion of the hip evoked the greatest emg activity of the sartorius muscle when only one joint was put through any movement. With combined movements of two joints, the greatest activity was recorded when hip flexion was accompanied by maximum knee flexion, although knee extension after the hip was flexed did show increased activity also (Johnson *et al.*, 1972).

Only when it was resisted did abduction of the hip produce slight-to-moderate activity. Medial rotation of the hip recruited little or no activity. There was only slight activity during lateral rotation of the hip while supine, and slight to moderate activity while sitting.

As expected, flexion of the knee was generally accompanied by more activity than any other motion confined to the knee, being most marked against resistance in a sitting position. Extension of the knee was accompanied by slight activity in most subjects, but in three it showed more activity than in flexion of the knee (Johnson *et al.*, 1972).

Variations in the insertion of the sartorius are well-known and may account for this. Sartorius is the most anterior muscle of the "pes anserinus" muscles and it is easy to see how the pull could occasionally be at, or anterior to, the knee axis; this condition would be enhanced when the knee is already in extension.

The sartorius may play a stabilizing rôle in strong knee extension as previously proposed (Houtz and Fisher, 1959). The rôle of the sartorius in walking appears to be that of a regulator. The swing phase of a normal gait is a low-energy phase. Once initiated the weight of the leg swings forward as a pendulum but its course is regulated by several muscles of the thigh and leg, one of which is the sartorius. Slight activity in this muscle occurs during the entire swing phase, being at its peak (moderate activity) about the middle of the swing phase. The hip is laterally rotating during the entire swing phase due to pelvic rotation in conjunction with the leg's forward controlled momentum. More hip and knee flexion is needed to clear the toe at the middle of the swing phase, perhaps explaining the rise in activity in the middle of the swing phase.

To clear the foot while ascending stairs requires more flexion of the thigh as well as dorsiflexion of the foot and appears to be accompanied by more lateral rotation of the thigh at the end of the swing phase. This would explain the increase in its activity during the latter part of this swing phase.

Vasti

The vasti, of course, are powerful extensors of the knee joint. Our experience agrees with that of many other investigators that the vasti are, however, generally quiescent during relaxed standing (Åkerblom, 1948; Kelton and Wright, 1949; Floyd and Silver, 1950; Portnoy and Morin, 1956; Joseph *et al.*, 1954 *et seq.*). Joseph and Nightingale (1956) found that when women wear high heels activity appears in the vasti in a substantial proportion of subjects. Professor Arienti (see p. 246) has demonstrated that during walking on a treadmill, the three vasti and rectus femoris do not act synchronously but have a phasic pattern.

Why are there four heads for a muscle that has most of its insertion on the restricted edges of the patella? Do the various parts have individual functions? Although Lieb and Perry (1968) answered this question in part by ingenious morphological and biomechanical studies of amputated limbs, EMG offers an added dimension to their findings by revealing the actual timing of activity and the interplay of function in the four parts.

Earlier emg studies by Pocock (1963) and Close (1964), limited in technique, offer provocative but useful ideas that require confirmation, expansion, and quantification. An incentive for gaining new and precise information arises from the growing interest in myoelectric assistive devices for the physically handicapped (including programmed muscle stimulators for recruiting of lower limb muscles in proper sequence). For these reasons, we undertook a detailed study of the interplay of activity in the three vasti and the rectus femoris in a group of normal adults (Basmajian, Harden and Regenos, 1972). Bipolar fine-wire electrodes injected in the middle of the belly of each head of quadriceps, provided excellent electromyographic response with no "crosstalk."

TIMING OF ONSET AND CESSATION. Figures 12.7, 12.8 and 12.9 summarize the tabulated results of the pooled data. In summary, the most striking features are: (1) the variation between the different heads; (2) the considerable standard deviation arising from individual differences; and (3) the late onset and early cessation of activity in rectus femoris during the squatting manoeuvre.

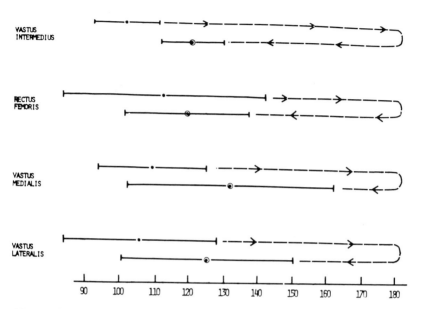

Fig. 12.7. Diagrammatic representation of onset and cessation of activity in the four heads of quadriceps during unweighted extension of the knee from 90° to 180° and flexion back to 90°.

Subject seated; no added load. Mean and standard deviation of the angle at which activity begins starts each line; the *broken arrows* indicate the direction of movement to full extension and back to the mean and standard deviation of the angle at which the activity ceases. (From Basmajian *et al.*, 1972.)

Grading of Activity. Figure 12.10 summarizes the findings. The most revealing findings are the similarities between muscles in the pooled data with the considerable variation in individuals at certain times (the most marked variation occurring at the end of straightening up from the squatting posture). There is no question that weighted extension recruits the greatest activity. Vastus medialis does appear to increase its activity more rapidly toward the end of unweighted extension but it is not greatly different from the other heads during the act of squatting.

It is now becoming widely recognized that vastus medialis acts through the whole range of extension, not just at its terminal phase. Also it is in this terminal phase that the muscle is supposedly the most active because it completes extension. However, another very real function of vastus medialis at the end of extension is its prevention of lateral dislocation of the patella.

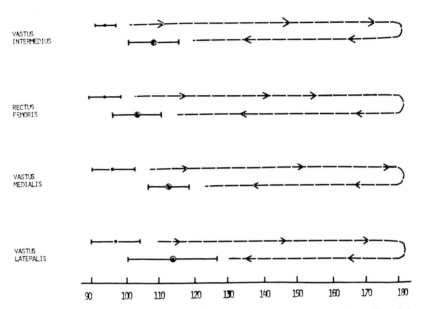

Fig. 12.8. Diagrammatic representation of onset and cessation of activity in the four heads of quadriceps during weighted extension of the knee from 90° to 180° and flexion back to 90°.

Subject seated; 30 pounds (13.6 kg) added load. Mean and standard deviation of the angle at which activity begins starts each line; the *broken arrows* indicate the direction of movement to full extension and back to the mean and standard deviation of the angle at which the activity ceases. (From Basmajian *et al.*, 1972.)

The special direction and insertion of the lowest fibres of the muscle point directly at this being the real rôle of that lowest part of the muscle which bulges so prominently here.

Most clinicians and functional anatomists believe that an unstable knee is due to an inability to produce the final 15° of extension. In turn, many clinicians continue to believe that this failure is primarily due to weakness of the medial head of the quadriceps. It is difficult to deny that the final 15° of extension is important. Indeed, the aim of quadriceps retraining is to enable a patient to extend his knee fully and to maintain that extension. Whether the fundamental part is vastus medialis itself is open to serious doubt. Lieb and Perry (1968, 1971) concluded that early atrophy of the vastus medialis prominence coupled with the loss of terminal extension is simply indicative of a *general* quadriceps weakness. Earlier Hallén and Lindahl (1967) reported

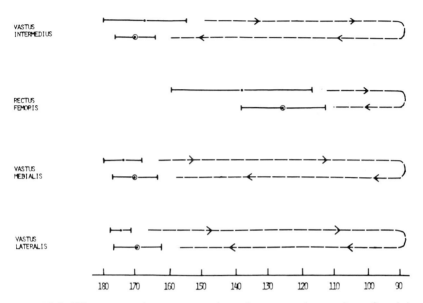

FIG. 12.9. Diagrammatic representation of onset and cessation of activity in the four heads of quadriceps during squatting to 90° from the fully erect posture and back up again.

Subject standing; no added load. Mean and standard deviation of the angle at which activity begins starts each line; the *broken arrows* indicate the direction of movement to 90° flexion (squatting) and back to the mean and standard deviation of the angle at which the activity ceases. (From Basmajian *et al.*, 1972.)

essentially the same findings. Both groups established the general rôle of vastus medialis rather than its widely touted terminal-extensor function. Hallén and Lindahl also stressed the importance of pain inhibition and adhesions in the limitation of extension in its last 10°.

The visible prominence of the vastus medialis is really related to the marked obliquity of the distal fibres of the muscle, its lowness of insertion and the thinness of fascial covering in that area. The extensor lag accompanying knee extension is a function of great loss in the mechanical advantage of the whole muscle during the final 15° of the extensor range because a 60% increase of force is needed to complete extension (Lieb and Perry, 1968). Thus, the only selective function attributable to vastus medialis is patellar alignment. We cannot deny the importance of the latter but it has no special part to play as a prime mover in the final extensor movement insofar as mechanical advantage is concerned; this appears to be confirmed by our

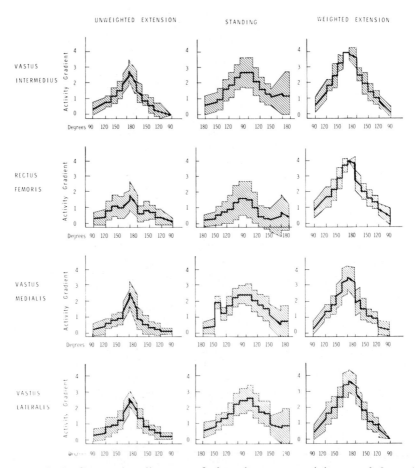

Fig. 12.10. Composite diagram of changing emg activity rated 0 to 4 during the movements of the knee reflecting mean and standard deviations at epochs of the movements shown on the horizontal axis. *Heavy lines* indicate the means; *shaded areas* the standard deviations. (From Basmajian *et al.*, 1972.)

findings. A surprising burst of activity during squatting as subjects pass from 150° to 135° remains unexplained.

Our detailed study supplements those of Pocock (1963), Close (1964), Lesage and Le Bars (1970) and Lieb and Perry (1971), and offers for the first time quantification in all four heads (including vastus intermedius). True, this does not greatly alter our understanding of the vastus intermedius which seems to have no discrete rôle. The question of "Why four heads?" is still unanswered except in a negative way: the muscles act in concert to achieve a common end. Minor differences (such as the burst of activity in vastus medialis during squatting) may alone explain the structural discreteness. The relatively short period of significant grades of activity in the rectus femoris (compared with the vasti), also noted by Pocock, may also have some importance which is not now apparent. Any judgments of the timing of the muscles based upon biomechanical probabilities appear to be especially fallible where the quadriceps femoris is concerned.

Pocock's finding that quadriceps activity is less during the down phase of squatting is contradicted by our findings which show a parallelism between the up and down stages. Some difference in technique may account for his conclusion.

An important additional finding in our study is the variation from subject to subject. This is particularly marked at the conclusion of standing up from the squatting exercise. Individual variations in responses of muscles must be taken into account in the design of myoelectric devices.

The study by Murphey, Blanton and Biggs (1971) of Dallas, Texas has special interest for they were concerned with the activity of the quadriceps during standing in subjects who habitually show *hyperextension* and *flexion* of the knee when standing up. In all instances, flexion subjects showed more activity than hyperextension subjects.

Ravaglia's (1957) investigation of quadriceps femoris is also of interest. In a series of normal subjects he recorded *via* surface and needle electrodes from the vastus medialis, vastus lateralis and rectus femoris simultaneously. During the movement of rising from the sitting to the standing position and *vice versa*, the activity in the three heads was not synchronized and equal. The vastus medialis was retarded and was not as active as the other two. In erect standing the activity in the three heads fell rapidly. The vastus medialis was more active than the other two muscles examined. Ravaglia demonstrated conclusively that the three heads acted in different ways in various phases of movement.

Wheatley and Jahnke (1951) found a greater activity in vastus medialis when the knee was held in extension with the hip joint flexed or the knee joint (tibia) laterally rotated. On the other hand, the vastus lateralis was more active in extension of the knee when the hip was flexed or the knee joint (tibia) medially rotated. During resisted extension of the knee, the various parts of quadriceps came into action at different phases of the movement (fig. 12.11).

In 1957, a group, consisting of Mrs. W. E. K. Brown, Miss Rita Harland and myself, performed a series of electromyographic examinations of quadriceps femoris, the results of which are as yet unpublished. I am grateful to

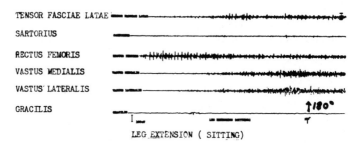

Fɪɢ. 12.11. EMGs of thigh muscles during extension of the knee from 90°
to 180° (subject seated, limb hanging). (From Wheatley and Jahnke, 1951.)

my physiotherapist colleagues for allowing me to report here some of the
more significant results. Our subjects included 11 young women in whom
simultaneous recordings were made from the vastus medialis, vastus lateralis
and rectus femoris using skin electrodes.

Our chief concern was the evaluation of a number of standard procedures
sporadically used in rehabilitation work ostensibly to help strengthen the
quadriceps. For example, associated movements of the toes have been advocated to augment the activity of the quadriceps. More than half of our
subjects showed no such augmentation. In those in whom augmentation of
quadriceps did occur, the actually effective toe movement was flexion in
some and extension in others.

Associated foot and ankle movements were somewhat more effective than
toe movements in causing augmentation of quadriceps activity; this was true
in most (but not all) of the subjects. However, there was no clear-cut
difference between the effects of any of the following: dorsiflexion, plantar
flexion, inversion and eversion. All seemed to augment in some subjects,
while one or other of the movements proved to be the most effective in others.

Medial or lateral rotation of the hip joint performed simultaneously with
contraction of quadriceps had essentially no effect on the emg activity of
quadriceps (except for some slight augmentation in one subject). Not only
did simultaneous hip flexion fail to augment the amount of activity in
quadriceps in most subjects, but it even decreased it in some.

The most effective technique for maximal motor unit activity was having
the subject actively perform extension of the knee against resistance—not in
a static position but during motion. Nonetheless, in many subjects, static
contractions were just as effective—or even more effective—and therefore
cannot be categorically condemned. The greatest activity during motion
occurred in the last half (i.e., 90°) of extension. With static contractions, the

position of the knee most effective for showing maximal activity in quadriceps was almost always the fully extended one.

In all our subjects, concentric actions caused more activity than eccentric ones, i.e., activity of the muscle during its primary action or shortening was considerably more than the activity while the muscle was acting as an antagonist or being forced to lengthen as it acted (negative work).

Flexion of the trunk or isolated contraction of the opposite quadriceps had little if any effect on the quadriceps under examination, suggesting that such techniques are practically useless for rehabilitation work. Simultaneous bilateral contraction of the quadriceps did not augment the activity in the one under study in some subjects while it did so in others. Finally, having the subject "push down" at the hip and knee regions (in an effort to hyperextend the knee joint actively) did not increase the activity of quadriceps, but actually diminished it in 3 of the 11 subjects. Instructing a patient to "push down your knee" is therefore not an acceptable procedure for producing maximal quadriceps activity.

The lesson to be learned by physiotherapists and rehabilitation specialists from these findings is one of healthy scepticism for many of the dogmatic teachings that bear on the best methods of evoking maximal quadriceps activity. Some of our findings appear to confirm the dogmas, others flatly contradict them, and still others show that different subjects react in different ways.

Popliteus

In a pioneer piece of electromyographic research with needle electrodes on this small, deeply set muscle lying behind the knee, Barnett and Richardson (1953) confirmed the classical teaching (often denied) that it is a medial rotator of the tibia. Its activity at the start of flexion of the knee is related to the unlocking of the knee joint. When a person stands in the semi-crouched knee-bent position, continuous motor unit activity of the muscle was demonstrated (see fig. 9.4 on p. 183). When the knee is bent the weight of the body tends to slide the femur downward and forward on the slope of the tibia. It seems that the continuous marked activity of popliteus aids the posterior cruciate ligament in preventing forward dislocation. Reis and Carvalho (1973) emphasized their findings based on equilibrium functions for the control of torsional forces.

Using improved fine-wire electrode techniques, we endeavored to delineate the functions of popliteus further (Basmajian and Lovejoy, 1971). We studied 20 normal persons. The data were recorded and stored on magnetic tape and analyzed after analogue-to-digital conversion by a LAB-8 computer. The digital information was normalized by making the greatest activity for each subject equal to one; lesser degrees of activity were then compared with this normalized maximum activity.

STATIC EXTENSION. Each subject was tested in the sitting position. Data were recorded with the leg and foot maintained by the subject in the neutral position, then in full medial rotation, and in lateral rotation, beginning with the knee joint at 90°, and then at approximately 120, 135, 160, 175 and 180° (full extension) (fig. 12.12). The emg activity in neutral and lateral rotation remained approximately the same in all degrees of extension (fig. 12.13). However, in medial rotation of the tibia on the femur there was marked increase in activity. The activity was greatest through the first 40° of extension and decreased as full extension was reached.

STATIC FLEXION. Each subject was tested in the prone position and data were recorded at the same degrees of flexion and rotation as previously (fig. 12.14). Again, the electromyographic activity remained essentially constant in neutral and lateral rotation of the tibia. There was a marked increase in activity through the first 20° of flexion with the leg medially rotated and it gradually decreased as 90° of flexion was reached.

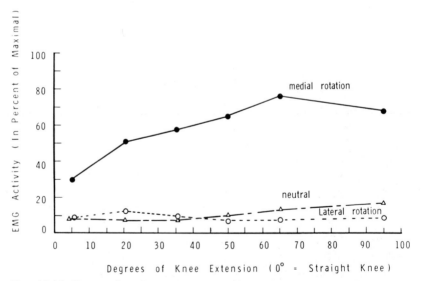

FIG. 12.12. Composite of mean emg results in popliteus with subject seated and unsupported knee held at different angles. (From Basmajian and Lovejoy, 1971.)

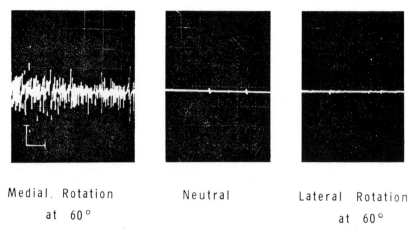

Medial. Rotation   Neutral   Lateral Rotation
      at 60°   at 60°

Fig. 12.13. Example of emg activity in popliteus during rotations of tibia.
(From Basmajian and Lovejoy, 1971.)

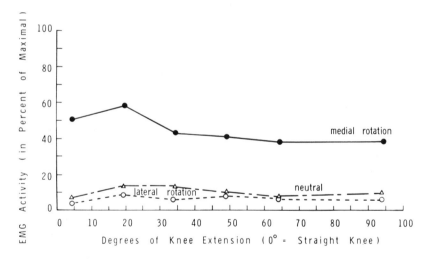

Fig. 12.14. Composite of mean emg results in popliteus with subject lying
prone and knee held at different angles. (From Basmajian and Lovejoy,
1971.)

MOVEMENTS OF KNEE. The emg activity was recorded beginning with the
knees locked and then with the subject flexing to 150°. This was repeated
with the right shoulder and trunk rotated forward and then with the left
shoulder and trunk rotated forward. In each instance recordings were made
with the feet in three positions: neutral, medial rotation and lateral rotation.

The emg activity in the right popliteus muscle remained constant with the right shoulder rotated anteriorly and the feet at neutral, in medial rotation, or in lateral rotation. With the left shoulder rotated anteriorly and the feet laterally, the activity increased with knee flexion. With the feet at neutral or in medial rotation, the activity was greatest at the initiation of flexion. The activity was persistently elevated above that obtained with the right shoulder rotated and with the feet in the same positions.

Gait was analyzed in all subjects using foot switches on the heel, toe and lateral margin of the shoe. The subject was tested on a treadmill at 1.6 and 3.2 km per hour with normal gait, toe-in and toe-out gaits. In general the activity was greatest at heel-strike and from foot-flat to toe-off, that is, during most of the weight-bearing phase of gait.

Leg Muscles

We have elsewhere reported in detail an electromyographic study with needle electrodes of three muscles of the leg (and three intrinsic muscles of the foot) (Basmajian and Bentzon, 1954). The muscles of the leg were tibialis anterior, peroneus longus and the lateral head of gastrocnemius. An outline of the findings will now be given.

Most of the subjects were tested in different postures and during a variety of movements. Often, other muscles were examined as well. The first part of this section concerns only the findings in the muscles named during the "relaxed standing at ease" position. In this regard we were especially interested in testing the validity of the popular theory that peroneus longus, tibialis anterior and the intrinsic muscles of the foot maintain the arches of the foot in ordinary standing. Special attention was also paid to the possible influence of certain variable factors, such as the sex of the subject, the type of feet and the wearing of high heels by women.

For each muscle, records were made while the subject was reclining (as a control), immediately upon assuming the standing barefooted posture and after 2 minutes of standing. Considerable testing showed that the initial activity invariably found on changing a position falls off rapidly to the resting level and that the 2-minute interval is adequate. Moreover, many of the subjects were tested at longer intervals (up to 15 minutes) with no change in the results.

Our "relaxed standing at ease" posture is the comfortable well-balanced stance with the feet several inches (about 8 cm) apart and bearing equal weight, and with the hands clasped loosely behind the back. The relative position of the feet would show minor individual variations. However, the slight changes that were necessary for some subjects to conform to the standard position made no difference in their comfort or in the findings.

In the women, additional similar records were made from the muscles of the leg while high heels were being worn. The heels were all 2½ inches (about 6 cm) high except in one case in which they were 3 inches (7.5 cm).

The lateral head of gastrocnemius was active in the majority of subjects. In the women, the group showing continuous activity was definitely smaller when the subjects were barefooted, but, when high heels were worn, almost the same as in the males. A quarter of both men and women showed no electromyographic activity in this muscle when the subject stood barefooted. Analysis of individual cases revealed that only one of the women continued to show no activity with high heels.

Almost half of the men and a quarter of the women showed no electromyographic activity in tibialis anterior when standing barefooted (fig. 12.15). But another quarter of both the men and the women showed pronounced activity; this could be abolished by leaning forward. Each of the women in this latter group exhibited at least the same degree of activity on standing in high heels. One woman showed moderate activity while standing with high heels but only slight activity while standing barefooted.

Only 1 of 16 men and 2 of 16 women showed continuous activity in peroneus longus while standing barefooted. An additional man and 5 women showed intermittent activity. Half of the men and a third of the women showed inactivity when barefooted. With high heels, half of the women showed continuous marked activity and none showed inactivity in peroneus longus.

There was no consistent significant relationship between the types of feet and types of activity in any muscle.

It now seems to be beyond controversy that the tibialis anterior and the peroneus longus (and, as we shall see, the intrinsic muscle of the foot) play no important active rôle in the normal static support of the long arches of the foot. Our figures show that these muscles are completely inactive electromyographically in many normal individuals while standing. Smith (1954) came to the same conclusion after examining six subjects who all showed electromyographic quiescence in the anterior crural muscles during standing. The same situation holds for both peroneus longus and peroneus brevis (Jonsson and Rundgren, 1971), and for tibialis anterior, the peronei, flexor and extensor digitorum longus, and abductor digiti quinti (Suzuki, 1956).

Although this investigation of predominantly normal feet showed no obvious relationship between types of feet and the types of activity in the muscles concerned during standing, no attempt is made to suggest that the muscles play no rôle in the abnormal flat foot. Furthermore, we are not

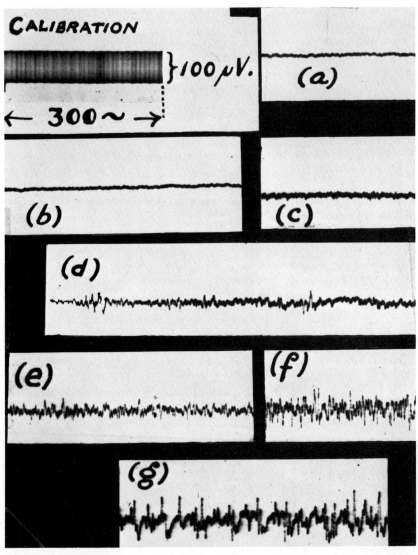

FIG. 12.15. Representative EMGs from leg muscles during standing: *a* and *b*, *nil* to negligible activity; *c*, slight; *d*, intermittent bursts; *e*, moderate continuous; *f* and *g*, marked continuous. (From Basmajian and Bentzon, 1954.)

dismissing the rôle of these muscles in the maintenance of the arch during locomotion. Indeed, the intrinsic muscles of the foot are always very active electromyographically when one rises on the toes to even the slightest degree (see below).

In so far as the muscles of the leg are concerned, the results showed a biological range of activity and were not in accord with some of the absolute findings of Smith, Joseph and Nightingale, and Åkerblom. We find that in the relaxed standing at ease position, there is activity in some individuals in the tibialis anterior and peroneus longus and that this can be abolished easily by unusual and varying stances or by the removal of weight from the limb. The disagreement probably stems largely from our use of the more sensitive needle electrodes rather than simple skin electrodes (fig. 12.16).

Joseph and Nightingale were concerned with soleus, gastrocnemius and tibialis anterior. In their subjects they found continued activity in every soleus and in some gastrocnemii, but none in tibialis anterior (fig. 12.17). Since their original paper they have added to their series with no change in their conclusions (Joseph, 1960). Granit's (1960) statement that, in general, mammalian soleus is tonic and gastrocnemius phasic may bear on this matter. However, Smith (1954) found that what postural activity there was in human legs during standing was intermittent and confined to gastrocnemius and not soleus. Levy (1963) finds that in man soleus produces greater reflex contraction than gastrocnemius during the ankle jerk. He suggests that a greater

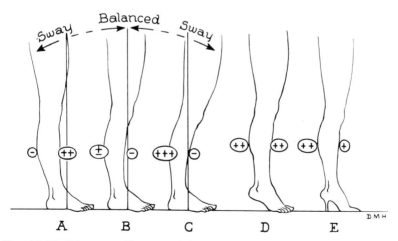

Fig. 12.16. Diagram of emg activity in anterior and posterior muscles of the leg under differing conditions.

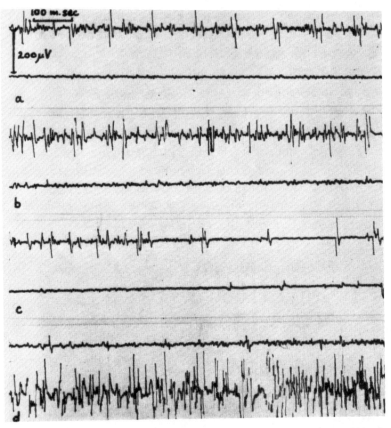

Fig. 12.17. EMGs from soleus (*upper trace*) and tibialis anterior (*lower*). *a*, Standing at ease; *b*, after swaying forwards; *c*, while swaying backwards; and *d*, after swaying backwards. (From Joseph, 1960.)

density of muscle spindles in soleus (now generally accepted) accounts for this. One might expect a greater sensitivity to stretch in soleus. Herman and Bragin (1967) further elaborated on the differences between gastrocnemius and soleus in man. In general, the former is more sensitive to conditions of length, strength and rate of contraction; soleus plays a more constant role and is more active at ankle dorsiflexion and in minimal contractions. Campbell *et al.* (1973) in Biggs' laboratory in the Baylor College of Dentistry were even more specific. Recently they showed with fine-wire electrodes that the stability of the foot support affects the activity of even selected parts of these two muscles. The medial part of soleus has distinct functions: it is both a strong mover of the foot on the leg and a stabilizer of the leg on the foot; however, the lateral part gives little power to moving the ankle and is largely a stabilizer—especially when the platform is unstable. The two heads of gastrocnemius are quiet until motion is required at the ankle, thus acting as an auxiliary plantar flexor. The main dynamic and static flexor, the medial part of the soleus, has never before had such a lime-light exposure.

Closely related to the soleus and gastrocnemius muscles is the much less obvious *plantaris* (p. 271). We found in a study of 12 normal subjects with fine-wire electrodes that its primary function is plantar flexion and inversion (Iida and Basmajian, 1973). Only in a loading situation does plantaris assist in knee flexion. Its function is not always the same as that of the closely related lateral head of gastrocnemius which always has a slight activity in eversion.

The increased activity in all three of the muscles of the leg when high heels are worn may seem to be due to the element of instability introduced by the posture (fig. 12.16). But it will be noted that gastrocnemius and peroneus longus are involved and that tibialis anterior is not affected to any great extent. In another series of experiments we found that peroneus longus is markedly active in plantar flexion of the ankle. It appears, then, that the wearing of high heels shifts the centre of gravity forward to a new position in many women with a resultant increase in the activity of gastrocnemius and peroneus longus.

Joseph and Nightingale (1956) confirmed our finding that the wearing of high heels caused an increase in activity of the calf muscles (specifically soleus) in most women (fig. 12.18). They investigated the line of gravity in 11

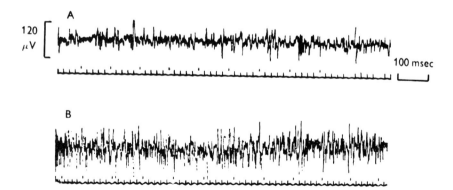

Fig. 12.18. EMGs of soleus in women. *A*, standing at ease; *B*, standing at ease with high heels. (From Joseph and Nightingale, 1956.)

women and found that it bore no constant relationship in spite of their reluctance to agree to the occurrence of intermittent activity of the leg muscles in standing. They also established beyond a doubt that the swaying forwards and backwards of a subject by as little as 5° produces reflex activity of the posterior and anterior leg muscles.

Portnoy and Morin (1956) tended to confirm our findings, reporting that 5 gastrocnemii showed intermittent activity while 9 (of 16) showed continuous activity during relaxed standing. Naponiello (1957) reported similar intermittent activity in tibialis anterior with or without high heels, as did Floyd and Silver (1950) who first drew special attention to the intermittent, unconscious, back-and-forth swaying which causes it.

Ferraz, de Moraes and Parolari (1958) have concluded from a study of seven subjects that the peroneus longus and peroneus brevis act intermittently as postural muscles, becoming very active in leaning forward and silent when leaning backward. Their activity is pronounced during the propulsive phase of normal walking, and the activity of the two muscles is synchronous.

Arch Support of Foot

The mechanism of arch support in the foot remains controversial despite years of investigation. According to one theory, the arches are maintained by the contraction of muscles; according to a second, by the strength of passive tissues; and according to a third, by the combination of both muscles and passive structures.

A century ago, Duchenne stated that by faradization of the peroneus longus in flat-footed children he was always able to produce the progressive formation of a normal plantar arch. Keith, in 1929, concluded that muscles are all-important in the support of the arch and that ligaments come into play only after the muscles have "failed." Morton (1935) disagreed. He concluded that the structural stability of the foot is not dependent on muscles. He claimed that appreciable muscle exertion is needed only when the centre of gravity of the body moves beyond the margins of structural stability, whereas only a slight controlling action by the muscles is required when the centre remains between those margins. In 1952, Morton and Fuller further showed that static strains upon the foot are relatively low in intensity, falling well within the capabilities of the ligaments. Their calculations showed that only acute, heavy, but transient forces, such as in the take-off phase of walking, require the dynamic action of muscles.

Thus the controversy continued and was kept alive by others including

Kaplan and Kaplan (1935) and Lake (1937). In 1941, R. L. Jones, using the method of palpation in the living and direct observation in cadavers, concluded that not more than 15 to 20% of the total tension stress on the foot is borne by the posterior tibial and peroneal muscles. Much the greater part of this stress is borne by the plantar ligaments of the foot; but the short plantar muscles, being in an advantageous position, also contribute to the support.

After World War II, Harris and Beath (1948) concluded from their extensive survey in the Canadian Army that both passive supporting structures and muscles are responsible for a normal arch. They frankly favored the rôle of the passive structures but admitted the readiness of the muscles to assume a rôle in arch support. Wood Jones (1949) agreed that maintenance of the normal arched form of the foot results from the dual control exerted by the passive elasticity of the ligaments and the active contractility of muscles. He concluded that the plantar aponeurosis and plantar tarsal ligaments hold the anterior and posterior pillars of the arch together and that the actively contracting intrinsic muscles between the aponeurosis and tarsal ligaments also play an important part.

From our general electromyographic study of the leg and foot with needle electrodes, we concluded that the tibialis anterior, peroneus longus and the intrinsic muscles of the foot play no important rôle in the normal static support of the long arches of the foot (Basmajian and Bentzon, 1954). As noted before, many if not most of these muscles showed inactivity during standing in a relaxed position. This was confirmed in general terms by Smith (1954) who used skin electrodes. Many standard textbooks, however, still over-emphasize the part played by muscles in the support of the arches of the foot. For example, *Gray's Anatomy* (Johnston, Davies and Davies, 1958) stated that the tibialis posterior is the most important factor when the foot is bearing weight, and that the peroneus longus, tibialis anterior, flexor hallucis longus, abductor hallucis and flexor digitorum brevis also contribute to the support.

In an attempt to settle the controversy, we have performed a special study of the muscular support of the loaded arch (Basmajian and Stecko, 1963). Because earlier studies ran the risk of confusing the muscle activity required for postural adjustment with that for the support of the arches, the subjects were seated and the leg and foot were loaded artificially in a special apparatus (fig. 12.19) that provides graded loads of up to 400 pounds (182 kg).

Six muscles of particular interest were chosen for this electromyographic study. They were: the tibialis anterior (since by its insertion it would appear to raise the summit of the medial longitudinal arch), the tibialis posterior and the peroneus longus (since by acting together, these two might provide a sling support), and the flexor hallucis longus, abductor hallucis and flexor

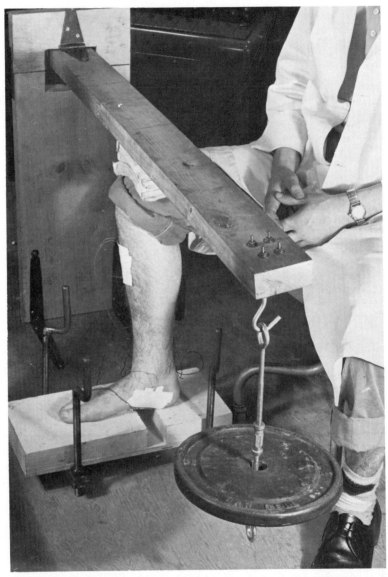

Fig. 12.19. Arrangement of a subject with the load applied by leverage and the foot on the adjustable platform (here set in the horizontal position). (From Basmajian and Stecko, 1963.)

digitorum brevis (since all three are in a position to act as longitudinal bowstrings). These muscles were studied with our special indwelling fine-wire electrodes, and simultaneous recordings were made with high-gain amplifiers.

The special load-applicator consisted of a lever made from an oak beam, fixed at one end to a heavy frame by a hinge to provide the fulcrum (fig. 12.19). The bent knee of the subject could be placed under the beam, and by use of leverage, loads of 100, 200 and 400 pounds could be applied through the vertical leg to the foot. These loads were chosen because 100 pounds approximates or exceeds the normal load on each foot in the upright bipedal stance, 200 pounds approximates or exceeds the load on the arch in upright unipedal stance and the 400 pound load is the maximum that can be applied without extreme discomfort at the knee. The system provides a convenient method of loading the arches while eliminating any postural effect that muscles might have on the leg and foot. To test the influence of various positions of the foot and ankle, the foot was supported on a specially constructed adjustable platform (fig. 12.19).

In these experiments, all six of the muscles, which are often considered to be important contributors to arch support, did not react to loads that actually surpassed those normally applied to the static plantigrade foot. One hundred pounds elicited little if any contraction. With loads of 200 pounds applied to one foot, a small number of the muscles showed some activity, but this was exceptional and varied with the muscle and the posture of the foot. The peroneus longus was dramatically quiescent except, perhaps, when 400 pounds was applied to the inverted foot. With this load, however, a substantial number of the other muscles also came into play (fig. 12.20).

An analysis of the forces on the arch by the method of Steindler reveals that 400 pounds does not exceed the normal forces imposed on the arch in the take-off position of walking. Our earliest study (Basmajian and Bentzon, 1954) showed a great deal of activity in the tibialis anterior, peroneus longus and intrinsic muscles of the foot when the subjects stood on tiptoe. From these earlier findings and those of the present study, one may conclude that in the standing-at-ease posture muscle activity is not required and the muscles are inactive; however, in positions in which excessive stresses are applied, as in the take-off phase of gait, the muscles do react. Without any question, the first line of defense is provided by the passive structures. During activity the muscles would appear to contribute to the normal maintenance of the longitudinal arches.

This brings us back to the conclusion of Harris and Beath that the normal foot is supported both by passive factors (bones and ligaments) and by active factors (muscles) and that these factors are reciprocal. They stated that, in

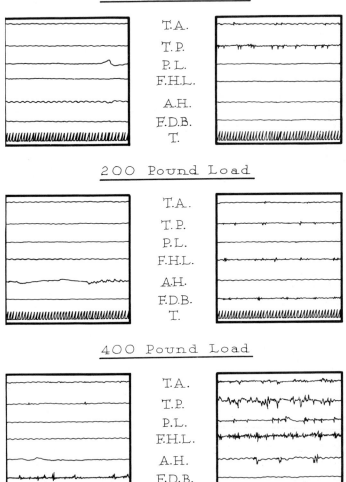

Fig. 12.20. Sample tracings of electromyographic recordings while the foot is on the horizontal platform to show range of activity from minimum (*left side*) to maximum (*right side*) activity with loads of 100, 200 and 400 pounds. T.A., T.P., P.L., F.H.L., A.H. and F.D.B. are abbreviations for tibialis anterior and posterior, peroneus longus, flexor hallucis longus, abductor hallucis and flexor digitorum brevis. T. is the time signal with a frequency of 100 cycles per second and an amplitude of 200 μv. (From Basmajian and Stecko, 1963.)

the average or strong foot, most of the support is provided by passive factors, little load being supported by the muscles. Much greater stability and strength for the foot can be provided by the passive support of a well designed skeleton than from the active support provided by muscles. However, they concluded that the muscles always play some part in maintaining balance and in supporting the load. This latter opinion is shown by the present experiments and those of Miyoshi (1966) to be incorrect for the static foot.

Harris and Beath stated further that the strong foot is one in which the tarsal bones are so articulated with each other that the weight of the body is borne without appreciable movement between them. With a strong foot, the muscles are used to maintain balance, to adjust the foot to uneven ground, and, of course, to propel the body in walking and running. The weak foot is one in which the tarsal bones are so shaped and are so disposed that they are unstable and shift in position when weight is superimposed. Only by increasing the support provided by the muscles can the normal shape of the foot be maintained and the body weight supported.

There is a limit to the contribution that muscles can make. They cannot function unremittingly, nor can they provide the powerful support furnished by the skeleton. The first line of defense of the arches is ligamentous. The muscles form a dynamic reserve, called upon reflexly by excessive loads including the take-off phase in walking.

Undoubtedly, a study of a series of subjects with flat feet, using the techniques of the present research would help to clarify the situation. Thus Gray and I (1968) found that more than half of the flat-footed subjects had activity in the muscles tested (tibialis anterior, tibialis posterior, peroneus longus and soleus). We concluded that the use of these muscles as a dynamic reserve in attempting to maintain the arch was real. We greatly expanded this study using fine-wire electrodes in tibialis anterior, tibialis posterior, flexor hallucis longus, peroneus longus, abductor hallucis and flexor digitorum brevis. Our findings lend further support to the theory that in the *imbalanced* foot muscular activity does occur, apparently reflexly. Gray (1969) elaborated further on these findings, showing that flat feet always are accompanied by activity in tibialis anterior, tibialis posterior and peroneus longus. The interesting finding of Novozamsky and Buchberger (1970) lends strong support to all that is written in the previous pages: 44 persons went on a 100-km hike; those who had normal feet maintained their arches, but those who had fallen arches before, had a further drop.

Plantaris

Clinically this small and sometimes absent muscle attracts the attention of surgeons after achilles-tendon rupture, when the plantaris tendon usually remains intact. Even with complete achilles-tendon ruptures, patients can plantar flex to some degree. Therefore, we determined the precise action of normal plantaris in 12 persons during voluntary movements of the knee and ankle and everyday activities of standing, walking and stair-climbing (Iida and Basmajian, 1975). To provide a reference, we studied the lateral head of gastrocnemius simultaneously.

The plantaris is very active when plantar-flexion occurs in full knee extension. As the angle of the knee decreases, the amplitude of activity falls progressively, apparently because the muscle becomes shorter and loses its mechanical advantage. Also, plantaris is a two-joint muscle (p. 156).

There are two types of inversion. In 'inversion 1,' there is no change of the ankle angle; in 'inversion 2' the foot is inverted while plantar flexing the ankle. In inversion 1, plantaris activity is slight and unaffected by knee angle; the foot is adducted and twisted in such a way that the medial border is raised and there is little or no plantar flexion; however, the plantaris may contract slightly to assist the twisting movement. In inversion 2, the plantaris muscle is strongly active although somewhat less than in plantar flexion, because plantar flexion always accompanies the medial rotation of the pendant foot so that the sole is directed inward.

In eversion, the plantaris muscle is slightly active only in full knee extension (in some subjects only) because knee flexion shortens the muscle. In flexion of knee in the prone position without resistance there is no activity until resistance is offered—when there is moderate activity. The moderate activity in some subjects during flexion of the knee in the standing position is related.

The moderate activity in climbing upstairs for all subjects—and in level walking and knee flexion in standing in some—suggests that the plantaris muscle assists the function of the knee in loading situations, although perhaps this is not a primary action of plantaris muscle. Slight activity in plantaris in some subjects while leaning forward occurs because the mechanical axis falls in front of the geometric center of the supporting bone. As the heel loses contact with the floor, strong contraction of the plantar flexors, including plantaris, quickly checks the forward movement of the trunk and the subject attains a new stable position on the balls of the foot.

Silence of plantaris in level walking, knee flexion in standing, leaning backward, left, and right, standing on forward and backward slopes, descending stairs and knee flexion in the prone position without resistance is pronounced. Apparently all these movements need neither plantar flexion nor inversion 2, nor strong flexion of the knee joint, all these being the chief actions of plantaris.

Difference of function between the plantaris and the lateral gastrocnemius emerges in eversion and in leaning right. The origin of plantaris muscle on the lateral femoral condyle is higher and thus closer to the axis of the femur than is the origin of lateral gastrocnemius; and the insertion is more central on the posterior aspect of the calcaneus. Thus, the momental line of plantaris is more vertical and parallel to the leg axis than that of the lateral gastrocnemius, in which the line is oblique from lateral to medial. Thus, in leaning right, when the body weight is shifted right, the mechanical axis falls lateral to the geometric center of supporting bone, and in order to hold the balanced standing position, the right foot must be everted, resulting in moderate activity in the lateral gastrocnemius muscle.

Free Movements of Ankle

Our own studies and those of O'Connell (1958) have confirmed the classical teaching in regard to the importance of tibialis anterior in dorsiflexion of the ankle with the assistance of extensor digitorum longus and extensor hallucis (fig. 12.21). The peronei are inactive during dorsiflexion but active during plantar flexion. The activity of the peronei seems to be transmitted chiefly to the transverse tarsal joint and not the ankle joint. Walmsley (1977) showed that the peroneus longus and brevis muscles act synchronously during the stance phase. Peak activity was at full-foot both on the level and on an incline.

O'Connell and I have each found independently that tibialis anterior is not very active in producing inversion *unless dorsiflexion occurs simultaneously.* I have only scattered experiments on the tibialis posterior but these indicate that it is a powerful invertor only when the ankle is simultaneously plantar flexed.

O'Connell proved the lack of a consistent pattern of activity in the two main peronei during eversion, sometimes the one showing activity first and sometimes the other. Moreover, the lateral part of the soleus appeared to become active during eversion while the medial part was active during inversion. This suggests a bipartite behavior of that muscle which requires further investigation. Arienti (1948) has also suggested patterns of activity of different parts of the triceps surae (i.e., gastrocnemius and soleus) during walking on a treadmill rather than unanimity of action. (In the upper limb, we have shown differences in the heads of biceps brachii—Basmajian and Latif, 1957.)

O'Connell and Mortensen (1957) reported on the activity of various muscles of the leg with the limb elevated. The action of tibialis anterior was variable during inversion (which, as seen above, must vary with the concurrent dorsiflexion or plantar flexion of the foot). It was strongly active during

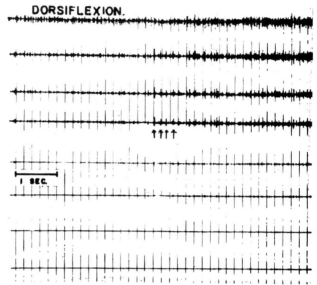

Fig. 12.21. EMGs of various leg muscles during dorsiflexion. (From O'Connell, 1958.)

forced eversion, a finding which remains unexplained and unconfirmed. One would question this result except for the knowledge of the integrity of the observers.

Houtz and Walsh (1959) compared the activity of the soleus and gastrocnemius in walking and in rising on tip toes. During the "stance phase" of walking, the activity was less in these muscles than in rising on tip toes. In other words, they seem to be stabilizers during walking. Apparently, rising up on the toes is not a normal part of the ordinary gait; this conclusion threatens the hallowed "push-off" concept of walking and might well be correct. Sheffield, Gersten and Mastellone (1956) showed in a simultaneous study of leg muscles during walking that the dorsiflexors (tibialis anterior and the two long extensors of the toes) act in unison during the swing phase, obviously to provide adequate clearance of the ground. Early in the stance phase there is a greater burst of activity in them, apparently to stabilize the foot on the ground. Sheffield's group appears to have shown that the plantar flexors are stabilizers during the stance phase. They also noted a paradoxical activity in soleus (with none in gastrocnemius) during the swing phase.

Houtz and Fischer (1959) found a surprising amount of activity in tibialis anterior during the pedalling of a bicycle. This must be due to the stabilizing function since the foot is already forced into dorsiflexion by the pedal itself. Gastrocnemius showed considerably less activity than tibialis anterior and, predictably, its occurrence was in the exactly opposite phase of pedalling (see fig. 12.5 on p. 243).

In an elaborate study of walking, Hirschberg and Nathanson (1952) described the patterns of activity in many of the lower limb muscles. In individuals, a consistent pattern of activity was found. Only one group, the calf muscles, started to contract in the middle of the stance phase, and these were the most active muscles. During the swing phase only the anterior tibial group contracted strongly. Frequently, a burst of activity occurred in gastrocnemius in the middle of the swing phase (see fig. 12.6 on p. 247). In a brief article, Richter (1966) reported similar findings.

Intrinsic Foot Muscles

Since the first edition of this book a number of new emg studies have been added to our first report (Basmajian and Bentzon, 1954). With needle electrodes, we studied abductor hallucis, flexor digitorum brevis and abductor digiti minimi in 12 men and 2 women. To test the muscles of the foot it was necessary to have the subject stand on a pair of raised blocks with a narrow interval between to accommodate the projecting electrode.

Generally, there was very little activity in these muscles while standing. This has been confirmed by Mann and Inman (1964). Almost all the abductors of the great toe and the short flexors of the toes showed electromyographic silence. A quarter of the abductors of the little toe showed no activity while more than half showed negligible or doubtful activity. In several cases there was marked activity in abductor hallucis and this was found to be due to "digging in" of the great toe. This activity was more or less abolished immediately when the subject straightened his toe.

When the subjects rose on tip toes there was a marked activity in the intrinsic muscles. This was confirmed in a later study by Oota (1956b). In the take-off stage of walking we found a similar marked activity and this was confirmed by Sheffield, Gersten and Mastellone (1956). (See also p. 181).

There was no consistent significant relationship between the types of feet and the types of activity in any muscle. However, Mann and Inman found that the pronated foot requires greater intrinsic muscle activity than does the normal foot, "to stabilize the transverse tarsal and subtalar joints."

As suggested before, it now seems to be beyond controversy that tibialis anterior, peroneus longus and the intrinsic muscles of the foot play no important active rôle in the normal static support of the long arches of the foot. Our figures show that these muscles are completely inactive electromyographically in many normal individuals while standing. Furthermore, when a subject suddenly lowers himself rapidly from a raised seated position to a direct relaxed standing position, there is little or no appearance of activity.

However, voluntary and visible efforts to increase the arch of the foot is accompanied by marked activity.

Mann and Inman (1964) in their excellent detailed study showed that the intrinsic muscles of the foot act as a group in many movements, especially the abductors of the great and little toes, the flexor hallucis and the flexor digitorum brevis. During walking on the level, they become active at or about the 35% mark of the whole walking cycle (fig. 12.22). (But activity is earlier with flat-footed subjects.) Activity always ceases just before toe-off. When walking on a downslope, the start of activity again is advanced, often occurring from the onset of the cycle (as the heel strikes). Mann and Inman relate the activity in intrinsic muscles to the progressive supination at the subtalar joint. They believe that an important rôle of the intrinsic muscles is the stabilization of the foot during propulsion, acting mainly at the subtalar and transverse tarsal joints. The pronated foot requires greater intrinsic muscle activity than does the normal foot.

Mann and Inman agree with us (Basmajian and Bentzon, 1954) that activity is not needed to support the arches of the fully loaded foot at rest. To fully investigate this, we investigated loading of the static foot while completely removing the factor of posture (Basmajian and Stecko, 1963); see above (p. 267).

Extensors of Toes

The activities of the extensor brevis of the little and great toes were compared by Carvalho, König and Vitti (1967) with needle electrodes in 20 persons. In half the cases the muscles did not act entirely synchronously. During walking they did not show a clear pattern. In different subjects the recruitment pattern showed different behaviour. The extensor digitorum brevis was often silent during gait. Carvalho and Vitti (1965) also described the EMG of the long extensor of the great toe which they found to be silent in ordinary stance but becoming active when the subjects swayed backwards and also during dorsiflexion of the ankle.

Movements of the Hallux

The great toe is not just a useless appendix. In human locomotion it is the last contact with the ground at takeoff. However, conscious control of its functions is usually primitive. Further, hallux valgus (with associated bunions) is a major surgical problem blamed on muscle imbalance. The opposite—hallux varus—also has been the object of surgical curiosity and intervention (Thompson, 1960). By implication or direct statements, authorities

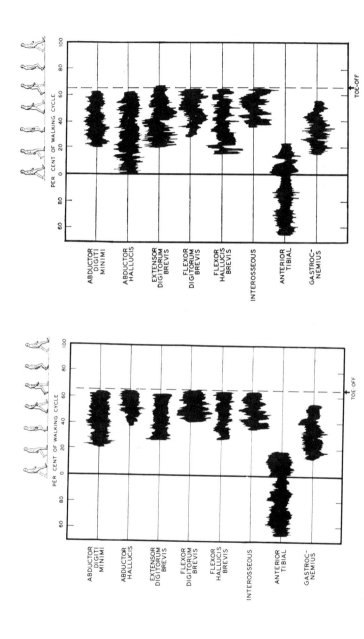

FIG. 12.22. Cycle of emg activity during walking in muscles of the leg and foot (composite of superimposed multiple records). *Left*, normal; *right*, flat-footed subjects. (From Mann and Inman, 1964.)

cite the rôles of contracture, imbalance, or predominance of activity in the adductor hallucis muscle over that in the abductor or *vice versa*. But no clear experimental proof of muscle imbalance exists. Therefore, we made an electromyographic investigation of 25 normal persons and ten patients with "idiopathic" hallux valgus to test the hypothesis of muscle imbalance (Iida and Basmajian, 1974). We chose muscles that insert at the MP joint of the great toe—abductor hallucis, adductor hallucis (both transverse and oblique heads) and flexor hallucis brevis. Although other muscles also act on the great toe, for simplification we have limited our enquiry to muscles acting on the MP joint of the great toe. Our experimental results indicate that the abductor force in normal subjects balances the adductor force at the MP joint. However, in hallux valgus, while the adductor force is markedly decreased, the abductor force actually falls to *nil* and so a weak adductor force becomes operative.

We found that in hallux valgus the activity of (1) adductor hallucis is markedly decreased in adduction; (2) flexor hallucis brevis is slightly decreased only in loading situations; (3) abductor hallucis has completely *nil* activity during abduction; but (4) adductor hallucis and abductor hallucis muscle show their strongest activity in flexion.

The changed patterns of activity can be explained as secondary to the following mechanical changes: with bony malalignment, the distance between origin and insertion of the intrinsic muscles is increased; so, the insertion of adductor hallucis moves toward the medioplantar side and the tendons of flexor hallucis brevis are displaced laterally; now, they will bowstring in the loading situation. As a result, the adductor hallucis and flexor hallucis brevis are stretched, with the adductor losing its force markedly and the flexor, slightly. The tendon of the abductor hallucis moves plantarwards in relation to the metatarsal head, and so it completely loses its abductor force, instead gaining flexor force. Thus, the imbalance is in the direction of adduction. But there also occurs a rotation of the proximal phalanx into an inverted position; this is due to imbalance caused by the changed function of the abductor muscle which is dragged into the orientation of a flexor. Now the long axis of the first ray is shortened by the pull of the adductor force—further enhancing the valgus tendency.

Morphology of Hallux Muscles

In an unpublished morphological study, J. W. Kerr and I made the following observations on 22 adult feet. These feet were dissected with particular emphasis on the insertion of the abductor hallucis muscle.

A great variation in the mode of insertion of the abductor hallucis tendon was revealed. In only one of the specimens did the tendon lie on the medial border of the foot and insert into the medial side of the base of the proximal phalanx in such a fashion as to be an obvious abductor.

At the other end of the scale, in several specimens the abductor hallucis and medial head of the flexor brevis had a common insertion into the base of the medial sesamoid, with the abductor tendon lying on the plantar aspect of the foot as an obvious flexor.

Between these two extremes, we found that in some specimens the abductor tendon was on the plantar aspect of the foot and, passing over the medial portion of the sesamoid without attachment, it inserted into the base of the proximal phalanx. In a quarter of the cases, a slip of insertion was given off the lateral side of the abductor to the medial sesamoid before its insertion on the phalanx. In another quarter, there was a common slip of insertion with the medial head of the short flexor into the sesamoid.

We concluded that in about one-fifth of our specimens the abductor hallucis was so placed as to be capable of true abduction. In most, the abductor must have acted at the metatarsophalangeal joint to flex the great toe. The abductor hallucis was always closely attached to the capsule of the metatarsophalangeal joint as it crossed it.

Great variation in the attachment of the medial head of the flexor brevis to the abductor tendon was found. In every case, there was an attachment between the two muscles proximal to the sesamoid. The insertion of the medial head of the short flexor was always into the ventral surface of the abductor hallucis tendon.

The Plantar Reflex

Landau and Clare (1959) analyzed the normal and abnormal plantar reflex (Babinski sign) electromyographically. The flexor response shows variable patterns of muscle contraction, while the abnormal extensor response shows both hyperexcitability and stereotypy. The unique feature of the abnormal extensor response is the recruitment of anterior crural muscles—extensor hallucis longus, tibialis anterior and extensor digitorum longus. Then there is an actual mechanical competition between the flexors and extensors of the great toe and it is the latter that triumph. If, perchance, the extensors are weak or denervated, flexion occurs when the Babinski extensor sign would be expected. Thus Landau and Clare concluded that the extensor reflex is really just a hyperactive flexor response with radiation to the extensors which, proving stronger as a rule, produce extension of the great toe (see fig. 4.11 on p. 99). However, caution must be exercised in accepting this conclusion, for Grimby (1963b) has shown that the plantar response is a complex phenomenon and not a simple reflex. It depends to a great deal on the exact area of skin stimulated (Grimby, 1963a; Engberg, 1964). While not disagreeing with this point of Grimby's, van Gijn (1975, 1976) more emphatically agrees with Landau and Clare and greatly stresses the powerful emg reaction of extensor hallucis longus during the Babinski response.

Chapter *13*

The back

THE muscle masses filling the space between transverse and spinous processes on both sides of the vertebral column are known as the deep, oblique or transversospinal muscle groups. Commonly they comprise (from superficial to deep): the semispinales, the multifidi and the rotatores muscles. Generally these are not considered to be part of the erector spinae. The superficial muscles of the transversospinal tract are accepted as the semispinales in the neck and thoracic region. Multifidi exist deep to the semispinales but are superficial in the lumbar area where there is no semispinalis. The next layer, the rotatores, are 11 pairs of small muscles of the thorax deep to the multifidi, best seen in the thoracic region.

Although others have studied the intrinsic muscles of the back, no names are better known in this field than those of Floyd (a physiologist) and Silver (an anatomist) of Middlesex Hospital Medical School in London. During the 1950's these two investigators broadened our understanding of the erector spinae, and this chapter will lean heavily on their reports. Further detailed studies are called for because the regional or local differences in the structure and function of the many muscles in this group must be explained. Furthermore, clinical conditions exist—the most important of which being scoliosis—that demand clarification.

The tentative effort of Riddle and Roaf (1955) was a step in the right direction. However, their conclusion that deep rotator muscle paralysis is the cause of idiopathic scoliosis cannot be substantiated by their published findings, though it may well be the truth (Basmajian, 1955b). Zuk (1960, 1962a, 1962b) of Warsaw demonstrated increased muscular activity on the convex side of the scoliotic curve. He believed it to be a secondary reaction of the body in an attempt to compensate for the curvature, the cause of which he blamed on "muscle imbalance." His series of patients examined electrically numbered some 250.

Hoogmartens (of Louvain) and I (1976), in our 1972 research at the Georgia Mental Health Institute in Atlanta, employed computer analysis of vibration-induced electromyography (VEMG) with 24 members of scoliosis families who had some measurable degree of scoliosis. Our hypothesis was that a concave-sided hypersensitivity of muscle spindles is responsible for idiopathic scoliosis. Since spindles are sensitive to vibration, we induced a VEMG bilaterally in the back muscles, using special apparatus. Of 23 thoracic curves, 15 showed vibration hypersensitivity of the spindle system on the concave side and 5 on the convex side.

Horn (1969) and Redford *et al.* (1969) have made valuable contributions to the EMG of scoliosis, while various authors have published clinical papers that reflect some light on spinal function (Pauly and Steele, 1966; Yamaji and Misu, 1968; Thomas, 1969).

Turning to the emg studies of the normal back, we must observe the early work of Lewer Allen in South Africa. In 1948, he reported a study with the emphasis on erector spinae. His chief conclusion was that the erector spinae is active during forward flexion of the vertebral column. Therefore, one main function is to control "paying out," which, in this muscle, is as important as the function of extension (fig. 13.1).

In very rapid flexion, little or no activity is required or in fact appears. As the slowly flexing trunk is lowered, the activity in erector spinae increases apace and then decreases to quiescence when full flexion is reached. If an attempt is made then to force flexion further, silence continues to prevail in the erector. In full flexion, then, the weight of the torso is borne by the posterior ligaments and fasciae—the posterior common ligament, the ligamentum flavum, the interspinous ligaments and the thick dorsal aponeurosis. The erector spinae again comes into action when the trunk is raised once more to the erect position.

In standing erect, Lewer Allen believed that activity in the erector spinae is not required except for forced extension. He seems to have concluded in this admittedly early and not completely elegant study that no activity in the erector spinae results during extension except with added resistance. This is a dubious conclusion which has never been either explained or confirmed.

Soon after, Floyd and Silver (1950) confirmed the main finding of Lewer Allen. They pointed out that Fick (1911), without the benefits of electromyography, had hypothesized the complete relaxation of the erector spinae in full flexion of the spine. Their experiments were, on the whole, more elegant than any others done on the back in the 1950's. With multiple surface electrodes on the skin over the muscle at the levels of T 10 and 12 and L 2 and 4 and the added use of needle electrodes for confirmation, they were able to show the activity of multifidus as well as erector spinae (fig. 13.2).

Their findings showed that in the initial stages of flexion of the trunk in bending forward the movement is controlled by the intrinsic muscles of the back. They agreed with Lewer Allen that the ligaments take over and were quite sufficient in the fully flexed position.

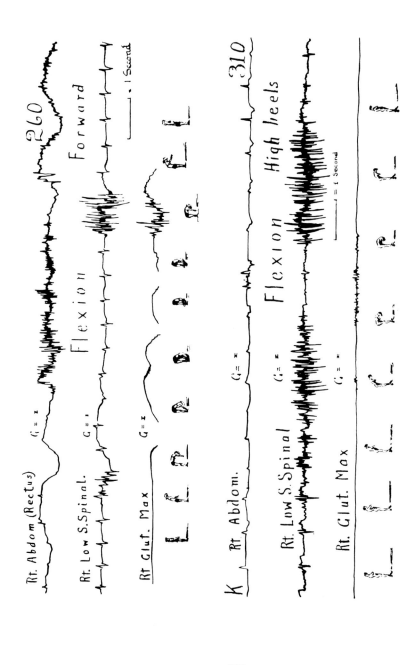

Fig. 13.1. EMGs of rectus abdominis, erector spinae (sacrospinalis) and gluteus maximus in two different subjects, one wearing high heels (*lower tracings*). Synchronized drawings to show phase of forward flexion of trunk. (Composite of parts of two illustrations from Allen, 1948; photograph retouched to improve engraving.)

283

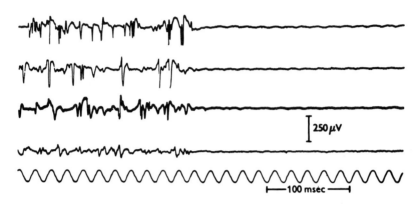

250 μV

├──100 msec──┤

Fig. 13.2. EMG of flexion-relaxation of erectores spinae. Recorded with needle electrodes at depths of 1, 2, 3 and 4 cm, at the level of L 3 vertebra. Similar EMGs were obtained at all depths. (From Floyd and Silver, 1955.)

They also showed that the position of full flexion while seated (usually considered by school teachers as a "bad" posture) is maintained comfortably for long periods and that during this time erector spinae remains relaxed. Quite correctly, they warned against jumping to conclusions, pointing out that "certain people experience backache if they sit in the fully flexed position for a sufficient time—e.g., patients sitting in bed with only the thoracic part of the vertebral column supported, motor-car drivers, etc." Floyd and Silver suggested, perhaps too cautiously in the light of our present knowledge, that a reflex inhibitory mechanism explains the complete relaxation of erector spinae in full flexion. Finally, they suggested that this relaxation of the muscle and the dependence on the ligaments, including the intervertebral disc, had implicit dangers including injuries to the disc.

In a later extensive series of investigations Floyd and Silver (1955) examined the function of the erector spinae in certain postures and movements and during weight-lifting. They used both surface and confirmatory needle electrodes for the thoracolumbar parts of the erector spinae. Posture was recorded by photography, direct measurements and radiography.

Most subjects standing in a relaxed erect posture showed a "low level of discharge" in the erector spinae. Small adjustments of the position of the head, shoulders or hands could be made which would abolish the activity of the muscle, i.e., an equilibrium or balance could be achieved.

From the easy upright posture, Floyd and Silver found that extension (hyperextension) of the trunk is initiated, as a rule, by a short burst of activity. Their findings during flexion of the trunk were described before (see p. 183 and fig. 9.5).

While standing upright, flexion of the trunk to one side is accompanied by activity of the erector spinae of the opposite side, i.e., the muscle is not a prime mover, but an "antagonist." However, if the back is already arched in extension (hyperextension) not even this sort of activity occurs.

Floyd and Silver state that erectores spinae contract (apparently vigorously) during coughing and straining. This occurs even in the midst of their normal silence whether the subject is erect or "full-flexed." The clinical implications of this last observation have not been, in my point of view, adequately explored by orthopedic specialists.

With the subject standing, the activity in erector spinae ceases earlier during forward bending than it does when he is seated. In some patients they found complete relaxation in the sitting, but not the standing posture.

Finally, Floyd and Silver reported that the erector spinae remained relaxed during the initial movement of lifting weights of up to 56 pounds (28.5 kg). They proved that it is movement at the hip joint that accounts for the earliest phase of apparent extension of the trunk. However, the ligaments of the back were required to carry the added weight without help from the adjacent muscles (fig. 13.3).

The studies on the erector spinae of Åkerblom (1948), Portnoy and Morin (1956) and of Joseph and McColl (reported by Joseph in his book *Man's Posture*, 1960), are especially concerned with posture and are discussed in that chapter on page 183. In essence they agree with Floyd and Silver.

In 1958, Friedebold of Berlin reported a study on the mode of action of erector spinae in a series of women who carried out a series of movements and postures of the trunk. In addition to confirming in general the earlier studies, this report enlarged upon the activity during lateral flexion. Most impressive is the recording of activity from both right and left erectores during bending to either side, though there seems to be a pattern of cooperative activity and not a simple simultaneous antagonism (fig. 13.4).

A new dimension in the EMG of intrinsic muscles of the back was added by Morris, Benner and Lucas (1962) and Waters and Morris (1972) of San Francisco. They investigated the activity of different layers and parts of the spinal musculature—iliocostalis in the thoracic and lumbar parts, longissimus and rotatores in the region of the 9th and 10th ribs and multifidus abreast of the 5th lumbar spine.

During the performance of various trunk movements, muscles showed patterns of activity that clearly showed two functions—sometimes they initiate movement and at other times they stabilize the trunk. Almost all the

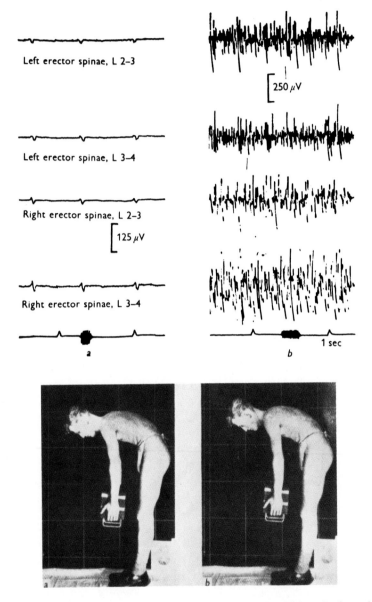

Fig. 13.3. EMGs of erectores spinae during weight-lifting. Left and right sides at levels L 2 to 3 and L 3 to 4, with corresponding photographs. The subject lifted a 28-pound weight from the ground without activity in erectores spinae until the trunk reached a position intermediate between those shown in photographs. (From Floyd and Silver, 1955.)

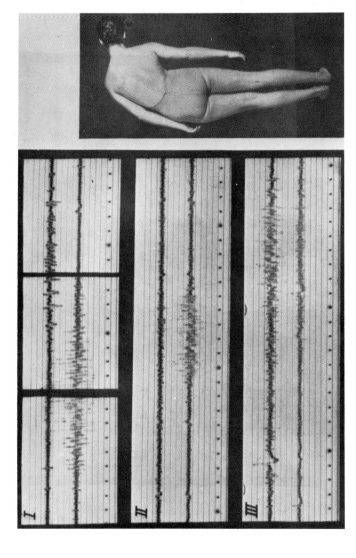

Fɪɢ. 13.4. EMGs of right and left erectores spinae during lateral bending. *I*, upper lumbar region; *II*, middle lumbar region and *III*, corresponding points at both levels reciprocally. (Surface electrodes used, but are not shown.) (From Friedebold, 1958.)

movements recruit all the muscles of the back in a variety of patterns, although the predominance of certain muscles is also obvious.

In compound movements, when subjects are not trying to relax, there is constantly more activity than when the movement is carried out deliberately and with conscious effort to avoid unnecessary activity of muscles. Complete relaxation and lower levels of contraction are the "ideal" rather than the rule for normal bending movements. Morris *et al.* found that muscles that might be expected to return the spine to the vertical position often remain quiet; they suggest that such factors as ligaments and passive muscle elasticity play an important role.

During easy standing, *longissimus* is slightly to moderately active; it can be relaxed by gentle ("relaxed") extension of the spine. During forced full extension, flexion, lateral flexion and rotation in different positions of the trunk, it is almost always prominently active.

A position of complete silence is easily found for *iliocostalis* in the erect position, but with slight forward swaying activity is instantly recruited. Forward flexion and rotation in the flexed position bring out its strongest contractions; but it is also fairly active in most movements of the spine.

Multifidus and rotatores have rather similar but not identical activity. With movements in the sagittal plane, they are active as they also are in contralateral rotary movements. Yet like all the other muscles, these too relax almost completely during full flexion, leaving the trunk practically hanging on its ligaments.

More recently, Pauly (1966) conducted a systematic exploration of the intrinsic muscle of the spinal column during various exercises widely advocated for physical fitness. At the same time he has clearly confirmed the earlier studies of the back and has revealed new aspects of the normal functioning of these muscles. Unlike some earlier workers, he finds that semispinalis capitis and cervicis apparently help to support the head by continuous activity during upright posture.

In almost all vigorous exercises performed from the orthograde position, Pauly finds that the most active muscle is spinalis; next in order is longissimus, and least is iliocostalis lumborum. Nevertheless all three muscles and the main mass of erector spinae act powerfully during strong arching of the back in the prone posture. During push-ups, there is considerable individual variation, but typically, the lower back muscles remain relaxed.

Simple side-bending exercises of the trunk do not recruit erector spinae so long as there is no concomitant backward or forward bending. This clearly refutes earlier opinions whose authors had ignored movements in the ventro-dorsal plane that do involve erector spinae. Much of this work has been confirmed by Jonsson (1970) and the technique has been adopted for ergonomics by Tichauer (1971).

Deep Muscles

Donisch and I (1972) tried to clarify the following question: what can we find out about the deep muscles close to the vertebral column by examining them bilaterally at different levels? Although it might have been interesting to duplicate the experimental conditions of the previous investigators, present-day regulations, improved equipment, and a different population of subjects ruled against a simple replication.

The deep layers of the transversospinal back muscles were studied in 25 healthy human subjects. Bipolar fine-wire electrodes were inserted bilaterally at the level of the sixth thoracic and the third lumbar spinous processes. Activity was registered simultaneously in sitting and standing, and during movements while in these positions. We found that the same muscle group displayed different patterns of activity in the thoracic compared to the lumbar level. Variations in the pattern of activity during forward flexion, extension and axial rotation suggest that the transversospinal muscles adjust the motion between individual vertebrae. The experimental evidence confirms the anatomical hypothesis that the multifidi are stabilizers rather than prime movers of the whole vertebral column.

Because of the complexity of the intervertebral joint mechanism and the impossibility of efficiently preventing motion between individual vertebrae, we preferred to study actual free movements rather than resisted ones in our experiments even at the risk of electrode displacement. While it was the purpose of this investigation to learn about the functions of these muscles during postures and movement, the interpretation of electrical activity presented some difficulties. Did a muscle showing activity produce the movement, prevent the movement, or was it contracting isometrically? We therefore emphasized the occurrence of electrical silence, knowing that the muscle tested was not taking part in the movement under observation. Decreasing or increasing activity during a movement also seemed to be functionally more important than unchanging activity.

Changes in the Lumbar Curvatures During Sitting and Standing

In most subjects the lumbar muscles were inactive during relaxed sitting but showed some activity in straight sitting and in the standing posture. This finding is in agreement with the results of most other workers, including Eastman and Kamon (1976) and Andersson *et al.* (1975). This last very productive group in Gothenburg, Sweden have found that disc pressure and myoelectric activity change together. When the back of a seated subject is supported, levels of both pressure and EMG fall. They also confirmed that intramuscular wire electrodes are superior to surface electrodes in the study of intrinsic back muscles (Andersson *et al.*, 1974).

Forward Flexion in Sitting and Standing Positions

As noted before, Floyd and Silver (1951) first showed that the back muscles became electromyographically inactive at a critical point during extreme flexion. Morris *et al.* (1962) found that flexion-relaxation can occur, but they felt that in normal bending movements the back muscles remained frequently active. We found spontaneous electrical silence of the lumbar muscles in extreme flexion in most of our subjects, but only half of them showed spontaneous inactivity of their thoracic muscles in both seated and standing postures.

During the Valsalva maneuver with increased intrathoracic and abdominal pressure while holding a sandbag of 11.25 kg, all thoracic and a number of lumbar muscles showed activity instead of electrical silence. This might be explained partly by the fact that most subjects were no longer in extreme flexion and might have reached the "critical point" of activity.

Extension from the Flexed to the Upright Posture

While inactivity of the back muscles during the last stage of flexion can be explained in that the muscles are no longer needed and ligaments are holding the vertebral column, we have no explanation why these muscles do not always become active immediately when extension is begun. Instead we frequently have short "bursts" of activity that occur (especially in the lumbar region) when the movement of extension is half completed. It therefore appears that in most persons the lumbar transversospinal muscles do not initiate extension from the fully flexed position.

Lifting weights with different mechanical advantages seemed to indicate that in most instances more energy is used in the lumbar and thoracic back muscles when the object lifted cannot be brought close to the line of gravity of the subject. Pauly (1966) and other investigators also noticed increased activity of the back muscles when the centre of gravity was shifted forward.

Axial Trunk Rotation

In our experiment less than half of the examined subjects showed this expected activity of the transversospinal muscles of the thoracic region, whereas more than half of our subjects showed the expected activity in the lumbar region. This finding is somewhat surprising considering that most of the actual rotatory movement occurs in the thoracic region. Paradoxical activity of the deep muscles was found in 5 subjects at the thoracic and in 3 persons at the lumbar level. Morris *et al.* (1962) also occasionally found activity of the rotatores and multifidi during ipsilateral rotation. These workers further stated that little activity was seen when their subjects returned to the original position.

In our experience all subjects showed bilateral activity in the thoracic level; in more than half of our subjects this activity did not appear to be related to the direction of rotatory movement (fig. 13.5). In the lumbar region the muscular activity seemed more often to support the theory of rotatory function. On the other hand, the position of articular facets in relation to the direction of muscle pull casts doubt on the anatomical feasibility of such a function.

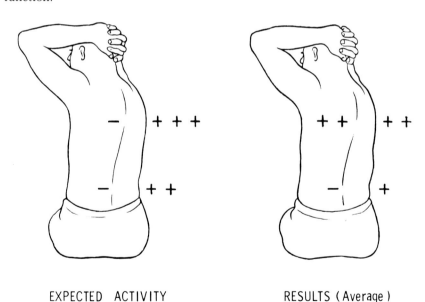

EXPECTED ACTIVITY   RESULTS (Average)

Fig. 13.5. EMG of deep back muscles during rotation. (From Donisch and Basmajian, 1972.)

Perhaps the designation of specific function is almost impossible in the back, where we have a complex arrangement of muscle bundles acting on a multitude of equally complex joints. Those who insist on finding prime movers, antagonists and synergists in the genuine musculature of the back, will be always disappointed. For example, in their wire-electrode study of the intrinsic muscles, the Tokyo orthopedics team of Iida *et al.* (1976) were unable to provide strong validation for re-education exercises widely advocated and used in France (Piette, 1974). Rather than abandon the exercises, they suggested that the transversospinal muscles are stabilizers and this function is important.

Standing

Joseph (1960) found continuous activity of the back muscles in the lower thoracic region during standing. He concluded that the activity of these muscles depended on their relation to the line of gravity. The segments of the vertebral column located further posterior to the line of gravity had the tendency to fall forward, a movement that was counteracted by the back muscles. Pauly's (1966) results were similar. In his experience the spinalis muscle showed more activity during the orthograde position than the erector spinae muscles at lower and more lateral levels. We found this greater activity of the thoracic transversospinal muscles not only in the orthograde posture but also frequently during other postures. The thoracic muscles showed a greater tendency to remain active, while the lumbar muscles acted with "bursts" of electrical potentials.

Both Joseph (1963) and Pauly (1966) stated that some muscles apparently contracted unnecessarily. These contractions were more often seen in women and untrained men. We found the same tendency in our subjects.

Asymmetry

There are some differences in activity of the transversospinal muscles at the same levels. This asymmetrical activity occurred during quiet sitting and standing, but was also noted with movements in the sagittal plane. Jonsson (1970, 1973) explained these differences of electrical activity by asymmetrical posture.

Our results in relation to the mechanical advantage, centre and line of gravity and the possible axis of movement, confirm the idea that the transversospinal muscles act as dynamic ligaments (MacConaill and Basmajian, 1969). These adjust small movements between individual vertebrae, while movements of the vertebral column probably are performed by muscles with better leverage and mechanical advantage (for details, see Donisch and Basmajian, 1972).

Conclusions

This chapter will be ended by emphasizing what must have become apparent to the informed reader. Electromyography has a great deal of practical value in this area and, aside from some general—but important—observations recorded above, much remains to be learned by this technique, especially about the fine functioning of various areas and depths of the intrinsic muscles of the back.

Chapter *14*

Human locomotion

O F all the electromyography done in recent years, studies on locomotion give the greatest promise of practical application. Yet the
sad fact is that such application is slow in coming. At least one
reason for this delay is a lack of synthesis of the various findings. These
remain relatively isolated and therefore meaningless to those who might use
the information. This chapter will be devoted to an attempt at such a
synthesis. Taken with the foregoing chapters on Posture, Limbs and Back, it
gathers in one place all that appears to be significant on the use of EMG in
studies of human locomotion. Locomotion and gait are not synonymous, but
relatively little emg research has gone beyond simple walking. Yet the title of
this chapter remains what it is in the hope that it will act as a goad for new
investigations.

The neurophysiology of locomotor automatism of terrestrial animals is
extensively considered in a review by Shik and Orlovsky (1976). In general,
many of the most important mechanisms—cortical, subthalamic, midbrain,
spinal and peripheral afferent—are still confusing or obscure. Cohen and
Gans (1975) revealed both constancies and variations in the limb-muscle
EMG of rats; however, temporal relationships between the onset of activity
in specific muscles and the gait cycle are remarkably constant. In view of
some similarities to human gait among the apes, readers may wish to refer to
papers by Ishida *et al.* (1974), Tuttle and Basmajian (1976) and Tuttle,
Basmajian and Ishida (1975).

Lower Limb in Gait

Although movements of the trunk and upper limb play a rôle in normal
walking (and will be discussed at the end of the chapter), the activity of the
lower limbs evokes the greatest interest. In study after study, my colleagues
and I have noted that walking elicits very slight emg activity in the thigh
and leg muscles compared to voluntary free movements. This has been
remarked on by others also (e.g., Koczocik-Przedpelska *et al.*, 1966).

A number of laboratories have devoted many years of study to the movements of the joints and accompanying muscular activity. Prominent among them was the Biomechanics Laboratory in the University of California at San Francisco, which has conducted a long series of studies since 1944. The early investigations there dealt with the problems of amputees, and their work remains unique. Normal body mechanics and gait have occupied a greater part of the Laboratory's time in the past decade. Particularly useful, but outside the scope of the present review, are other California studies on the energy cost of various types of gait (reviewed by Ralston, 1964, and added to by Delhez *et al.*, 1969). The emg studies of Liberson and the biomechanical studies of Elftman are also especially important and will be drawn upon below.

In any study of gait, EMG by itself would be of limited value. If quantitative evaluations are to be made, it must be supplemented with other biomechanical techniques. Photographic methods particularly high-frequency cinematography, have been used since the classic studies of Muybridge (1887). Marey (1885) in France, Braune and Fischer (1895) in Germany, and Bernstein (1935, 1940) in Russia, all greatly improved the techniques of photographic cyclograms. Liberson (1936) was the first to combine these methods with electronic accelerometers while Schwartz, Trautman and Heath (1936) introduced the recording of contacts of various parts of the foot with the ground (with an instrument called an electrobasograph). This later led to the use of walking on force-plates in which multiple, electronic, force and displacement transducers are incorporated. The latter technique has been used intensively in several centres, but its expensiveness in equipment and time has proved forbidding for most laboratories. Nevertheless, Carlsöö (1962, 1973) and Carlsöö and Skoglund (1969) have designed a fairly simple apparatus, as has the group at Purdue University (Ismail *et al.*, 1965). Magora and his team (1970) at Hadassah University Hospital in Jerusalem and Grundy *et al.* (1975) in Manchester have also developed excellent apparatus.

A number of investigators have been exploring the use of telemetering for gait studies (Joseph, 1968; Rainaut, 1971). Rainaut has also described a workable pneumographic foot switch. Brandell *et al.* (1968) developed a miniaturized portable EMG tape recorder device which is carried by running subjects and free of all external connections. Winter and Quanbury (1975) and Dubo *et al.* (1975) described an excellent working technique developed by a large team of workers in Winnipeg, Manitoba, led until the mid-70's by David A. Winter, now of the University of Waterloo in Ontario. Much of their work employed multifactorial computer analysis (see below). In Italy, the work on telemetering by Casarin *et al.* (1974) and Pedotti (1977) is worthy of note.

Accelerometers appear to offer additional useful and tidy results when combined with multichannel EMG. While our early experience tended to confirm this, difficulties in calibration and jiggling of the transducers on the subjects led us away—temporarily, I believe—from these devices. At this time, however, Liberson and his colleagues (1962, 1965) of Chicago (now in Miami) and Smidt's group at Iowa University (Smidt *et al.*, 1971; Smidt, 1974) have the longest and most fruitful experience with accelerographs applied to gait. *In vivo* recording of tendon strain during walking has been reported in sheep by Kear and Smith (1975); some day it may be applicable to man.

The results of the San Francisco studies led Saunders, Inman and Eberhart (1953) to define the six major determinants of human gait as: (1) pelvic rotation, (2) pelvic tilt, (3) knee flexion, (4) hip flexion, (5) knee and ankle interaction and (6) lateral pelvic displacement. Actually, the phenomenon of walking is much more complex; yet these are the components that provide the unifying principles. Locomotion is "the translation of the center of gravity through space along a path requiring the least expenditure of energy." Pathological gait "may be viewed as an attempt to preserve as low a level of energy consumption as possible by exaggerations of the motions at the unaffected levels." When a person loses one determinant from the above six, compensation is reasonably effective. Loss of the determinant at the knee proves the most costly, according to Saunders *et al.* Loss of two determinants makes effective compensation impossible; the cost in terms of energy consumption triples and apparently discourages the patient to the point of his admitting defeat.

Multifactorial Analysis with Computer

The use of computer analysis for gait is gaining acceptance as improved techniques of pattern recognition are developed in many research centres; its use with indwelling fine-wire electrodes was first reported by my colleagues and myself from the National Research Council of Canada in Ottawa (Milner *et al.*, 1971; Kasvand *et al.*, 1971). Our study was designed to provide some insight into the precise timing of vastus lateralis, the long head of biceps femoris (the lateral hamstring), tibialis anterior and lateral gastrocnemius, in the course of normal walking. There were two sets of experimental conditions: (1) level walking at various speeds in the range 2.2 to 7.5 feet per second

(0.67 to 2.28 m per second); and (2) level walking at a speed nominally fixed at 4.5 feet per second (1.37 m per second) but with pace periods set by metronome beats in the range 0.2 to 0.8 beats per second. As suggested by Eberhard *et al.* (1954), are average electromyographic values dependent upon walking speed and pace frequency?

Emg signals from indwelling electrodes and footswitch data together with other control signals (including footswitches) to facilitate automatic processing of the data were collected on five channels of a seven-channel FM tape recorder. The tape was input to the A/D converters of an SDS 920 computer via signal conditioners. A series of computer programs enabled the automatic processing of the data which were first digitized. Next the footswitch timing characteristics were analyzed and averages obtained. Using these analyzed data the emg data were subdivided and a mean wave form for each muscle at each step computed and various data output in typewritten and graphic form. From the processed data average emg characteristics were plotted as functions of speed and pace periods; phasic activities for these various conditions were plotted for 6 normal adult male subjects.

For details, see our lengthy paper (Milner *et al.*, 1971) with its many illustrations. Only the highlights are given here. The average EMG in each of the several muscles showed clearly that there is a strong dependence in each instance of the emg values upon speed. Perhaps the most interesting result was exhibition of a minimal value of emg activity in tibialis anterior. Since emg value is a measure of muscular energy, we concluded that a minimal energy condition can be found for most subjects in the range of 3.00 to 5.00 feet per second (0.91 to 1.52 m per second). A generalization about comfortable, optimal walking speeds seems probable when the emg speed characteristics of the other muscle groups are examined in this speed range. The most comfortable walking speed lies close to 3.00 feet per second (0.91 m per second) to minimize "energy consumption" and permit a reasonable propulsion speed. The "energy cost" would appear to increase considerably as speeds in excess of 6 feet per second (1.82 m per second) are attained, at least under experimental conditions.

When a subject is permitted to walk without the imposition of a pace frequency constraint, he selects a walking pace for the set speed in such a manner as to allow a minimum of muscular activity. During the course of such a walk, random variations in emg activity seem to be more marked. This gives an indication of some measure of ongoing adaptive control. The technique of averaging emg records was designed primarily to eliminate the differences between successive steps and to determine a meaningful average value (Milner *et al.*, 1971).

More recent studies have polished up our findings and extended them (Bajd *et al.*, 1974; Richards and Knutsson, 1974; Wait *et al.*, 1974; Winter *et al.*, 1974; Cheng *et al.*, 1974; Cappozzo *et al.*, 1975; van der Straaten *et al.*, 1975; Kasvand *et al.*, 1976; Hershler and Milner, 1976; Takebe and Basmajian, 1976). Pearson (1976) drew special attention to studies of neural mechanisms in his semi-popular article in *Scientific American*.

Phases in Walking Cycle

Traditionally, human gait is composed of two phases: (1) stance, beginning when the heel strikes the ground, and (2) swing, beginning with toeing-off. The fairly precise division of the whole cycle into 10% segments is illustrated in figure 14.1. Carlsöö (1966) showed that the initiation of walking from a stance posture consists of the body losing its balance as a result of cessation of activity in postural muscles (including erector spinae and certain thigh and leg muscles). The various torques of the body weight displace the line of gravity, first laterally and dorsally, and then ventrally, to a position in which the propulsive muscles are able to contribute to and complete the first step.

Radcliffe's (1962) diagram (fig. 14.1) illustrates the interaction between the knee and ankle joints and the phasic action of the major muscle groups recorded electromyographically. (The terms "knee moment" and "ankle moment" refer to the action of muscles about the knee or ankle which tend to change the angle of these joints towards either flexion or extension.)

In figure 14.1, one should not miss the following features:

As the heel strikes the ground the hamstrings and pretibial muscles reach their peak of activity.

Thereafter, the quadriceps increases in activity as the torso is carried forward over the limb, apparently in maintaining knee stability.

At heel-off the calf group of muscles build up a crescendo of activity which ceases with the toe-off. Before and during toe-off, quadriceps and sometimes the hamstrings reach another (but smaller) peak of activity.

The pretibial muscles maintain some activity all through the cycle, rising to a peak at heel-contact and a smaller peak at toe-off. This has been often confirmed (e.g., by Dubo *et al.*, 1976).

The complex phasing of these normal actions has led the San Francisco investigators to studies of amputees and prostheses (Radcliffe, 1962). Similar work on the muscle functions in lower amputees is in progress in Poland under the direction of J. Tomaszewska (1964) of Poznan, and M. Weiss (1959, 1966) of Warsaw. There is little doubt that all such work will stimulate

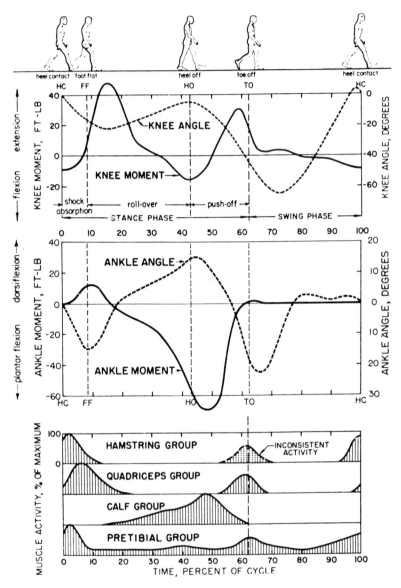

Fig. 14.1. Normal walking: knee and ankle moments (in foot-pounds) compared with muscular activity during one cycle of walking (right heel to right heel contact) on level ground. (From Radcliffe, 1962.)

improvements in prosthetic appliances of both the conventional and myo-electrically-controlled types.

In a sober review, Elftman (1966), a pioneer in the multifactorial approach for the study of gait, gives the apt warning that electromyography requires the addition of other criteria for the monitoring of tension in non-isometric contractions. Studies of the calf and foot during walking have in the past two decades employed such adjuncts with notable success. Eberhart, Inman and Bresler (1954) showed that the function of the calf muscles during walking is limited to the push-off. They lift the body against gravity on the forepart of the foot.

Radcliffe (1962) postulated that during the stance phase the stabilizing function of the ankle plantar-flexors at the knee is most important. This was confirmed by Sutherland (1966) by means of combined EMGs and motion pictures of gait. The period of activity in the calf muscles and of knee extension and dorsiflexion of the foot corresponded. Only at the end of plantar flexion of the ankle did plantar flexion of the foot occur. A bizarre finding was that knee extension occurred after quadriceps activity had ceased. This is related to the fact that full extension of the knee never occurs during walking in the way that it does in standing (Murray *et al.*, 1964).

Sutherland believes that knee extension in the stance phase is brought about by the force of the plantar flexors of the ankle resisting the dorsiflexion of the ankle; this dorsiflexion is in turn the resultant of extrinsic forces—kinetic forces, gravity and the reaction of the floor. Because the resultant of extrinsic forces proves to be greater, increased dorsiflexion of the foot continues until heel-off begins. The restraining function of the ankle plantar flexors in decelerating forward rotation of the tibia on the talus proves to be the key to their stabilizing action.

Using similar techniques, Houtz and Fischer (1961) have produced evidence that a movement of the torso and hip region that shifts their position over the feet initiates the movements of each foot during walking. Movements initiated in the trunk lead automatically to changes in the position of the leg and foot. Houtz and Walsh (1959), by showing that soleus functions to stabilize and adjust the tibia on the talus, gave additional evidence for the view that movements of the ankle during walking occur as a reaction to muscular forces far removed from the foot. (p. 274).

Liberson (1965b), combining the techniques of motion picture photography, accelerograms, electrogoniograms, myograms and EMGs, has reported the following correlation of activity (figs. 14.2, 14.3, 14.4).

1. Contraction of the triceps surae is followed by that of gluteus maximus on the opposite side.

Fig. 14.2. Diagram of typical multifactorial gait-recording, showing on the *left* a motion picture frame and on the *right* oscillograms, the terminal points of which correspond to the instant the picture of the walking subject was taken. *A*, angular accelerometer on left leg; *B* and *C*, vertical and horizontal accelerometers, respectively; *D*, Lissner strain gauge tensiometer on the left gastrocnemius muscle; *E*, electrodes in the left gastrocnemius muscle. *A'*, accelerogram of the left leg; *B'*, vertical accelerogram; *C'*, horizontal accelerogram; *D'*, tensiogram from left gastrocnemius muscle; and *E'*, electromyogram from left gastrocnemius muscle. Note that emg activity precedes the major tensiogram deflection and the relationship of the latter to the accelerograms. (From Liberson, 1965b.)

2. Contraction of iliopsoas occurs simultaneously with that of gluteus maximus of the opposite side.

3. Dorsiflexion of the foot begins at the time of maximum acceleration of the lower leg.

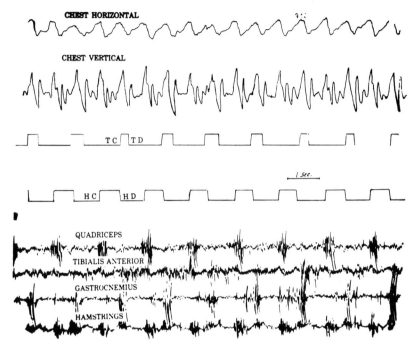

Fig. 14.3. Series of multifactorial gait-tracings from a normal subject when walking; listed from top down: horizontal and vertical accelerograms; right toe switch (TC, toe contact and TD, toe off); right heel switch (HC, heel strike and HD, heel off); and four electromyograms from muscles in the right lower extremity. (From Liberson, 1965b.)

4. Extension of the knee begins at the time of maximum velocity of the leg.

5. Contraction of the triceps surae corresponds to the first hump of the vertical accelerogram.

6. Contraction of the gluteus maximus on the opposite side corresponds to the second hump of the vertical accelerogram.

7. In many cases, two-joint muscles show an increase of tension without emg potentials because they act as simple ligaments during the contraction of the antagonists.

Two comprehensive emg studies of gait have been completed by my students and myself: on muscles in the region of the leg and foot—Gray and Basmajian (1968); and in the region of the hip joint—Basmajian and Greenlaw (1968) and Greenlaw (1973). Other multifactorial studies are now in progress.

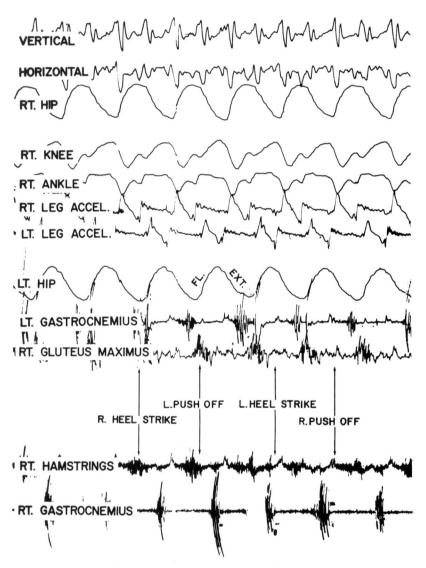

FIG. 14.4. Series of multifactorial gait-tracings from normal subject during walking, listed from top down: vertical and horizontal accelerograms; goniograms from right hip, knee and ankle; angular accelerograms from right and left legs; left hip goniogram; and electromyograms from left gastrocnemius, right gluteus maximus, right hamstrings and right gastrocnemius muscles. (From Liberson, 1965b.)

Leg and Foot Muscles

Tibialis Anterior

This muscle has been a favored object of attention, and so it is commonly accepted that peak emg activity occurs in it at heel-strike of the stance phase. Our movies show the foot to be inverted and dorsiflexed at this time.

Notwithstanding the above, there has been no general agreement as to the function of tibialis anterior at heel-strike. Without offering direct evidence, some suggest only that it counteracts forces applied to the heel by the ground, while others propose that the tibialis anterior decelerates the foot at heel-strike and lowers it to the ground by gradual lengthening (eccentric contraction). Perhaps the clinical condition known as "drop-foot" due to paralysis of the tibialis anterior forces this conclusion.

During the more central moments of the stance phase (full-foot, mid-stance and heel-off) modern techniques reveal no tibialis anterior activity in "normal" subjects. Our flatfooted subjects and those of Battye and Joseph (1966) are like "normals" except for extended activity into full-foot. Curiously, the movies of our flatfooted subjects show the foot staying inverted during full-foot, maintaining inversion in order to distribute the body weight along its lateral border.

A peak of emg activity that occurs at toe-off of the stance phase is apparently related to dorsiflexion of the ankle, presumably to permit the toes to clear the floor.

Although earlier workers believed that there is a slight fall in the activity of tibialis anterior at mid-swing, there is, in fact, a period of electrical *silence* at mid-swing. The explanation emerges from our movies which show the foot everting at the end of "acceleration" and remaining everted through mid-swing. This allows for adequate clearance, while the inactivity of the invertor fits the concept of reciprocal inhibition of antagonists. We conclude that the brief period of electrical silence of tibialis anterior is essential.

The peak of activity at toe-off tapers to a slight-to-moderate level of activity during acceleration of the swing phase. Conversely, prior activity in deceleration of the swing phase builds up to a peak of activity at heel-strike. Thus, the pattern of activity of tibialis anterior is biphasic. Apparently, tibialis anterior is in part responsible for dorsiflexion during acceleration and for inversion of the foot during deceleration of the swing phase.

The pattern of activity of tibialis anterior suggests that it does not lend itself to direct support of the arches during walking. At heel-strike, when the muscle shows its greatest activity, the pressure of body weight is negligible. Conversely, during maximum weight-bearing at mid-stance when all the body weight is balanced on one foot, the tibialis anterior is silent. When the activity resumes at toe-off, the weight-bearing of the involved foot is minimal

Tibialis Posterior

During ordinary walking tibialis posterior shows activity at mid-stance of the stance phase. Our movies show the foot remaining inverted throughout full-foot and turning to a neutral position (between inversion and eversion) just before mid-stance. First, the fourth and fifth metatarsal heads make contact; then, as the foot everts increasingly toward neutral, more of the ball of the foot makes contact at mid-stance until the tibialis posterior acts. It is an invertor in non-weight-bearing movements of the foot, but its role at "mid-stance" appears to be a restraining one to prevent the foot from everting past the neutral position.

R. L. Jones (1941, 1945) showed in human cadaveric preparations that the tibialis posterior distributes body weight among the heads of the metatarsals. In living subjects he showed that a lateral torque on the tibia results in an increase or shift of body weight onto all but the first metatarsal head; a medial torque has the opposite effect. He concluded that by inverting the instep of the foot the tibialis posterior increases the proportion of body weight borne by the lateral side of the foot. Sutherland (1966) concluded that the plantar flexors, including the tibialis posterior, have a restraining function, to control or decelerate medial rotation of the leg and thigh observed at mid-stance; by controlling the eversion of the foot at mid-stance, the tibialis posterior provides an appropriate placement of the foot. In our flatfooted subjects, the emg activity of tibialis posterior in the early stance phase is consistent with the maintenance of an inverted position during full-foot. By maintaining inversion the foot is supported in order to keep the body weight on the lateral border of the sole.

The foot must be inverted to accomplish lateral weight-bearing in the early "moments" of the stance phase. This of course is because the middle part of the medial border of the foot does not bear body weight in "normal" subjects; the lateral border with its strong plantar ligaments is well-equipped to bear the stresses of body weight in walking (Napier, 1957).

Although tibialis posterior is often considered to be a plantar flexor of the ankle, during level walking with an accustomed foot position it shows *nil* activity at heel-off (when plantar flexion of the ankle takes place to raise the heel). (This is not to deny that tibialis posterior may be a plantar flexor of the ankle when more powerful contractions are needed.)

Flexor Hallucis Longus

At mid-stance, when the entire body weight is concentrated on one foot, flexor hallucis longus shows its greatest activity. Flexing the big toe apparently positions and stabilizes it during mid-stance. During heel-off, our movies show the big toes hyperextended. Although Napier (1957) felt that the flexor hallucis longus helps maintain overall balance and prevents instability induced by excessive extension of the big toe, our emg observations support this only for the flatfooted subjects and then only weakly. There is a slight activity during heel-off which may be related to preventing overextension and so giving a better balance. In contrast, the "normal" subjects show negligible activity. Consequently, one may conclude that the flexor hallucis longus is not needed in most "normal" subjects to play this rôle.

Peroneus Longus

The pattern of activity of the peroneus longus confirms the findings of many who have suggested that the peroneus longus helps to stabilize the leg and foot during mid-stance. Our movies and electromyograms show how the peroneus longus and tibialis posterior, working in concert, control the shift from inversion during full-foot to neutral at mid-stance. Thus, the opinion of R. L. Jones is again confirmed; from static studies, he inferred that peroneus longus is related to eversion of the foot at mid-stance during level walking. Sutherland further concluded that peroneus longus, like tibialis posterior, is involved in controlling rotary movements at the ankle and foot. We found that eversion of the foot and medial rotation of the lower limb occur together. One may conclude that the peroneus longus is in part responsible for returning the foot to, and maintaining it in, a neutral position at mid-stance. As noted before, Walmsley (1977) found peroneus brevis to act synchronously with the peroneus longus during ordinary walking.

Throughout most of the stance phase, peroneus longus is generally more active in flatfooted subjects than in "normal" subjects. This appears to be a compensatory mechanism called forth by faulty architecture.

During heel-off, our movies showed some inversion while peroneus longus, an evertor, is active, and the invertors are relaxed. Duchenne (1867) first suggested that the inversion is caused by triceps surae. We believe the activity in peroneus longus affords stability by preventing excessive inversion, thus maintaining appropriate contact with the ground.

In flatfooted subjects, the interplay of activity between peroneus longus and tibialis posterior appears to play a special rôle in stabilizing the foot during mid-stance and heel-off. At mid-stance the tibialis posterior is notably more active, but at heel-off the emphasis shifts to peroneus longus.

Abductor Hallucis and Flexor Digitorum Brevis

These two muscles become active at mid-stance and continue through to toe-off in "normal" subjects; in flatfooted subjects most show activity from heel-strike to toe-off. Perhaps the flexor digitorum brevis and abductor try to grip the ground since they are flexors of the toes. Although others are not opposed to this idea, they believe that the muscles are also in an ideal location to help support the arches. Our findings tend to confirm this opinion only for flatfooted subjects because they showed higher mean levels of muscular activity.

Triceps Surae

Dubo *et al.* (1976) found a single high peak of activity during the push-off of the gait cycle. At heel contact, mean muscle activity was a fifth of maximum and this continued thus through foot flat to mid-stance. As the foot enters push-off the gastrocnemius sharply increases its activity. Following this there is a rapid drop that continues as the swing starts and reaches zero at mid-swing. Then as the transition to stance begins, activity once more increases.

Hip and Thigh Muscles

Our definitive locomotion studies of the muscles that cross the hip joint (Greenlaw and Basmajian) have appeared extensively only in the form of a Ph.D. thesis by Greenlaw (1973) as well as being given at meetings (Basmajian and Greenlaw, 1968). The essence of our findings is given below.

Gluteus Medius et Minimus

In the *anterior fibres of gluteus medius* there is moderate activity at heel-contact that persists through to mid-stance. There is also a brief burst at toe-off and another just before heel-contact. The posterior fibres are rather (but not exactly) similar. *Gluteus minimus* has only a biphasic response (at heel-contact to 40% and at mid-swing).

Tensor Fasciae Latae

Its pattern is biphasic with a peak during early stance through mid-swing and another short smaller peak during toe-off.

Gluteus Maximus

Upper, middle and lower parts were all tested simultaneously. *The upper part* shows a clearly biphasic pattern with a small peak at heel-strike and one near the end of swing phase. *The middle part* is more triphasic with an additional high peak just-before to just-after toe-off. *The lower part* is biphasic, rather like the upper fibres.

Hamstrings

Semitendinosus has a triphasic pattern with an initial low peak at heel-contact, a second peak at 50% of the cycle and a small third one just before the end (90%) of the cycle. *Semimembranosus* is biphasic, lacking the peak in the middle. *Biceps femoris* is also biphasic, but more crisply so.

Sartorius and Rectus Femoris

Rectus femoris is generally biphasic or triphasic depending on cadence. Sartorius really shows only one peak, immediately during toe-off.

Iliopsoas

Iliacus acts continuously through the walking cycle with some rises and falls. The highest rise is during the swing phase but there is another in mid-stance.

Psoas is triphasic (except with slow cadence); the main peaks correspond to those of iliacus, with a third peak at 50% of the cycle.

Adductors

Adductor Magnus is really two muscles. The upper (horizontal) part is active nearly continuously; it reaches *nil* only at mid-swing. The lower (vertical) part acts like a biphasic hamstring.

Adductor Brevis varies with the speed of walking. At moderate speed it is biphasic—at 40% and 90% of the cycle. *Adductor Longus* and *Gracilis* also have a main peak of activity at toe-off and additional peaks at late-stance and early-swing phases.

In a brief preliminary report, Joseph (1965) described his findings of telemetered EMGs from a number of muscles used in gait. Generally, his findings were similar to those reported in the '40's and '50's by the San Francisco group. In the swing phase, the hamstrings were inactive (even though knee flexion occurred) and the tibialis anterior was also inactive—but only for a brief period. In the supporting phase, activity occurred early in the calf muscles, hamstrings and gluteus maximus but ceased toward the end. Two periods of activity were found in the sacrospinalis (at level L 3): one during the swinging and one during the supporting phase.

Later, Battye and Joseph (1966) reported details of a study which, because of its clarity, will be heavily drawn upon below, supplemented by the findings of Greenlaw and myself:

In general, they found more similarities than differences in the walking patterns of 14 persons (8 men and 6 women). They also emphasized the importance of the inertial forces as factors in producing certain movements.

Soleus begins to contract before it lifts the heel from the ground; it stops before the great toe leaves the ground. Apparently these are supportive rather than propulsive functions.

Quadriceps femoris contracts as extension of the knee is being completed, not during the earlier part of extension when the action is probably a passive swing. Quadriceps continues to act during the early part of the supporting phase (when the knee is flexed and the centre of gravity falls behind it). Quadriceps activity occurs at the end of the supporting phase to fix the knee in extension, probably counteracting the tendency toward flexion imported by gastrocnemius. Similar findings have been reported (for vastus lateralis) by Dubo *et al.*, 1976.

The *hamstrings* contract at the end of flexion and during the early extension of the thigh apparently to prevent flexion of the thigh before the heel is on the ground and to assist the movement of the body over the supporting limb. In some persons, the hamstrings also contract a second time in the cycle during the end of the supporting phase; this may prevent hip flexion.

Gluteus medius and *gluteus minimus* are active at the time that one would predict, i.e., during the supporting phase; however, we found with wire electrodes that gluteus medius is actually triphasic while minimus is biphasic during each walking cycle (Greenlaw, 1973).

Gluteus maximus shows activity at the end of the swing and at the beginning of the supporting phase. This is contrary to the general belief that its activity is not needed for ordinary walking (Battye and Joseph, 1966; Greenlaw,

1973). Perhaps gluteus maximus contracts to prevent or to control flexion at the hip joint.

The same 6 female subjects that provided the above findings were also studied while they wore high heels (Joseph, 1968). There were comparatively few differences from the findings previously noted. These differences were: tibialis anterior contracted less strongly but more continuously; soleus was more active; quadriceps femoris contracted (either continuously or intermittently) during the stance phase; and gluteus maximus contracted during swing phase when walking with high heels.

The gait pattern of women also interested Finley *et al.* (1969) who compared elderly women with young ones. Rectus femoris, tibialis anterior, peroneus longus and gastrocnemius were recorded through surface electrodes along with a multijoint electrogoniometer complex. The elderly women showed greater activity in tibialis anterior, peroneus longus and gastrocnemius. Except for rectus femoris, the muscles of the elderly women showed considerably greater activity during foot-flat. The other differences do not appear to be functionally significant.

Comparing changes in integrated EMG of the quadriceps with those of the calf muscles under increased degrees of stress (increased slope and/or speed), Brandell (1977) reported that the vasti responded relatively more vigorously to the stress. The peak activity for the calf occurs when the extending activity of the knee stretches the gastrocnemius across the back of the knee joint, thereby helping the calf muscles as a group to lift the heel (plantar flexion), the "most essential actions for producing the push-off and thrust in the normal walking cycle."

Ascending and Descending Stairs

Although a number of investigators have included stair-climbing in their work, Joseph and Watson (1967) give the best formal results. During both ascent and descent, each limb has a support and swing phase. In walking up, the body is raised by contraction of soleus, quadriceps femoris, hamstrings and gluteus maximus (along with gluteus medius activity for hip stability). During the swing phase, tibialis anterior and the hamstrings are active as would be expected.

Walking downstairs recruits the same muscles in lowering the body, with the exception of gluteus maximus. Tibialis anterior inverts the foot at the beginning of the support phase and dorsiflexes it during the swing. The hamstrings are also active during the swing.

Both erector spinae muscles contract twice in each step, apparently in controlling bending which occurs in walking both up and down stairs.

Gait of Children

The most systematic emg studies of children's gait have been carried out by several investigators in Japan who continue with this valuable area of investigation. Okamoto and Kumamoto (1972) made long-term serial multichannel EMGs (combined with foot contact switches) in 2 infants learning to walk and in 30 subjects (ranging from infant to adult). The muscles investigated were tibialis anterior, gastrocnemius, vastus lateralis, rectus femoris, biceps femoris, gluteus maximus, sacrospinalis and deltoid.

Okamoto and Kumamoto found a varying time in development at which these muscles acquire their adult patterns. Tibialis anterior acted in an adult fashion within 1 month after infants learned to walk but the others were delayed to the middle of 2 years of age (vastus lateralis) or even later: thus, in gastrocnemius, biceps femoris, gluteus maximus and sacrospinalis adult patterns appeared at the end of 2 years of age, in the deltoid at 4 years of age.

During the 10 days after learning to walk (about a dozen steps), there is a continuous discharge in tibialis anterior during the stance phase but it drops off and by day 15, it has become adult in its pattern. Between the 50th and 85th day after learning to walk, the reciprocal discharge pattern of tibialis anterior and gastrocnemius is clearly established. Tsurumi (1969) independently found that in children, EMG of gastrocnemius appears during the push-off at 6 years of age. Although his findings agreed with those of Okamoto and Kumamoto for gastrocnemius, he found that tibialis anterior does not assume the adult role until the 5th year. Differences in technique could account for the discrepancy.

While the general adult mode of gait and its EMG patterns are recognizable from 2 years of age on, the child must reach its 7th birthday before the walking pattern truly becomes adult in form and function (Okamoto and Kumamoto, 1972; Okamoto, 1973; Kazai *et al.*, 1976). Similar results have been reported by Marciniak (1975) of Poznan, Poland.

The Human Bicycle

Human walking on a flat surface is best described as bicycling, the most economical form of mechanical transportation (Basmajian, 1976). The reason we have failed to grasp the idea of the human bicycle for so long is rooted in our failure to realize that each foot is a short segment of an invisible or imaginary wheel—unseen by us until we take a good look (fig. 14.5). Think then of two wheels set (as in a bicycle) one in advance of the other and yet

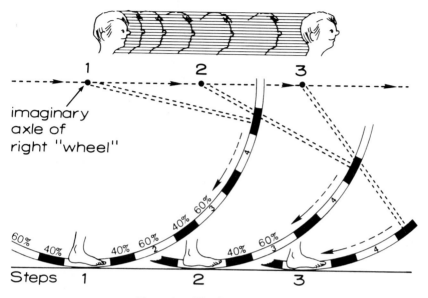

Fig. 14.5. The human bicycle

offset by the width of the footsteps on the ground (not the width of the pelvis, as one may first think). After walking starts, each wheel rolls smoothly along and remains about the same distance in front or behind the other. The lateral off-setting provides the stabilization along with complex neuromotor inputs—perhaps the same ones required during bicycle riding. Generally it is conceded that the lateral oscillations of the front wheel account for a major part of the stabilization; with a fixed front wheel the bicycle cannot be kept erect by most riders, for the wobble is a necessity.

Carlsöö showed that the initiation of walking from a stance posture consists of the body's losing its balance as a result of cessation of activity in postural muscles (including the erector spinae and certain thigh and leg muscles). The various torques of the body weight displace the line of gravity, first laterally and dorsally, and then ventrally, to a position in which the propulsive muscles are able to contribute to and complete the first step. Then the bicycle starts rolling.

Running

Increasingly more thorough papers have appeared on emg patterns during running. Hoshikawa *et al.* (1973) revealed a non-linear correlation between speed and muscular activity. Elliott and Blanksby (1976) found a high reproducibility of results from one day to the next when surface electrodes and automated analysis techniques are used.

Trunk Muscles During Gait

Erector spinae shows two periods of activity, according to Battye and Joseph (as noted before by other investigators). They occur "at intervals of half a stride when the limb is fully flexed and fully extended at the hip at the beginning and end of the supporting phase." Battye and Joseph's explanation is that the bilateral activity of the erectores spinae prevents falling forward of the body and also rotation and lateral flexion of the trunk.

More recently, and using intramuscular wire electrodes, Waters and Morris (1972) reported striking periodic electrical activity in the trunk muscles of all subjects during walking at two different speeds (4.39 and 5.29 km per hour). It occurred faithfully with each cycle and was very constant in relationship to parts of the cycle from subject to subject. The muscles included various paravertebral (erector spinae) muscles (iliocostalis in thoracic and lumbar parts, longissimus, multifidus and rotatores) as well as quadratus lumborum, obliqui externus et internus and rectus abdominis (figs. 14.6, 14.7).

ABDOMINAL MUSCLES. In addition to the work of Waters and Morris noted above, Sheffield (1962) performed a study of abdominal muscles which differs surprisingly: the abdominal muscles were reported as inactive in level walking which the wire-electrode investigations of Waters and Morris negate. Very little investigation of gait has been done otherwise in this part of the body.

Swinging of the Arm During Gait

Fernandez-Ballesteros, Buchthal and Rosenfalck (1964) of Copenhagen recorded the activity in the arms of normal subjects during walking. Buchthal and Fernandez-Ballesteros (1965) applied the same techniques to a study of patients with Parkinson's disease. The first emg abnormality to appear in the early stages of the disease is a non-rhythmical pattern with random or wrongly timed activity. Later, swinging movements disappear and this becomes quite obvious clinically. In fact, it is characteristic of Parkinson's disease.

In normal persons, the posterior and middle parts of deltoid begin to show activity slightly before the arm starts its backward swing, and this continues throughout the backward swing. The upper part of latissimus dorsi and the

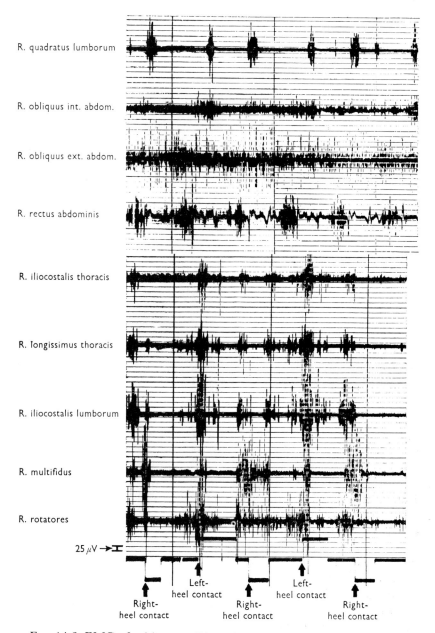

R. quadratus lumborum

R. obliquus int. abdom.

R. obliquus ext. abdom.

R. rectus abdominis

R. iliocostalis thoracis

R. longissimus thoracis

R. iliocostalis lumborum

R. multifidus

R. rotatores

25 μV

Left-
heel contact

Left-
heel contact

Right-
heel contact

Right-
heel contact

Right-
heel contact

FIG. 14.6. EMG of subjects walking. (From Waters and Morris, 1972.)

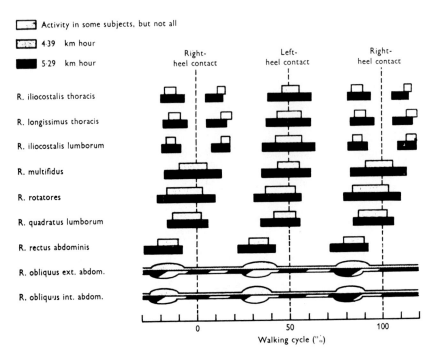

FIG. 14.7. Average durations of EMG in trunk muscles (10 subjects) during walking. (From Waters and Morris, 1972.)

teres major act from the onset of the backward swing until the arm reaches the line of the body. Similar results were reported by Hogue (1969) who also showed that the activity of scapular muscles and the muscles of the arm and forearm are silent. The following summary of the work of Fernandez-Ballesteros *et al.* does not agree with Hogue in some details however.

During forward swing of the arm, activity is confined to some of the medial rotators (subscapularis, upper part of latissimus dorsi and teres major); the main flexors are strikingly silent.

In half of their subjects, they found activity in both forward and backward swing in the rhomboids and infraspinatus. This was most marked in persons who walk with a stoop.

Apart from brief silent periods in the extreme positions of swing, trapezius is active in both phases to maintain elevation of the shoulder. Similar activity occurs in supraspinatus; this obviously is related to the prevention of downward dislocation (p. 196). However, Fernandez-Ballesteros *et al.* ascribe it (unconvincingly) to abduction of the shoulder to allow the arm to bypass the trunk. Trapezius falls silent immediately after a subject comes to a halt, contrary to earlier opinions, but corroborating Bearn's (1961b) finding.

Primate Evolution: EMG Studies

Our emg studies on the flexor muscles in the forearm of a gorilla suggest that future comparative morphological studies on the wrists of African apes may reveal special bony features related to certain close-packed positions imperative to knuckle-walking (Tuttle and Basmajian, 1972). These features may then be employed to discern evidence of knuckle-walking heritage in the wrists of other extant hominoids and to trace the history of knuckle-walking in available fossils.

The fact that the flexor digitorum profundus muscle, which constitutes approximately 44% of total forearm musculature in the gorilla, is relatively inactive during many knuckle-walking behaviours indicates that special close-packed positioning mechanisms may be operant in the metacarpophalangeal joints of digits II to V. But these mechanisms probably are not exclusive of muscle activity since the flexor digitorum superficialis and perhaps also the lumbrical and interosseous muscles may participate severally in knuckle-walking episodes.

The relative inactivity of the extensor carpi ulnaris muscle during knuckle-walking is probably related to the fact that the same basic posture of the wrist is maintained in the swing and stance phases of most slow and moderately paced progressions. During swing phase, when activity of the wrist extensors might be anticipated, elbow flexion elevates the hand clear of the floor and shoulder movements are probably chiefly responsible for its placement anteriorly.

With Russell Tuttle, and assisted by others, we conducted a series of extensive emg studies in gorilla, orang and chimpanzee from 1970 to 1975 to determine the function of both forelimb and hindlimb muscles in posture and locomotion. These studies, carried out at the Yerkes Regional Primate Research Center of Emory University, are being reported in a series of papers with Tuttle as the senior author (Tuttle and Basmajian, 1972 *et seq.*; Tuttle *et al.*, 1975).

Chapter 15

Anterior abdominal wall and perineum

S INCE 1948 a considerable number of papers have appeared which deal with the actions of either specific abdominal muscles, e.g., rectus abdominis, or in connection with specific functions, e.g., posture of the vertebral column or breathing. A few have dealt with the abdominal wall in a more general way and this method of approach will be our first concern. Though I have not published any data concerning the abdominal wall, I have had the occasional opportunity (usually during clinical emg examinations with needle electrodes) of confirming almost all of the systematic observations now to be noted.

Abdominal Wall in General

Floyd and Silver (1950) were the first to make an extensive emg study of the abdominal musculature in normal people. With a grid of paired multiple electrodes on the anterior abdominal wall (fig. 15.1), they recorded simultaneously from various parts of the rectus abdominis, the external oblique and the internal oblique where it is not covered by external oblique, i.e., in the triangular area bounded by the lateral edge of the rectus sheath, the inguinal ligament and the line joining the anterior superior iliac spine to the umbilicus. Here in this triangle, the external oblique is represented only by its aponeurosis and, quite fortunately, it is this very part of internal oblique which is of greatest interest. (They did not try to study the transversus abdominis because of its depth, admitting the possibility of a pickup from it through the electrodes for internal oblique.)

Floyd and Silver frequently found some difference between the right and left side of the abdominal musculature even when electrodes were carefully matched. They ascribed this to a basic asymmetry in function and found individual variations in the amount of difference.

319

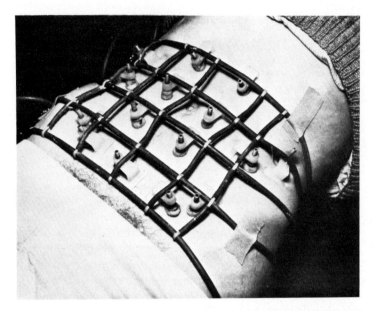

Fig. 15.1. Photograph from Floyd and Silver, 1950, to show their electrode grid for recording of abdominal EMGs.

With the subjects lying supine and resting, slight activity was found in some nervous subjects, but none was found with relaxed comfortable persons (fig. 15.2). Even in nervous subjects the activity could be reduced or abolished by proper positioning, e.g., by propping up the upper part of the trunk. Lowering of the trunk caused an exaggeration of activity.

With the "head raising" movement used commonly as an exercise for strengthening abdominal muscles, the recti were powerfully active while external oblique and the part of internal oblique that was studied were only slightly active at first. Even with increased effort they become only moderately active (fig. 15.2). These findings were confirmed later by Campbell (1952) using needle electrodes, and by Ono (1958) of Hirosaki University. One might conclude from their finding that only the rectus is benefited maximally by the head raising exercise, in spite of this exercise being advocated to increase the general "tone" of the abdominal wall. In contrast to the head-raising exercise, the bilateral leg-raising exercise brought all the abdominal musculature into activity to steady the pelvis. One-sided leg-raising was much less effective, calling upon activity predominantly on the same side of the abdomen.

In the relaxed standing position, all the electrodes except those over the lower part of the internal oblique picked up no activity. Internal oblique apparently is on constant guard over the inguinal region.

When the subjects (whether recumbent or erect) were made to strain or to "bear down" with the breath held, the external obliques and the internal obliques (lower parts) contracted to a degree that was directly related to the effort; but rectus abdominis, in contrast, was very quiet (fig. 15.3). This was later confirmed by Ono (1958). Surely it is surprising that physiotherapists have not seized upon these findings for application in the strengthening of weak or stretched obliques. Perhaps they are not dramatic enough!

Floyd and Silver found no inspiratory or expiratory activity in the abdominal muscles during quiet breathing, a finding that was later enlarged upon by Campbell (1952) (see p. 326) and by Ono (1958), Carman *et al.* (1971), de

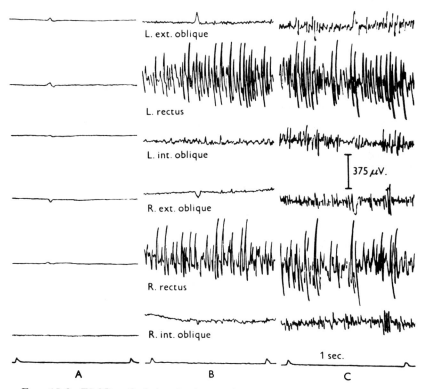

Fig. 15.2. EMGs of abdominal muscles. *A*, supine; *B*, raising head; *C*, greater effort in head-raising. (From Floyd and Silver, 1950.)

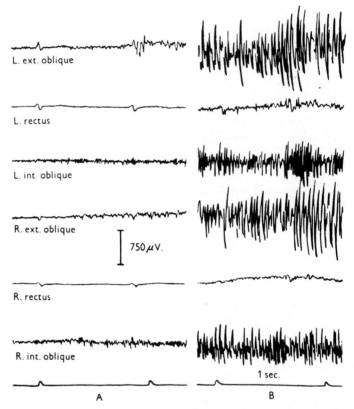

L. ext. oblique

L. rectus

L. int. oblique

R. ext. oblique

750 μV.

R. rectus

R. int. oblique

1 sec.

A   B

FIG. 15.3. Abdominal EMGs of "straining." *A*, during start of straining, and *B*, 4 seconds later (maximal effort). (From Floyd and Silver 1950.)

Sousa and Furlani (1974) and Rideau *et al.* (1975). With forced expiration, with coughing and with singing, the pattern was similar to that in straining, i.e., marked activity in the obliques and none in the rectus.

Most investigators quite correctly emphasize the importance of the rectus sheath in protecting the abdominal area occupied by the rectus during all these physiological functions which are *not* accompanied by contraction of the rectus. Therefore, they point out, repairs of the sheath and maintenance of its integrity during surgical closures is vital. It has been my own experience that surgeons erroneously think abdominal hernia is actively prevented by the activity of rectus abdominis in the region it covers. Floyd and Silver have proved conclusively that the apparent hardening of the recti on straining is usually only a passive bulging of the muscles and their sheaths. One can only hope that this knowledge, available now for a decade, will soon reach the practising surgeon. But de Sousa and Furlani (1974), while they agreed in general with all previous workers about the rectus abdominis, disagreed violently with the lack of activity reported by others in the recti *during coughing*. In every one of 20 subjects studied with needle electrodes (instead of the usual surface electrodes), they found exuberant activity in the recti bilaterally during a cough. So there is a difference, apparently, between the reactions of the recti to the increase of intra-abdominal pressure from "bearing down" and the sharp, short increase of coughing.

Walters and Partridge in 1957 (then at the University of Illinois) reported on the electromyography of the abdominal muscles during exercise and later added further information (Partridge and Walters, 1959). They found that in movements of the trunk performed without resistance in either the sitting or the standing posture the obliques and recti remained quiescent. However, lateral bending of the trunk does produce activity in the more posterolateral fibres of external oblique (a fact that is also mentioned by Campbell, 1952). Inclining the trunk backwards gives activity in all the muscles, but forward bending was, as Floyd and Silver also found, unaccompanied by activity. They also confirmed, in general, the findings of Floyd and Silver concerning forced expiration and coughing, i.e., the recti remain relaxed while the obliques contract. During forced trunk-twisting exercises the internal obliques of the side to which the twisting occurred were greatly active while both external obliques showed some slight activity and the recti showed none (unless the subject violently flexed the trunk simultaneously).

Because Partridge and Walters were concerned about exercises in the bedridden patient, they also studied the abdominal muscles during exercises in the supine position (fig. 15.4). They found that all portions of the external oblique and rectus abdominis were activated best by "a lateral bend of the trunk, pelvic tilt, straight trunk curl, and trunk curl executed with rotation." They state, in the naive jargon of physiotherapy, that "rotation of the pelvis on the thorax (hip roll) and the reverse curl are excellent activators of the internal oblique in this (supine) position."

Concerned more with athletic training, Flint and Gudgell (1965) put a series of subjects through vigorous exercises while recording EMGs from the rectus abdominis and external oblique. Their most effective exercises for bringing out the greatest activity were: the "V-sit," "basket hang" on the horizontal bar, "side-lying trunk raise," backward leaning and "curl-up." (Detailed definitions of these and other technical but explicit terms are given in their article.) Less effective were "chin-up," "pull-up," "pelvic tilt" in the supine position, isometric contraction of the abdominal wall, "low bicycle

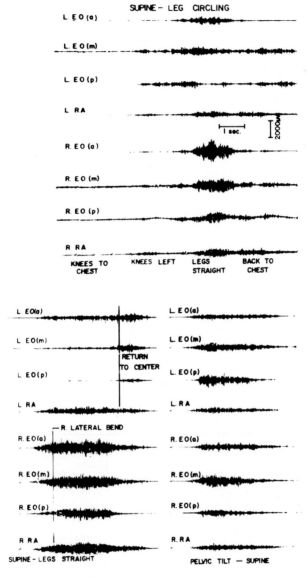

FIG. 15.4. EMGs of abdominal muscles during various activities while subject supine. (From Partridge and Walters, 1959.)

with pelvic tilt," vertical jumping and straight leg-raising in the supine position. Least effective, to the point of being useless, were "full waist circling" and vertical reach in the standing position, controlled leg extension in the supine position and "hip-roll" while lying on the mat.

Flint (1965a,b) showed further that the upper and lower parts of the rectus abdominis vary in response to different movements. Most of the activity in the recti during trunk flexion from the supine position occurs during the first half of the movement. Trunk raising elicits more activity than trunk lowering. Maximal abdominal muscle activity occurs during hook-lying (knees bent) unsupported slow and fast sit-ups (Godfrey *et al.*, 1977).

Postural Rôle of Abdominal Muscles

Lewer Allen was the first to prove conclusively with electromyography that the rectus abdominis does not draw a resistant spine forwards, but that gravity does this. Only in full-flexion does rectus show activity, apparently in an effort to force the trunk further downwards against resistance. In hyper-extension (at the other end of the range of motion) the rectus abdominis shows activity while being stretched; this apparently steadies the torso. Floyd and Silver (1950), Ono (1958) and Partridge and Walters (1959) have confirmed these findings which were predicted many years ago by Duchenne and others. Campbell (1952 *et seq.*) and Campbell and Green (1953 *et seq.*) using needle electrodes found some activity in quiet standing which might be missed by surface electrodes; but this activity was never very marked.

Morris, Lucas and Bresler (1961), Waters and Morris (1970) and Bearn (1961a) have stressed the importance of the abdominal muscles in the developing of positive pressure in the abdomen. This is an important adjunct to the vertebral column in stabilizing the trunk.

As was mentioned above, in ordinary standing the only muscle to show important continuous activity was the internal oblique, but this activity was not related to maintenance of the general posture; it will be now considered below.

Control of Inguinal Canal

In the consideration of the abdominal wall posture, we must consider in particular the lower part or region of threatening hernia. Here the inguinal canal tunnels through the muscular layers of the abdominal wall and so provides a weak spot. Through this area excessive intra-abdominal pressures (particularly while the person is standing) may force a hernia. Since, in the male, the opening transmits the ductus (or vas) deferens, it must be protected without, however, causing complete occlusion. This delicate, but dynamic, function is performed by the internal oblique and transversus abdominis—in particular by their lowest fibres which arise from the inguinal ligament. These fibres arch over the inguinal canal and insert medially on the pubic bone. One would imagine that they must be in constant contraction during standing. The work of Floyd and Silver (described above) gives ample evidence to prove this long-held opinion of anatomists and surgeons. Furthermore, one would imagine that, regardless of a person's position, straining and coughing would require increased activity in the muscular protection of the canal. Indeed, the evidence now is overwhelmingly favourable to this view (fig. 15.3).

Respiratory Rôle of Abdominal Muscles

Following upon the work of Floyd and Silver, their graduate students at Middlesex Hospital Medical School expanded that part which dealt with respiration. Campbell, joined in parts of his research by Green, performed both extensive and intensive studies, combining electromyography of the abdominal muscles and direct and accessory respiratory muscles with various other techniques, such as spirometry. The respiratory muscles will be considered in the next chapter, but here we should consider the findings for the abdominal musculature only.

Campbell's first paper (1952) confirmed and underlined the work of the earlier electromyographers who found that there was no activity in the external oblique and rectus abdominis of supine normal subjects breathing quietly. The new dimension added by Campbell was his use of needle electrodes. He showed that with maximal voluntary expiration these muscles contracted as they also did towards the end of maximal voluntary inspiration (fig. 15.5). Yet they did not contract under the latter condition when the breathing was increased by imposing asphyxia. In contrast, the activity in maximal expiration was increased further when the volume of breathing was increased by asphyxia.

Campbell reported that the great activity during expiration with hyperpnea appeared first towards the end; it was never prominent at the beginning. He concluded that abdominal contraction was a factor in limiting voluntary inspiration, but in the presence of very rapid deep breathing due to asphyxia it was inhibited. It would seem, then, that contractions of the abdominal

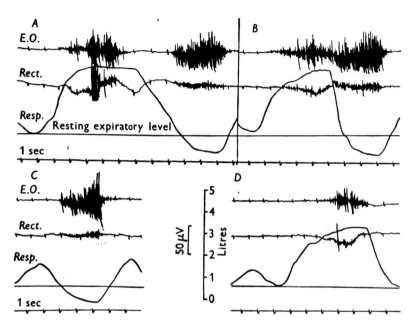

Fig. 15.5. EMGs of external oblique and rectus abdominis and spirometry of maximal inspiration and expiration. "Resp." is the spirometry trace (inspiration upward). *A*, inspiration held for 4 seconds and released, then followed by maximal inspiration; *B*, forced expiration within a second of attaining full inspiration; *C*, maximal expiration followed by normal inspiration; *D*, maximal inspiration followed by relaxation to the resting respiratory level. (Note: superimposed ECG spikes should be ignored.) (From Campbell, 1952.)

muscles to aid expiration only occurs in severe cases of greatly increased pulmonary ventilation under stress. In any case, they do not initiate the expiratory phase, but rather they help to complete it quickly. Campbell showed that a pulmonary ventilation of more than 40 litres per minute was required for the abdominal muscles to play their accessory respiratory rôle (fig. 15.6).

Campbell and Green (1955) showed that the early findings were essentially true both in the supine and in the erect posture, but that normally in the erect posture there is some continuous activity in the muscles. This activity can be abolished by certain postures and is not particularly related to respiration.

In order to summarize the present knowledge of the rôle of abdominal

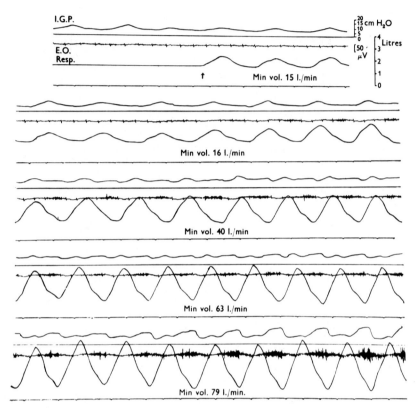

Fig. 15.6. Simultaneous EMG (E.O.), spirometry (Resp.) and recording of intra-abdominal (intragastric) pressure (I.G.P.) showing effects of progressively increasing pulmonary ventilation on the abdominal muscles. Continuous recording. (From Campbell and Green, 1955.)

muscles in respiration, it seems wise to present below the essence of the paper by Campbell (1955c):

The abdominal muscles are the most important and the only indisputable muscles of expiration in man. The obliques and transversus are much more important than the rectus abdominis.

Vigorous contraction occurs in all voluntary expiratory manoeuvres (such as coughing, straining, vomiting, etc.).

The abdominal muscles (almost exclusively the obliques) contract at the end of maximum inspiration to help limit its depth, but, in normal persons, they do not contract in hyperpneic asphyxia, apparently being inhibited by central mechanisms.

Hyperpnea with a ventilation rate of greater than 40 litres of air per minute calls upon activity of these muscles at the end—and only at the end—of expiration.

The finding by Fink (1960) of phasic expiratory activity in patients while being anesthetized does not (as Fink himself believes) invalidate Campbell's conclusion for normal subjects. Fink has, however, described a practical use of the integrated EMG of the abdominal wall muscles. He advocates it for monitoring the relaxation of the abdomen during operations in which the muscle-relaxant succinylcholine is used. I believe that the activity occurring during expiration under these circumstances is real and that the EMG provides an excellent tool for determining relaxation. Bishop (1964) has shown that continuous positive pressure breathing initiates expiratory activity of the abdominal muscles in cats. Youmans *et al.* (1974) further reviewed the control of "involuntary" or tonic responses of abdominal muscles, demonstrating in their studies that the tonus is under neural inhibitory centres which react to many influences including intragastric and intracardiac pressure, respiration, etc.

Perineum

The muscles of the human perineal region that have been investigated are the external anal sphincter, the striated sphincter of the male urethra and the striated muscles of the pelvic floor and urethra in normal women and patients with "Manchester repairs" for prolapse. These studies will be considered in that order, following which, a brief account of reflex control of micturition will be given.

Sphincter Ani Externus

Aside from Beck's early electromyographic study (1930) of the anal sphincter almost exclusively in dogs, only Floyd and Walls (1953) have reported on the electromyography of this important muscle. Beck's results are now more provocative than useful; but those of Floyd and Walls are extremely interesting and practical. I have had occasion to confirm some of them coincidentally during a study of the urethral sphincter (see below).

Floyd and Walls found that the anal sphincter is in a state of tonic contraction (fig. 15.7). The degree of this tone varies with posture and the subject's alertness, falling to a very low level during sleep. Presumably, the internal sphincter is the main agency for keeping the rectum closed during sleep. They found that the subjects can voluntarily produce an outburst of

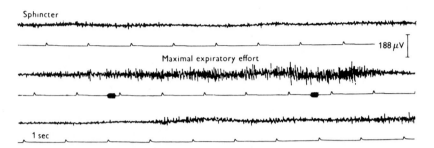

Fig. 15.7. EMG of external anal sphincter, showing resting tone and (between signal marks) the increase in tone during a maximal expiratory effort. The record is in three serial horizontal strips and reads continuously from left to right. (From Floyd and Walls, 1953.)

activity in the anal sphincter. W. B. Spring and I (1954, unpublished study) found that the contraction of sphincter ani externus is not isolated but is accompanied by general contraction of the perineal muscles, especially the sphincter urethrae. Since these muscles are of common origin from the cloacal musculature, these findings are not surprising.

With increased intra-abdominal pressure produced by straining, speaking, coughing, laughing or weight-lifting, Floyd and Walls found increased sphincteric activity related in amount to the degree of pressure. This has been confirmed by Cardus *et al.* (1963) and Scott *et al.* (1964). However, actual efforts to defecate were usually (but not always) accompanied by relaxation of the sphincter ani.

Duthie and Watts (1965) found that electrical activity in the striated external sphincter, though greatly reduced, persisted even under general anesthesia. In response to rectal distension, the sphincter showed an increased activity as the maximal rate of diminution in pressure occurred. Thus relaxation in the anal canal is independent of the action of this sphincter which contributes to pressure only when a bolus is present.

Porter (1960) also has shown that the external sphincter and the puborectalis show continuous activity at rest, heightened activity with effort and coughing, and inhibition with defecation and micturition. A critical volume brings about the desire for a bowel movement with sphincteric inhibition. After evacuation, tonic activity is restored in the external anal sphincter and puborectalis at the same time as the internal (smooth-muscle) sphincter returns to its normal contracted state (Ihre, 1974).

A growing clinical interest in electromyography of the external anal sphincter has led to the appearance of special papers on the subject. Some of these reflect normal function and may be of interest to readers (e.g., Archibald and Goldsmith, 1967; Haskell and Rovner, 1967; Bailey *et al.*, 1970; Jesel *et al.*, 1970; Ruskin, 1970; Waylonis and Aseff, 1973; Chantraine, 1974; Ihre, 1974; Blank and Magora, 1975; Schuster, 1975; Kiviat *et al.*, 1975; Lane, 1975; Frenckner and Euler, 1975; Shafik, 1975; Kiesswetter, 1976; van Gool *et al.*, 1976).

Striated Male Sphincter Urethrae

Our brief report (Basmajian and Spring, 1955) dealing with this almost inaccessible muscle, after years of virtual eclipse, has been referred to so often in recent years that it can be repeated here to advantage. Various authors have generously mentioned our early findings, which happily have been borne out by subsequent, more elaborate studies. We inserted very fine, self-retaining wire electrodes through the perineum into, or very near, the sphincter urethrae in six men during cystoscopic examination. Action potentials of the muscles were obtained and recorded on our Stanley Cox 6-channel electromyograph.

We discovered that when the bladder is empty, there are a few, occasional, small potentials in the sphincter urethrae with long periods of inactivity. As the bladder is filled slowly through the cystoscope, the action potentials increase in number. There is a continuous low level of activity in the striated muscle surrounding the membranous urethra as long as the bladder contains fluid. When the subject is instructed to micturate after the cystoscope is removed, the potentials disappear as micturition begins, and remain absent during the whole period of micturition and remain absent if the bladder is empty (fig. 15.8). Sudden voluntary stopping of micturition before the bladder is empty is accompanied by a marked outburst of potentials, the frequency of which then falls off rapidly to "the resting level."

Nesbit and Lapides (1959) in a study of male patients conclude that the striated sphincter is necessary for sudden interruption of micturition and for maintenance of continence when the vesical neck is incapacitated. Lapides is emphatic in his belief that micturition may be initiated and terminated consciously by voluntary effort without the use of any striated muscles and that urinary continence is normally maintained by the internal vesical sphincter, not by striated muscles. (See Lapides *et al.*, 1957, 1960.) On the other hand, Susset, Rabinovitch and MacKinnon (1965) clearly show that under conditions of stress, such as coughing or straining, the external sphincter is necessary and very active. A full bladder even results in minimal opening of the bladder neck on coughing, throwing added responsibility on the

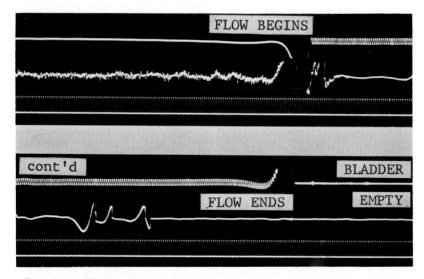

Fig. 15.8. EMG of male striated sphincter urethrae with partly filled bladder (continuous activity), during micturition (relaxation of sphincter) and with bladder empty (relaxation continues). Uppermost trace is a recording of urine flow ("flow begins—flow ends"); middle trace is EMG; and lowest trace is 10-msec time-marker.

striated muscles. Susset *et al.* have also proved the importance of striated muscle spasticity in patients with upper motor neuron lesions; in such cases the external sphincter takes over the total function, and being spastic, obstructs micturition.

Meyer Emanuel's (1965) review of control of the bladder outlet appears to rationalize the findings of many types of investigation. In men, the urinary sphincter system consists of a tubular extension of the bladder containing elastic tissue with a collar of striated muscle at the urogenital diaphragm; in women this system accounts for the whole length of the urethra. Incontinence occurs only if both ends of the sphincteric system become damaged. The striated muscle is important for interrupting micturition, but it does not maintain sustained contraction indefinitely.

Petersén and Franksson of Stockholm (1955) reported the electromyography of the male striated urethral sphincter and the bulbocavernosus muscle in 10 and 11 patients, respectively. When their patients were asked to contract their muscles to stop micturition, there was a sudden burst of activity in the sphincter urethrae, just as in our findings, and similar activity in bulbocavernosus (fig. 15.9). This agrees, in general, with my opinion mentioned above, namely, that the cloaca-derived musculature contracts simultaneously and indiscriminately when one or the other muscle is suddenly contracted. It also accounts for Muellner's (1958) erroneous belief that levatores ani and the pubococcygeus are the "primary muscles used to stop the urinary stream voluntarily." Indeed, these muscles do also contract reflexly when the sphincter urethrae contracts, but they have other functions to perform.

Scott, Quesada and Cardus (1964) have found such a close association in the activities of cloacal musculature that they use electromyography of sphincter ani externus for the routine indication of activity in the striated urethral sphincter. Study of their results reveals that they are indistinguishable from direct recordings from the urethral muscles.

To elaborate this point further, we studied the electromyography and the morphology of the external anal sphincter and external urethral sphincter in rabbits (Basmajian and J. R. Asunción, 1964, unpublished; Basmajian, Sharon McKay and Ron Hons, 1964, unpublished). The outstanding finding by both techniques was the unseparable nature of the two sphincters. The morphology is intricately linked; bundles of one sphincter intermingle in a complex fashion with bundles of the other. In rabbits (as in some other mammals) the structure is quite primitively cloacal in nature when compared with that in man.

Franksson and Petersén (1955) have also reported emg studies of patients with various disorders in micturition, convincingly demonstrating neurogenic sphincteral disturbances in some. Such studies must, of necessity, be expanded by urologists. Work along this line has been reported by Hutch and Elliott (1968).

Giovine (1959) of Milan has also presented some electromyographic results from the striated sphincter. In addition, he reported histological findings

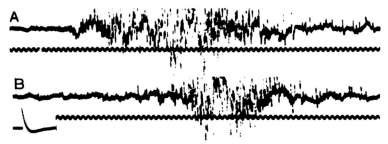

FIG. 15.9. EMG from bulbocavernosus, *A*, in voluntary contraction, and *B*, in coughing. Time in milliseconds. Calibration (*in lower left corner*), 100 μv. (From Petersén and Franksson, 1955.)

from which he concluded that the muscle is made up of red fibres to a large extent. The significance of red and white fibres not being firmly established, this matter must await further study. Meanwhile, Vereecken *et al.* (1975) have demonstrated that fatigue ensues rapidly during sustained voluntary contraction (even for 1 minute); thus the intensity of contraction waxes and wanes rather than being maintained steadily. Fatigue in the levator ani is less pronounced; women have a high voluntary control of the activity in those pelvic floor muscles (Vereecken *et al.*, 1975).

Pelvic Floor and Urethra in Women

The pelvic floor or pelvic diaphragm is mostly muscular, very important in parturition, and generally misunderstood. Electromyography, one might suppose, would have been invaluable in clearing up misunderstanding, but only one group has actually studied the musculature of the female pelvic floor—Petersén, Franksson and Danielsson (1955). This is the same Stockholm group mentioned before with the addition of a gynecologist, who then assumed senior authorship of a second paper on abnormalities of the pelvic floor (Danielsson, Franksson and Petersén, 1956).

In the first study, Petersén *et al.*, using needle electrodes and without general anesthesia, explored: (1) the pubococcygeus which is the medial or most important part of the levator ani and (2) the urethral sphincter. The electrodes were inserted through the vaginal wall in 24 normal women (about half of whom had borne children). They concluded that some subjects were able to relax the sphincter urethrae completely, while others were unable to relax it. However, none could relax the pubococcygeal part of levator ani even though they were in the "lithotomy position."

Diminution or complete cessation of activity in the sphincter urethrae at micturition (or attempted micturition) agrees with our findings in men (fig. 15.10). Furthermore, their finding that voluntary efforts to contract the one muscle automatically recruits the contraction of the others agrees with my impression, already noted above, that individual contraction in the perineum is difficult if not impossible.

More recently, Petersén *et al.* (1962) proved in women that voluntary complete relaxation of the external sphincter was possible even with a partially filled bladder. Voluntary interruption of micturition results in a rapid closing of the striated external sphincter; only afterwards does the posterior urethra empty relatively slowly in a proximal direction. Considerable variations were found in normal persons.

Petersén *et al.* (1955) reported regional differences in the activity of the sphincter, and they stated that the pattern was more or less related to whether the subject had borne children. Nulliparous subjects showed little difference in the response of the whole circumference of the sphincter while the multiparous subjects had much less or even no activity in the dorsal part of the sphincter. This dorsal part, related as closely as it is to the vagina, might have been destroyed by laceration in the course of childbirth, according to these investigators. Pubococcygeus, however, showed no difference attributable to parity. As noted earlier, women have a high voluntary control of levator ani (Vereecken *et al.*, 1975).

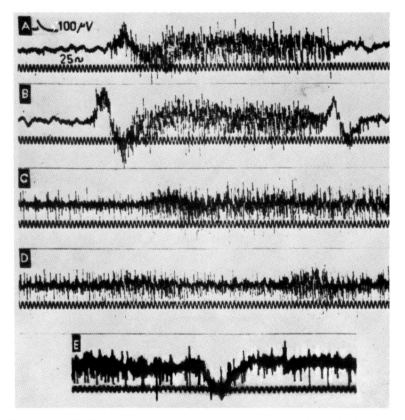

Fig. 15.10. EMGs from normal female sphincter urethrae during *A*, voluntary contraction, and *B*, a cough; and from pubococcygeus during *C*, voluntary contraction, *D*, a cough, and *E*, an attempt to micturate. (From Petersén *et al.*, 1955.)

Danielsson *et al.* performed electromyographic explorations with similar techniques on women who had recovered from the Manchester reparative operation for prolapse but who were still complaining of urinary stress incontinence. In all of these women there were no electromyographic potentials obtainable from the dorsal part of the urethra, suggesting again that the part of the sphincter urethrae next to the vagina had been torn by parturition (fig. 15.11). Lesions of the pudendal nerves, either at childbirth or at the time of the Manchester operation, may have also played a rôle.

Spinal Reflex Activity from the Bladder

Bors and Blinn (1957) of Long Beach, California, have presented convincing evidence that demonstrates the importance of the bladder mucosa in influencing the striated musculature of the pelvic floor. Thus, the sphincter ani and the sphincter urethrae contract in response to mucosal stimuli of the bladder wall. Our own finding that filling of the bladder with more than a few cubic centimetres of water starts up activity in the male sphincter urethrae is in agreement with this. The further findings of Bors and Blinn (1957, 1959) and of Pierce *et al.* (1960), demonstrate that all these reflexes

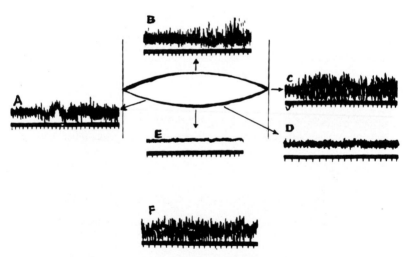

Fig. 15.11. EMG of abnormal sphincter urethrae of woman who had borne three children and had had a Manchester operation. *B,* ventral part of sphincter; *E,* dorsal part (nearest vagina); *A, C* and *D,* from sides; *F,* from normal pubococcygeus. (From Danielsson *et al.,* 1956.)

may be grouped simply under one generic term, "the bulbocavernosus reflex"; this reflex may be elicited from the stimulation of any genitourinary mucosal surface including the glans penis. (See also Nakaarai, 1968; Aranda and Jara, 1969; Allert, 1969; Ertekin and Reel, 1976; but see the contrary and less convincing view of Rattner *et al.*, 1958.)

Detrusor Muscle of Bladder

Having developed a valid technique with fine-wire electrodes for recording emg activity from the musculature of the bladder wall, La Joie, Cosgrove and Jones (1976) proved that the spike potentials were of local origin. Simultaneous emg recordings were made from the detrusor muscle and the overlying part of the rectus abdominis, showing disparity in their actions. The normal detrusor muscle is electrically silent when the bladder is empty; it shows increasing activity with filling. If the filling is interrupted, accommodation occurs. When the bladder is completely full, emg activity remains high; it gets even higher at the initiation of voiding. Then, progressively, emg activity falls and reaches zero as the bladder empties.

Relationship of Abdominal and Perineal Muscles

Bors and Blinn (1965) investigated the activity of the rectus abdominis in relation to micturition and to contractions of perineal muscles. Before or at the onset of micturition, whether *on desire* or *on volition*, the EMG of rectus abdominis usually remained unchanged; also at the normal cessation of micturition there was generally little or no reaction in most subjects. However, when either micturition was suddenly interrupted or the external anal sphincter was consciously contracted, activity in rectus abdominis usually increased. There would seem to be a related contraction of abdominal and perineal muscles, but the exact relationship is still obscure. Bors and Blinn's view is "that phasic contractions of the pelvic floor cast their shadow upon the abdominal muscles."

EMG of Ejaculation and of Penile Muscles

In a preliminary study, Kollberg, Petersén and Stener (1962) reported the train of events recorded electromyographically from the striated external sphincter urethrae and adjacent striated muscles. Some seconds before ejaculation occurs there is a lively contraction in muscles of the u.g. diaphragm.

The cause or effect of this remains obscure. It may play some role in penile engorgement just before ejaculation. On ejaculation, the sphincter (and probably its neighbours) contract rhythmically for 15 to 20 times in about 25 seconds. These contractions, which must have some part in propelling the semen, appear as salvoes of action potentials alternating with quiet intervals. Some reciprocity is also noted, i.e., activity in the muscles alternate.

The preliminary study was enlarged and reported by Petersén and Stener (1970) and Kadefors and Petersén (1970). The rhythmic emg activity in the striated urethral sphincter is split up into two parts. The first is characterized by a comparatively high-frequency content. The later part has fewer high-frequency components. Perhaps, these enterprising Swedish investigators will also succeed in recording the state of contraction of the smooth muscle at the neck of the bladder which is so often believed to prevent reflux of sperm into the bladder.

Hart and Kitchell (1966), using special bipolar needles, recorded the emg activity from the penile muscles of dogs—ischio-urethralis, bulbocavernosus and ischiocavernosus. Different patterns of reflex activity were obtained on light stimulation of different parts of the penis. Rubbing behind the bulbus glandis elicited tonic contractions of the ischio-urethralis, rhythmic contractions of the other two muscles and rapid penile detumescence. This reflex would seem to be related to normal responses in ejaculation. Penile tumescence was then obtained by application of pressure behind the bulbus glandis and rubbing the urethral process; the tumescence was accompanied by the same pattern of activity in the three muscles. Stimulation of the corona glandis resulted in tonic contraction of the bulbo- and ischiocavernosus muscles, again accompanied by rapid detumescence; but this reflex is not as amenable to explanation.

Chapter 16

Muscles of respiration

THE muscles usually considered to be the primary muscles of respiration are the diaphragm and the intercostals, but the following also have been implicated as either primary or accessory in respiratory function: the scalenes and the sternomastoid in the neck, the musculature of the shoulder region including the pectoral muscles and serratus anterior, and the anterior abdominal muscles. These "accessory" muscles are discussed, but only briefly, at the end of this chapter, and reference is also made in appropriate places elsewhere to the broader aspects of each group of muscles.

Since the earliest recorded medical history respiration and its mechanical production have been the subject of inquiry. Both before and after Galen, theories waxed and waned. Galen in the second century was perhaps the first to direct attention to the action of the intercostals though he did not belittle the rôle of the diaphragm in breathing. Furthermore, he was aware of the two layers of intercostals—external and internal. His assignment of inspiration to the former and expiration to the latter still reaches down to the present day, causing renewed controversy. Such illustrious names as Willis, Hamburger and Magendie continue to appear in any history of the respiratory function, but a review of their work here would serve no particular use.

Toward the end of the nineteenth century, Martin and Hartwell (1879), from crude experiments in anesthetized cats and dogs, made conclusions that were not much different than those of Galen. The history of our knowledge of diaphragmatic function unfolded until the turn of the century. Newer techniques and medical advancement have increased our knowledge of the function of the muscles of respiration, but have not lessened the arguments.

In the past decade, electromyography has provided a tool which promises to remove much of the uncertainty. Already there has been definite progress which will be reported in some detail in this chapter. The names of Jones, Pauly and Beargie, then of Chicago; Campbell, first in England and now at McMaster University in Canada; and Koepke, Smith, Murphy, Rae and Dickinson of Ann Arbor, Michigan, all stand out prominently in any review of this subject. Contributions have also been made by others; these will be mentioned in the appropriate places. Studies on the diaphragm have been particularly voluminous.

It is convenient first to consider the movements of the ribs and the muscles that produce these movements.

Costal Respiration

Jones, Beargie and Pauly (1953) were the first to make a substantial electromyographic contribution to the knowledge of costal respiration. With surface electrodes over the upper four internal and external intercostals, the scalene muscles in the neck, and the abdominal muscles, they put their subjects through various tests.

The usual concept of normal quiet breathing is that the scalenes anchor or fix the first rib while the external intercostals elevate the remaining ribs towards the first—this in spite of radiographs showing no approximation of the ribs. Jones and his colleagues showed that both sets of intercostals in man were slightly active *constantly* during quiet breathing and showed no *rhythmic* increase and decrease. In contrast with this, the scalenes did show a rhythmic increase during inspiration (fig. 16.1). Their rôle in quiet breathing was confirmed with needle electrodes by Raper *et al.* (1966).

With forced inspiration, the scalenes, sternomastoid, and internal and external intercostals showed marked activity (Jones *et al.*, 1953; Raper *et al.*, 1966). In contrast, with forced expiration, the scalenes were quiescent while the intercostals were still active. Attention was focussed anew on the scalenes as fundamental muscles of inspiration.

These findings led Jones, Beargie and Pauly to the conclusion that the function of the intercostal muscles in respiration is to supply the tension necessary "to keep the ribs at a constant distance from each other while the chest is expanded from above and contracted from below." They insisted that passive membranes between the ribs instead of muscles would be inadequate because they would be sucked in and blown out during respiration. Therefore, they would not provide the constant fine control of the rib positions. Moreover, the intercostals were shown to function in flexion of the trunk (as in sitting up from the supine position) and, probably for the first time, they were suggested as being postural muscles.

This group of investigators emphasized the belief that the main rôle in human respiration is performed by the diaphragm, while the intercostals were necessary for markedly increasing the intrathoracic pressure. Thus they were agreeing with Hoover (1922) who showed that a person with paralyzed intercostals had a sharp reduction of sucking and blowing power with comparatively much less embarrassment of quiet respiration. In a later paper, Jones and Pauly (1957) state that perhaps the intercostal muscles are "used

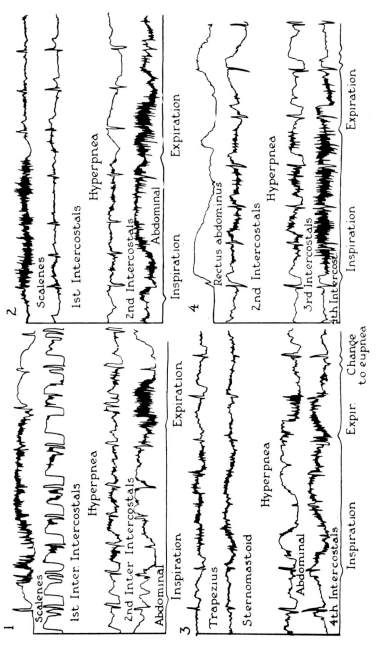

Fig. 16.1. Four EMGs of scalenes and intercostals and other muscles during four different respiratory cycles. (From Jones, Beargie and Pauly, 1953.)

in nonrespiratory activity more than in ordinary respiration." (See also Pauly, 1957).

Campbell (1955a) did not agree completely with these American workers. Using interspaces (6th, 7th and 8th) lower than those that they had used for recording, he concluded that in quiet breathing the intercostals contract during inspiration and completely relax during expiration. He went further in his disagreement by stating that considerable hyperpnea (forced breathing) still did not recruit intercostals in expiration (fig. 16.2) (see also Campbell's 1958 monograph. *Respiratory Muscles*, for further details).

From the work that we have done, it appears that the findings of both groups may well be correct—the apparent disagreement arising from the use of different interspaces for the two studies. Koepke *et al.* (1958) showed electrical activity in the 1st, 2nd and 3rd interspaces in quiet breathing but none in the 4th, 5th and 7th. These inherent differences that have been revealed have not yet been explained, but in their rationalization may lie the final solution of the rôle of the intercostals in breathing. In spite of this

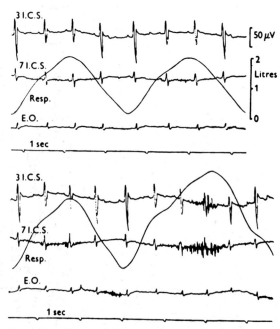

Fig. 16.2.   EMGs of intercostal muscles (3rd and 7th intercostal spaces—3 I.C.S. and 7 I.C.S.) and of external oblique of the abdomen (E.O.) along with spirometry (Resp.). (From Campbell, 1955a.)

indecision, the following cannot be ignored: (1) the vital rôle of the scalenes as ordinary respiratory muscles, (2) the possible rôle of the intercostals as primarily postural muscles, (3) the marked differences in activity which are found between different interspaces and (4) the primary rôle of the diaphragm in respiration.

In deeper breathing, Koepke, Smith, Murphy and Dickinson (1958) showed that the lower intercostals that were so noticeably silent in quiet breathing became progressively recruited in descending order until even the lowest became active on very deep inspiration. This pattern of recruitment is further discussed below (p. 353). Hirschberg's (1957) studies on patients with partial respiratory paralysis tended to confirm the findings for normal patients. Generally the intercostals are recruited during inspiration but not during expiration. Morosova and Shik (1957) of Moscow, from an electromyographic study of respiratory muscles, have concluded that in patients with various respiratory deficiencies extreme augmentation of the electrical activity occurs. This suggests that there is some form of compensatory stimulation *via* the respiratory centre of the central nervous system.

A clear direct relationship has been shown by Viljanen (1967) of Helsinki between the electrical and the mechanical recording of human intercostal muscles during voluntary inspiration.

Draper *et al.* (1957) and then Taylor (1960) re-awakened the Galenic teaching that the external and internal intercostals had different functions. Although many other workers could find no evidence for this with standard needle electrodes, Taylor succeeded in a long series of human subjects by probing with very fine needle electrodes. He demonstrated that there are two functionally distinct layers of intercostal muscle everywhere *except* anteriorly in the interchondral region and posteriorly in the areas medial to the costal angles.

Where there are two functional layers, the superficial one (external intercostal) acts only during inspiration and the deeper (internal intercostal) during expiration. Thus, Taylor seems to have proved in man what has been common experience in experimental animals for centuries.

Taylor found that in quiet breathing what emg activity that occurs is limited to the parasternal region during inspiration—here there is only one functional layer and it is exclusively internal intercostal. This explains why other workers, who usually study internal intercostal in this very place because here it is not covered by the external intercostal, have generally considered internal intercostal to be inspiratory. Indeed, this part of it *is* inspiratory, according to Taylor. Yet, to confuse the issue, Nieporent (1956) found this same part to be active in expiration and Boyd (1968, 1969) found and emphasized that in rabbits both external and internal layers acted only in inspiration. (See also the discussion of work of Hirschberg *et al.* on p. 356.)

During expiration, in quiet breathing, Taylor found emg activity limited to the lower lateral part of the thoracic cage and coming from the internal intercostals. Apparently he found no external intercostal activity whatsoever in quiet breathing.

His finding of intercostal expiratory activity even in quiet breathing, if true, shakes the old theory that expiration is entirely passive. Probing deeper, Taylor found that the transversus thoracis is purely expiratory in function—including its parts known as sternocostalis, intercostales intimi and subcostalis.

More vigorous respiratory effort brings the layers of muscular activity into reciprocal action all over the chest wall. Now the external layer is entirely inspiratory and the internal layer expiratory. Taylor suggests that these two layers exert opposite rotational forces on individual ribs around their long axes. At this point, obscurity creeps into the interpretation that one might put on the rôle of such possible rotation; no purpose is served in speculating further here.

EMG of Intercostals during Phonation

Since the early and unsubstantiated work of Stetson (1933) no great progress was made in this type of study until the '60's. Hoshiko (1960, 1962) has investigated the sequence of activity during phonation from the intercostals and rectus abdominis. Whatever the speech material and rates of utterance may be, the internal intercostals are always recruited first, followed by the rectus abdominis and then the external intercostals. Action potentials disappear at the end of phonation in the external intercostals.

Hoshiko believes that the internal and external intercostals cooperate in releasing a syllable, i.e., initiating the simple pulse associated with syllabication. Contrary to Stetson's teaching, the external intercostals are not involved in terminating the syllable movement.

Extending this type of work to *connected speech* in native English-speakers, Munro and Adams (1971, 1973) found similar emg patterns in the internal intercostal muscles for the utterances of rhymes, fabricated sentences with rhythm similar to rhymes, and prose passages. Thus, a preliminary burst of activity before each phrase was followed after an interval by increasing activity in the utterance of the phrase. Activity increased further through each phrase of the sentence. However, phasic activity was *not* apparent in the external intercostals. The localized burst of activity in the internal layer seems to contribute to the increased intrathoracic pressure necessary to overcome the resistance of the glottis and later—during the utterance—it may help to maintain an adequate intrathoracic pressure despite diminution of lung volume.

Diaphragm

As indicated above, the main respiratory muscle, at least in man, is the diaphragm. Most emg studies of this vital muscle—except the most recent ones—have been indirect. Admittedly, studies of the diaphragm in experimental animals under anesthesia are not rare, but, being conducted under highly unphysiological conditions, they can hardly give definitive answers to a vital problem. For example, a recent emg study by Di Benedetto *et al.* (1959), which appears to be as technically acceptable as most such studies, used dogs that were actually hepatectomized in addition to their bellies being left wide open during the tests. To be fair to these workers, we must note that they were not attempting to categorize the actions of the diaphragm in detail. However, many of the statements on diaphragmatic function in the textbooks are based on exactly this type of observations, with or without electromyography.

The function of the diaphragm has been of longstanding interest. Recognition of its importance dates back to Hippocrates, but he attributed no movement to the diaphragm. During the past 100 years, various comprehensive studies (see Boyd and Basmajian, 1963) showed that the the diaphragm—at least when stimulated electrically—expands the base of the thorax by moving the ribs upward and outward. Other studies combining fluoroscopy and spirometry showed that a change from the erect to the supine position causes marked alteration in the pattern of diaphragmatic movement in man.

J. C. Briscoe (1920) and G. Briscoe (1920) postulated that the diaphragm is a tripartite organ consisting of the right and left costal parts and the crura forming one part. This view was supported by the fact that the costal and the crural parts develop from different muscular sheets.

Research on anesthetized rabbits by Wachholder and McKinley (1929) showed that the diaphragm was almost continuously active during quiet breathing with only brief relaxation during expiration. Campbell's (1958) attempts to confirm these findings in normal human subjects with surface electrodes met with disappointing results because of the intervening mass of bone, cartilage and lung. Studies of diaphragmatic activity in many species, ascribed to Chennells by Campbell (1958), apparently showed that there is no activity during expiration. In human subjects Nieporent (1956) and Draper *et al.* (1957) found that diaphragmatic activity could be obtained with needle electrodes during inspiration, but there was none during expiration. However, other workers, also using needle electrodes, showed that diaphragmatic activity in man could be recorded during both inspiration and expiration (Koepke *et al.*, 1958; Murphy *et al.*, 1959; Petit *et al.*, 1960). Bahoric and Chernick (1975) showed with EMG that in fetal sheep diaphragmatic activity is present though minimal.

In recent years, then, little if any advance has been made in our understanding of the total function of the diaphragm. This fact led to our detailed investigation in which recordings were made simultaneously from indwelling multiple electrodes (up to 16) along with spirometric tracings in conscious rabbits. The correlated results provided the first complete account of the activity of the whole diaphragm in quiet breathing (Boyd and Basmajian, 1963).

A series of 25 adult male rabbits had multiple clip electrodes implanted in their diaphragms at open operation. Following postoperative recovery, the wires from these electrodes were connected to an electromyograph and records were made along with spirometry under normal physiological conditions. A colour motion picture demonstrates our techniques (Basmajian and Boyd, 1960).

The rabbit's diaphragm, like man's is divided clearly into eight left and eight right muscular slips or digits. Each side of the diaphragm has one sternal slip, six costal slips and a lumbar slip from the aponeurotic arches and crus. The muscular slips, may be numbered anteroposteriorly: sternal slip—1; costal slips—2 to 7; and lumbar slip, including fibres from the crus—8.

The question as to whether the diaphragm is functionally a single muscle, or two functional halves, or a tripartite organ with a lumbar or crural portion and two costal portions was answered conclusively: all portions of the diaphragm contract simultaneously. Almost certainly, the whole diaphragm normally functions as a unit. This is true also in the cat, as shown by Sant'Ambrogio *et al.* (1963) and Grassino *et al.* (1976). Most likely it is true in all mammals.

Phases of Respiration

One of our main findings was that quiet respiration includes not only the two simple opposite phases of inspiration and expiration but also a static phase before each. These we have called pre-inspiration and pre-expiration.

In duration, they are much shorter than the air-moving phases; nonetheless, the static phases make up an appreciable part of the respiratory cycle. They would appear to correspond to the well-known inspiratory and expiratory "pauses" in man.

Pre-inspiration

This static phase, which occurs prior to inspiratory air movement, lasts from 20 to 120 msec. During pre-inspiration, the activity is rarely greater than 1 +, or slight.

Other workers noted activity occurring in the diaphragm before the onset of active inspiration without recognizing a statis phase. With needle electrodes, Koepke *et al.* (1958) determined from spirometric tracings that the onset of contraction in human subjects occurred as much as one-fourth of a second before the onset of inspiration. This finding was confirmed by Taylor (1960), potentials began immediately before inspiratory airflow, increased rapidly to a maximum, and died away in the first half of the expiratory phase. Similar findings were reported by Petit *et al.* (1960) using esophageal electrodes in man.

Inspiration

Inspiration is an increase in the volume of the thorax with an actual inward flow of air. The diaphragm contracts and increases the cavity in a caudal direction and perhaps in the ventrodorsal and lateral directions as well. The motor units of the diaphragm increase their rate of firing at the onset of inspiration and as this phase proceeds, new units are added or "recruited" so that the inspiratory effect gains force as it proceeds (Bergström and Kertula, 1961; Lourenço *et al.*, 1966). As inspiration continues, the individual motor units in the diaphragm accelerate in rate, resulting in a progressive increment in the strength of contraction of each unit. With an increase in the number of active units and by augmentation in the strength or frequency of each unit, inspiration reaches its peak. In all but an insignificant number of our records, activity was recorded throughout the entire inspiratory phase. Indeed, activity was continuous throughout both pre-inspiration and inspiration. However, the level of activity fluctuates over these phases.

The entire inspiratory phase in rabbits last from 300 to 550 msec (and rarely longer). This inspiratory phase may be divided for convenience into four quarters. Of course, the length of these varies with the length of each phase, which in turn depends on the respiratory rate.

The peak of the inspiratory motor unit activity terminates some 30 to 40 msec before the end of inspiration. In most instances, the peak is followed by a rapid decline—and in some instances by a very sharp drop—to the baseline, where varying degrees of activity continue to the end of the fourth quarter of the inspiratory phase. A similar finding was reported by Koepke *et al.*, indicating that the diaphragmatic voltage pattern in man consistently reached its greatest amplitude at or slightly before the end of inspiration; then it tapered to its stopping point in what was considered to be expiration. No instance was found during which activity stopped at the end of inspiration in man; but we found electromyographic silence following inspiration in 44% of the recordings in rabbits.

At the onset of inspiration, almost all of our recordings (98%) showed a carry-over of pre-inspiratory activity into the first quarter of inspiration. Then, in all of the recordings, activity was present throughout the second and third quarters of inspiration. During the first quarter of inspiration, in general, the diaphragm is only slightly active. The motor units of the diaphragm do not begin to increase their rate of firing during quiet breathing until the second quarter of the inspiratory phase is reached; most of the recordings (74%) show moderate activity by that time. One might postulate that the intercostals play the main role in the first quarter.

The third quarter of inspiration forms the peak for inspiratory activity during quiet normal respiration with 76% of our recordings showing marked activity. Nonetheless, in a substantial number (12%) of the activity remains only slight during the third quarter.

The activity in the third quarter of inspiration, whatever its character, is carried into the fourth quarter. In 76% of the recordings the great activity recorded during the third quarter continued into the fourth quarter and stopped some 20 to 40 msec before the termination of the inspiration. However, the great activity at the peak of the inspiration does not continue throughout the entire fourth quarter. The fourth quarter thus can be divided into two unequal parts, the line of demarcation being the terminal point at which the muscular activity ceases. Thus, during the first part of the last quarter, 76% were markedly active and the remainder moderately or slightly active (12% each).

Great activity when it did occur at the end of inspiration (7%) was never carried into pre-expiration in our series.

Pre-expiration

The static phase of pre-expiration usually lasts from 20 to 50 msec (and somewhat longer in some cases). Electromyographic activity is slight or absent (44% and 56%, respectively) in this phase.

Pre-expiration is a regular precursor of expiration. Studies on the human diaphragm by Murphy *et al.* during passive expiration seemed to show that electrical activity occurred even though each subject had been instructed to relax after taking in a breath. They found that an increase in the duration of activity during expiration was directly related to the depth of the preceding inspiration. These findings could not be confirmed in our study, for most of the recordings were taken during quiet respiration and, or course, they were in non-human subjects.

Expiration

Expiration, the last phase in the respiratory cycle, consists of a decrease in the volume of the thorax with air moving outward. One way that this might be accomplished actively is be the contraction of the abdominal muscles forcing the diaphragm up into the thorax. During quiet breathing, however, expiration is generally regarded as passive. Nonetheless, activity in the diaphragm is recorded during expiration. In almost every instance, the (slight) activity that we recorded during expiration lasted for a longer period of time than did the greater activity that occurred in inspiration. Previous to our findings, Murphy *et al.* reported that the activity continued through as much as 98% of expiration; Agostoni *et al.* (1960) found activity persisting into the early part of expiration in human diaphragms.

Many varied opinions exist on the activity of the diaphragm during expiration. The literature includes only one study specifically in the rabbit (Wachholder and McKinley, 1929). This report indicates that the diaphragm is almost continuously active during quiet breathing with only a very brief period of non-activity during expiration. In contrast, using needle electrodes, Nieporent (1956), Campbell (1958) and Draper *et al.* (1957) found no diaphragmatic discharge during the expiratory phase in man or in animals.

Murphy *et al.* (1959) found no activity originating in the human diaphragm during forced expiration; they reported that activity occasionally carried over from deep inspiration but always stopped abruptly at the onset of a forced expiratory effort. In patients with transverse myelitis, irrespective of the size of the preceding inspiration, diaphragmatic voltages were not found during a forced expiratory effort. During passive expiration, activity always continued from inspiration into expiration, and this pattern was like that in normal subjects.

During quiet breathing in man (Petit *et al.*, 1960) activity started in the diaphragm at the beginning of inspiratory flow, increased in intensity through inspiration, and persisted after the onset of expiration, but with decreasing intensity. A similar pattern occurred during increased ventilation, with the difference that the activity started before the beginning of inspiration. Some workers tend to agree that there is a carry-over of inspiratory activity into the expiratory phase, but none indicates either how far into the expiratory phase this activity extends or the degree of activity carried into expiration.

In most of our recordings (63%), some activity occurred throughout the entire expiratory phase. Complete absence occurred in only 14%. Almost always the diaphragm is active during the last quarter of expiration. One possible reason why activity carries into or continues throughout expiration is that activity in the diaphragm during passive expiration is a braking action to oppose the normal elastic recoil of the lungs (Murphy *et al.*, 1959). In effect, it is not a true expiratory effort. Agostoni and Torri (1962) hold like views from like findings in man. However, they implicate reflexes to balance the antagonist activity of the abdominal wall muscles. Delhez *et al.* (1963 *et seq.*) discount the importance and even question the occurrence of activity at the end of forced expiration.

The anatomical structure of the diaphragm supports this view because all of its fibres are arranged in radiating fasciculi inserting into the central tendon. Shortening of the fibres can only cause a flattening of the dome and so actually resist the production of an expiratory force. Furthermore, no electrical activity could be recorded during forced expiration in human subjects.

According to Campbell, the rate of airflow at the onset of expiration does not rise rapidly to a maximum (as would be expected if the muscles of inspiration relaxed immediately), indicating that during the early part of expiration the muscles of inspiration decrease their force of contraction only gradually. He further states that there is a persistence of intercostal activity during the early part of expiration. Although no direct reference to the diaphragm is made, he reports that measurements of the work of breathing based on pressure-volume diagrams suggest that the muscles of inspiration may exert considerable force in opposing the elastic recoil of the lungs during expiration. One may conclude, then, that the slight activity recorded from the diaphragm during expiration is a braking action to oppose the normal elastic recoil of the lungs.

Koepke *et al.* (1958) at Ann Arbor, Michigan, have also been making a concerted attack on problems of the respiratory muscles including the diaphragm. They have used needle electrodes in the diaphragm in several human subjects. These electrodes were inserted through the 11th intercostal space into the muscular digit of the diaphragm that arises from the 12th rib We must keep in mind the limitations imposed by this localization of pickup (in a muscle which—in our own findings on rabbits—may act somewhat differently in its various parts), With ordinary inspiration, the diaphragm became active before any of the intercostals and before the flow of air by as much as a quarter of a second. During quiet breathing the diaphragm never failed to act in inspiration although some of the intercostals were only recruited with deeper breathing (see above).

The same group, with Murphy as senior author (1959), reported on the emg activity during expiration. They insisted that the diaphragm always shows electrical activity as a carry-over into the early stage of passive expiration (fig. 16.3). The diaphragm consistently showed its greatest activity at or slightly before the end of inspiration; this tapered off to silence in early expiration. The duration of activity in the diaphragm during expiration was directly related to the depth of the previous inspiration. In some instances, the activity continued through as much as 98% of expiration. None of the intercostal muscles showed the same degree of carry-over of activity as did the diaphragm. During quiet breathing *only* the diaphragm showed such expiratory activity.

Murphy *et al.* suggest that the activity of the diaphragm during passive expiration may be a braking action "to oppose the normal elastic forces of the lungs rather than the exertion of a true expiratory effort." Observing that the fibres of the diaphragm radiate from the central tendon, they too believe that activity of the muscle can only flatten and lower the dome. Furthermore, no activity in the diaphragm was seen during forced expiration, which appears to confirm their thesis.

A large and vigorous group of investigators in Liège, Belgium (Petit, Milic-Emili and Delhez, 1960) devised a novel technique for diaphragmatic elec-

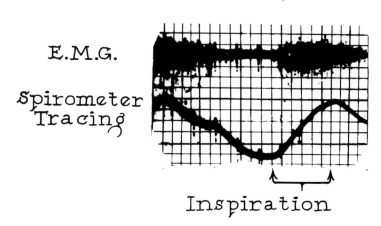

FIG. 16.3. EMG of diaphragm. (From Murphy *et al.*, 1959.)

tromyography in conscious man. The electrical activity was detected by means of electrodes passed down the esophagus to the level of the diaphragmatic esophageal hiatus. In four normal persons they found the activity to be synchronous with the respiratory variations of intra-abdominal and intrathoracic pressures. Potentials occurred from the onset of inspiration and increased in intensity. They continued into expiration for a varying length of time with decreasing intensity. During increased ventilation, potentials began immediately before inspiration rather than just at its onset. For further details of the work of the Liège group, see Delhez and Petit (1966), Delhez, Bottin-Thonon and Petit (1968), and Petit *et al.* (1968).

Admittedly, the intra-esophageal electrodes pick up potentials only from the crural fibres of the diaphragm, but this does not invalidate the results. Though in the future we may show some variations between various parts of the diaphragm, no major differences are likely to be revealed. In any case, the approach of Petit's group to diaphragmatic electromyography is a promising one.

Similar (and successful) techniques were used by Miranda and Lourenco (1968) in New Jersey and Hixon *et al.* (1969) at the University of Wisconsin for technical investigations beyond our present concerns. Somewhat related to the transesophageal emg technique for the diaphragm, are the esoteric emg studies of the esophagus itself, although they present a quandary as to where to fit these interesting studies in this book. Interested readers should consult Coman and Car (1970) for a description of swallowing as it affects esophageal reflex contractions produced by stimulation of the vagus and superior laryngeal nerves, and Tokita *et al.* (1970) for a study of normal and abnormal functions.

Breath-holding and Snoring

Agostoni (1963) has reported a curious diaphragmatic activity during human breath-holding. After an initial silence, a marked discharge occurs and repeats at progressively higher rates until the breaking point. These are ineffectual respiratory efforts that cause a fall of intra-thoracic pressure apparently because the glottis is clamped shut by effort. Agostoni *et al.* (1960) found a rather similar fluctuation during the very brief episode of either coughing or laughing. Readers interested in intercostal EMG during snoring should consult the work of Lugaresi *et al.* (1975).

Quadratus Lumborum as an Accessory

In a later study using our earlier techniques, Boyd, Blincoe and Hayner (1965) showed that quadratus lumborum acted simultaneously with the diaphragm to stabilize the rib which might otherwise be elevated. Thus quadratus acts in concert with the diaphragm as a respiratory muscle; further, its activity coincides with the diaphragm to produce the normal braking action during expiration noted on p. 351.

Diaphragm-Intercostal Interrelationship

Although many generalized statements have been made in the literature (both of respiration and of electromyography) about the interplay of dia-phragmatic and intercostal activity, only the work of the above-mentioned Ann Arbor group (Koepke *et al.*, 1958; Murphy *et al.*, 1959) and our own studies appear to cast any direct light on their relationship (Boyd and Basmajian, 1963; Boyd, 1969).

In quiet respiration, the diaphragm became active a quarter of a second before the onset of inspiratory air flow as measured by spirometry. Further-more, this occurred before any activity in any intercostal muscle. The intercostals that were recruited earliest were the first topographical pair (i.e., the first intercostal muscles). Then, the recruitment proceeded progressively to include lower and lower intercostals. With quiet breathing the diaphragm was always active in inspiration, the 1st intercostals usually were, and the 2nd intercostals occasionally were. All the others were inactive. During inspirations deep enough to call upon all the intercostals, the onset of activity advanced progressively in successively lower intercostals.

During natural passive expiration (as distinct from forced expiration) these workers found that a carry-over of activity was also present in the intercostals. Among them, the lowest intercostals were more important than the highest, the duration decreasing sequentially from the 11th pair upwards.

Though no activity was found in the diaphragm during forced expiration, intercostal activity was almost always present. The likelihood of such activity was related inversely to the volume of air in the lungs at the end of inspiration. When intercostal activity was present during forced expiration the recruit-ment again was progressively upwards from the lowest intercostals to the highest.

Murphy and his colleagues postulated that when the intercostals are recruited sequentially from below upwards during forced expiration the lower portion of the thorax becomes relatively smaller than the upper to produce a desirable pressure gradient within the chest to empty the lungs. This remains pure hypothesis and is not, as yet, particularly convincing.

During active or REM sleep in the supine position, a sharp drop occurs in intercostal muscle activity and rib-cage movements compared with quiet sleep and the awake upright posture—which are about the same (Tusiewicz *et al.*, 1977).

Other Respiratory EMG Studies

Fink *et al.* (1960) used the integrated emg recording of patients during general anesthesia to study the threshold of the respiratory centre. Though their study revealed nothing about the normal functioning of the diaphragm, it did show the feasibility of human diaphragmatic EMGs. They inserted unipolar needle electrodes into the diaphragm through the 8th, 9th or 10th intercostal space and successfully obtained recordings. Their main findings relating to the onset of apnea revealed—as indicated by cessation of diaphragmatic potentials—that it occurs when the average alveolar carbon dioxide tension or pressure falls to 38 mm Hg. Diaphragmatic activity reappears as the CO_2 tension rises above this value, the stimulus acting through the CO_2-sensitive respiratory centre.

The studies of Björk and Wåhlin (1960) on the effects of the muscle-relaxant drug succinylcholine upon the cat diaphragm would be of interest to a very few readers and are mentioned here for completeness only. The effect of the drug is to disturb the synchronization of parts of the motor unit and to cause a progressive decay of motor unit potentials down to individual fibre potentials. This effect is peripheral rather than central.

Di Benedetto *et al.* (1959) used electromyography combined with phrenic nerve stimulation to investigate the innervation of the diaphragm in dogs. They found that, contrary to widely held opinion, the muscle mass to the right of the esophageal hiatus was commonly innervated by both phrenic nerves, i.e., the right crus was bilaterally innervated in about one third of cases. However, they found no instance in which left crus was bilaterally innervated.

Jefferson *et al.* (1949), who also worked with dogs, found complete paralysis of the left hemidiaphragm with left phrenicotomy. This (1) confirms the above findings and (2) confirms the teaching (which has been often challenged without adequate evidence) that the only innervation of the diaphragm is the phrenic nerve.

Sant'Ambrogio and Widdicombe (1965) have studied respiratory reflexes from the diaphragm and intercostals in rabbits to assess their strength. They used single unit EMGs as a direct and quantitative measure of activity before and after vagotomy. There is little doubt from their results that proprioception plays an important part in driving the respiratory muscles. Guttmann and Silver (1965) approaching the problem quite differently, i.e., by studying the reflex activity in the intercostals of tetraplegics, confirm the rôle of stretch reflexes in respiratory muscular activity.

Youmans and his group at the University of Wisconsin have been studying the "abdominal compression reaction" by means of emg records from the diaphragm and intercostal muscles of anesthetized dogs (Briggs *et al.*, 1960). This reaction is initiated by procedures which cause a decrease in central blood volume and it consists of a steady state of activity of abdominal muscles rhythmically interrupted by breathing. In some instances, the intercostal muscles show no action currents while in others they show a burst of activity during inspiration and again during the abdominal compression reaction. The steady-state contraction of the external oblique abdominal muscles commonly begins while expiration is in progress and reaches a maximum after completion of expiration. When a strong abdominal compression reaction is present, the initial phase of inspiration is the movement of the diaphragm caudally, related to sudden inhibition of the abdominal compression reaction and a corresponding decrease in intra-abdominal pressure. The diaphragm begins to move caudally because of less pressure on the abdominal side (and not because of motor activity) and it continues to move as a result of its contraction, according to Youmans and his colleaques (1963).

Delhez and various colleagues in Liège, Belgium, have made an important series of contributions to the literature of respiration and diaphragmatic function during the past several years (1963 *et seq.*). The amount of activity in the diaphragm in human subjects was found to comply proportionately with the ventilation up to 50 or 60 litres per minute. Above that, emg activity rises more rapidly than the ventilation, apparently to counteract elastic forces and antagonistic activity (Delhez, 1964; Delhez *et al.*, 1964a, b, c, d). Their insistence that strong diaphragmatic activity occurs during forced expiration has created considerable attention and controversy without satisfactory resolution as yet. Their views, expanded upon in a long review written by Petit, Delhez and Troquet (1965), are interesting; but fuller consideration is out of place here. Also interesting to some readers will be the emg work of Kelsen *et al.* (1976) on the neural responses of respiratory muscles during elastic loading and that of Sharp *et al.* (1976) and Lopata *et al.* (1976) in evaluation of respiratory regulation.

Motor Units in Diaphragm

Even in the highly developed human diaphragm, the muscle is quite thin and likely there is a great lateral spread of fibres belonging to one motor unit. To test this hypothesis, Krnjević and Miledi (1958) at the Australian National University in Canberra, investigated the distribution of single motor units in the rat diaphragm electromyographically using a "phrenic-hemidiaphragm" preparation. They found the fibres of one motor unit irregularly scattered over an area of several millimetres. They also found that the motor units were considerably intermingled (see also p. 9).

Accessory Respiratory Muscles

The muscles which are usually considered to be accessory respiratory muscles are the following: the muscles of the vocal cords (discussed on p. 369); sternomastoid and scalenes in the neck p. 340 and 398); abdominal muscles (p. 326); pectoral muscles (p. 190); serratus anterior (p. 191); and trapezius (p. 189). As I have pointed out in the appropriate sections of this book, the scalenes should be considered primary respiratory muscles, the abdominal muscles are certainly accessory respiratory muscles, and, generally, the upper limb muscles (including serratus anterior) take no part in quiet or even laboured respiration—except under highly abnormal conditions, diseases and postures (Grønbaek and Skouby, 1960).

Nieporent (1956) found that there was no activity in pectoralis major during quiet breathing, but during maximal inspiration some slight to moderate activity appeared. During dyspnea, pectoralis muscle functions primarily in inspiration as an accessory muscle. Campbell (1954, 1958) found some activity only during very deep inspirations in trapezius (upper part), latissimus dorsi, pectoralis major and minor, and serratus anterior. On the other hand, Tokizane *et al.* (1954), reported activity with ordinary breathing in some of these muscles, but, being unconfirmed by later studies, their findings are subject to serious doubt.

In patients with complete diaphragmatic paralysis, Hirschberg *et al.* (1962) found that most could breath quietly by using their accessory muscles. Surprisingly, the intercostals (whose viability could be proved by other tests) were the least active during quiet breathing; the abdominal muscles were the most active. Accessory muscles in the neck (although they were partially paralyzed in the patients with poliomyelitis) were also quite active. Of course, readers of p. 340 would expect this. Guttmann and Silver's (1965) finding that the intercostals of tetraplegics resumed reflex activity is of some impor-

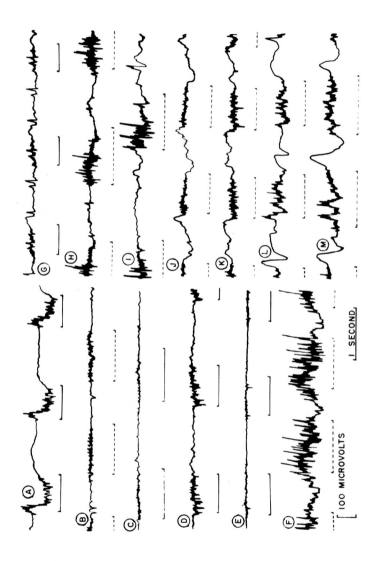

FIG. 16.4. EMGs of various canine muscles during respiration. Solid marker line—inspiration; dashed line—expiration. *A*, dilator naris; *B*, mylohyoid; *C*, sternohyoid; *D*, sternothyroid; *E*, hyothyroid; *F*, scalenus anterior; *G*, scalenus posterior; *H*, external intercostal; *I*, internal intercostal; *J*, rectus absominis; *K*, external oblique; *L*, internal oblique; *M*, transversus abdominis. (From Ogawa *et al.*, 1960.)

tance. This developed in response to stretch reflexes initiated by the diaphragm or the "accessory" muscles of the neck (mainly sternomastoid).

Ogawa *et al.* (1960) at Michael Reese Hospital in Chicago found in dogs that the following muscles had no respiratory function: digastric, masseter, levator nasolabialis, scutularis, cervicoauricularis, splenius, brachiocephalis, trapezius, rhomboideus, supraspinalis, infraspinalis, deltoid, semispinalis, serratus anterior, pectoralis superficialis and profundus, serratus posterior superior and inferior, and psoas. In contrast, they consistently found respiratory activity in the following canine muscles: nostril, intrinsic laryngeal, scalenus anterior, intercostals, rectus abdominis, external and internal obliques, and transversus abdominis (fig. 16.4).

Chapter *17*

Mouth, pharynx and larynx

U NTIL 1958, electromyography of the mouth and pharynx was virtually an unexplored frontier, and even now the tongue—though
it is an obvious and accessible muscular mass—has not been
adequately explored. Indeed, only one systematic study had been done on
the functions of the intrinsic or extrinsic muscles of the tongue before the
second edition of this book appeared in 1967. Furthermore, the easily
accessible musculature of the floor of the mouth (i.e., mylohyoid, etc.) also
has been badly neglected until recently. On the other hand, the palate and
the pharyngeal constrictors have now been well explored by several groups.

Insofar as the palate and pharynx are concerned, this chapter will deal
chiefly with our own findings, supplemented by those of clinical electromyographers. Also to be discussed at the end are the proliferating reports on
the EMG of the larynx.

Tongue and Mouth

Bole (1965), the first to use our inserted, fine-wire electrodes for the tongue,
performed definitive studies of the actions of genioglossus muscle. He found
that the right and left muscles act together with approximately similar emg
response during many general movements of the tongue—even lateral shifts.
The greatest emg activity appeared when the tongue met resistance. Surprisingly, little activity accompanied protrusion of the tongue unless it was
against the back of the incisor teeth.

Bole showed that there may be several different patterns of glossal movement during swallowing. The duration of activity in genioglossus ranged
from 1.42 to 2.74 seconds and appeared in two or more bursts. Generally the
bursts were at the onset of swallowing and after the substance had left the
tongue. Bole's pioneer work on the human tongue, because it gives great

promise of revealing the true functions of this vital organ, have been partially followed up by my colleagues and myself and by scattered investigators elsewhere. Thus, in our emg investigation of the activity of the paired genioglossus and geniohyoid muscles of 26 human subjects during deglutition, a general pattern of muscular activity was revealed involving an initial build-up, gradual summation and tapering of electrical potentials during swallowing of both saliva and water. There is an observable difference in the pattern of swallowing of individuals within a group and among the individual swallows of a single subject (see fig. 17.1).

There are longer periods of electrical activity during a saliva swallow than during a water swallow. The type of bolus also seems to affect the pattern of activity in the individual muscles as well as the length of time that they are working. The geniohyoid muscles do not appear to begin their activity with the genioglossus muscles but rather lag behind and they do not appear to be active for as long. Both pairs of muscles appear to remain active during and after the time that the bolus has passed the area of the laryngopharynx. A

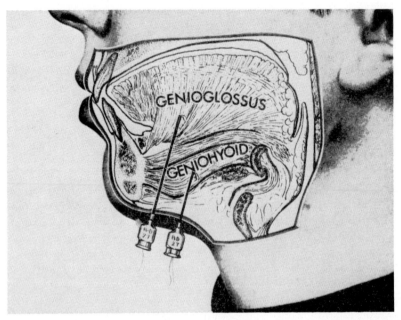

FIG. 17.1. Schematic drawing of genioglossus and geniohyoid muscles showing the position of the needle cannulae before their withdrawal. Small hooks at end of fine wires will hold them in place. (From Cunningham and Basmajian, 1969.)

period of electrical silence occurs prior to the characteristic burst of activity associated with a swallow. This appears to be the result of an active inhibition (Cunningham and Basmajian, 1969).

The "tongue-safety" function of the genioglossus muscle has been the special interest of Sauerland in Galveston, Texas (Sauerland and Mitchell, 1975; Sauerland and Harper, 1976). More active during inspiration both during working and sleeping hours, the muscle prevents obstruction of the airway. The glossopharyngeal part of the superior pharyngeal constrictor ("myloglossus muscle") acts as an antagonist (Jüde *et al.*, 1975).

Following our early studies, we expanded them by investigating muscles in the floor of the mouth related to the tongue. The geniohyoid, anterior belly of digastric, mylohyoid and genioglossus muscles of 20 human subjects were studied electromyographically to determine the temporal relationships of their activities during the act of swallowing. Although the firing order of the four muscles varied within the same subject, the best estimate of the "true" firing sequence was established for each of the 18 subjects who provided statistically significant data. However, no definite universal pattern could be established for the four muscles because there was great inter-subject variability in both the duration and the sequence of activity. Therefore, at least with respect to these four muscles, each individual has his own swallowing pattern, but different people may swallow quite differently (Hrycyshyn and Basmajian, 1972; Vitti, Basmajian *et al.*, 1975).

The type of bolus (saliva *vs* water) may influence the duration of the muscles' activity. On the other hand, posture (semi-reclined *vs* sitting) did not seem to have any influence. There was no evidence to indicate that posture and/or the type of bolus are correlated with the sequence of muscular activity.

The anterior belly of the digastric muscle was not active in one-quarter of the swallows studied. When active during deglutition, all muscles had a general emg pattern of one to many summations of activity separated by relatively quiet periods before and after each swallow (Hrycyshyn and Basmajian, 1972).

The mylohyoid muscle has also been studied by Lehr, Blanton and Biggs (1971) to determine its activity relative to isolated movements of the tongue and mandible and during various functions involving multiple parts of the oral apparatus. Data were obtained from 20 subjects. Using bipolar fine-wire electrodes, the anterior fibres were found to be more active than the posterolateral fibres in a majority of activities performed. Tongue movements produced slightly more activity in the posterolateral fibres; the anterior fibres were more active during mandible movements. During mastication, deglutition, sucking and blowing, both the anterior and posterolateral fibres were markedly active. Other emg work on mylohyoid and geniohyoid, among other muscles, has been reported by Miller (1972a, b) and by Car (1970) who were concerned with the swallowing reflex induced by peripheral nerve and brainstem stimulation.

Other studies of the tongue have been concerned with speech production and disturbances and use rather global pick-ups from electrodes applied to the tongue's surface (MacNeilage *et al.*, 1967; Shankweiler *et al.*, 1968; Huntington *et al.*, 1968). Fine-wire electrodes have been used in such studies only sparingly (Harris, 1971; Borden and Harris, 1973; Raphael, 1975; Raphael and Bell-Berti, 1975). As techniques improve—and fine-wire electrodes simply must gain acceptance—these studies should flourish.

Buccinator

Buccinator emg research is increasing. This facial muscle which has sunk deep and become the lining of the mucosa of the cheeks has many roles to play. Its part in facial expression will be described with facial muscles on p. 392. Blanton, Biggs and Perkins (1970) have made an analysis of the buccinator muscle using indwelling, fine-wire electrodes. Electromyograms were made from 22 subjects with normal dentition during various oral activities. The buccinator muscle was found to be markedly and consistently active during swallowing, blowing, sucking, masticating and various lip and mandibular movements. Its activity during mandibular movements was not believed to be for the sake of direct propulsion but rather as an "expression of an effort to perform the movement." The studies of de Sousa and Vitti (1965) and by my own group are related chiefly to mobilizing the face and will be described on p. 392.

Palate

Cho-luh Li and Arne Lundervold (1958), while working in Montreal, had the opportunity of examining electromyographically a series of normal human palates as controls for a broader study of cleft palates. They were able to record separately from the tensor palati (which becomes aponeurotic as it turns around the pterygoid hamulus into the palate) and the various fleshy muscles in the palate itself. In our laboratory, we have performed examinations of the soft palate in a series of rabbits (Basmajian and Dutta, 1961a) and compared these with examinations in a series of normal human volunteers (Basmajian and Dutta, 1961b).

[Extensive investigations by Doty and Bosma (1956) on dogs, cats and monkeys were concerned chiefly with swallowing. These investigators did not report on the soft palate as such (but see below). Broadbent and Swinyard in 1959 reported on their emg studies following human cleft palatal repairs in which they had used a "dynamic pharyngeal flap." These studies are mentioned here for the sake of completeness.]

Our emg experiments on the palate of rabbits were of two types. In the first type, rabbits under general anesthesia were tested for reflex swallowing with a bipolar needle electrode inserted into the part of the palate that contains the bulk of levator palati. In the second type of experiment, EMGs were recorded by two different techniques in a series of conscious rabbits made to swallow "normally" by placing water on the tongue. Our experiments on human subjects were on conscious normal volunteers in whom recordings were made with special long, fine, bipolar electrodes inserted under direct vision into the levator palati.

In conscious human subjects, we always found, as did Li and Lundervold, a burst of activity upon inserting the electrode (fig. 17.2). This lasted for several seconds, but it never lasted more than 10 seconds in either our investigations or those of Li and Lundervold. Then complete relaxation of the levator palati and tensor palati ensued and continued for as long as the subject remained at rest. This was confirmed by Fritzell (1963) for both muscles by direct recordings. When our subjects sucked water through a

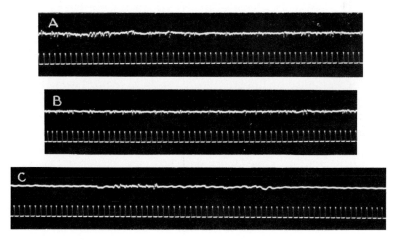

Fig. 17.2. EMG of human palate. *A,* "insertion" potentials; *B,* activity during sucking water through a straw; and *C,* swallowing. Calibration signal: 10-msec intervals; amplitude, ca. 500 μv.

straw, the levator palati became slightly active and remained thus as long as the water was held in the mouth (fig. 17.2).

During swallowing, potentials came as a burst lasting about 1/3 second and followed by complete relaxation (fig. 17.2). Li and Lundervold comment only on the normal appearance of the potentials obtained during voluntary swallowing, but one gets the impression that their findings were similar to ours.

The results of our experiments on the soft palate of conscious rabbits were similar to those on the palates of human beings. In contrast, the most striking finding in anesthetized rabbits was that reflex swallowing (caused by prodding the pharyngeal mucosa) showed very little activity in the palate (see fig. 17.5 on p. 368).

Palatal Activity During Speech

Fritzell (1963) found that activity started simultaneously in tensor and levator palati just before speech begins. This "acoustically silent" period of palatal activity varied in different subjects. The potentials diminish and usually disappear before an utterance is finished.

When words beginning with a nasal sound are spoken, action potentials precede the microphone signal. But when nasal sounds appear within a word the potentials disappear or diminish only to return when oral sounds are made. "The production of nasal is regularly announced in the electromyograms before the sound appears in the microphone record," according to Fritzell.

The hypothesis of a very simple organization of muscular activity (for example, two discrete levels) underlying a more complex pattern of velar positioning was tested by Lubker (1968). The data obtained in his study were not consistent with this hypothesis. Palatal emg activity during speech appeared to vary in a relatively continuous manner and was positively correlated with velopharyngeal positioning.

A majority of Lubker's subjects demonstrated little or no velar movement during production of the sustained, detached nasal consonant. A relatively marked positive correlation was obtained between tongue and palatal positioning during speech production, thus suggesting that a high tongue position is associated with a high palatal position. The Moll-Shriner hypothesis predicts such a relationship and suggests that it is due to anatomical interconnections between tongue and palate. An alternative is supported by the data obtained in Lubker's research: greater palatal elevation may acompany vowels with high tongue position simply because such elevation is needed to prevent the vowel from being detected as nasal in quality. The latter explanation may be more consistent with the high correlation observed between palatal position and emg activity. Further emg studies of palato-glossus muscle by Bell-Berti and Hirose (1973) suggest that during speech this muscle is associated more with glossal rather than palatal movements. Ushijima and Hirose (1974) and Bell-Berti (1974, 1976) have greatly ex-panded knowledge of palatal muscles during speech using fine-wire electrodes.

Lubker joined Fritzell and Lindqvist in Sweden to expand this work with fine-wire electrodes (Lubker *et al.*, 1970). This work suggests that palatoglossus does not contract maximally during speech in contrast to its vigorous action in swallowing and nasal breathing. More recently Lubker and May (1973) using bipolar fine-wire electrodes found that palatoglossus functions in a variable fashion during speech, assisting to move the tongue or reacting to tongue movements when the levator palati is contracting, and lowering the palate when the levator is inactive. Fritzell and Kotby (1976) in turn compared the emg activity of the levator palati with that of thyroaryten-oideus. They are clearly different. During the rest position the levator palati remains silent but thyroarytenoideus (as part of its respiratory function) is constantly active. Although both are intimately involved in speech, their latencies before utterances differ significantly depending on the characteristics of the sounds produced. Lubker (1975) further enlarged upon studies of the palatal muscles with fine-wire electrodes. Of special interest were his findings of a temporal organization of speech, with considerable controlled variations among the various muscles involved.

Clinical application of palatal EMG is promising. Thus, Chaco and Yules (1969) proposed that a candidate for tonsillo-adenoidectomy should have EMG of the soft palate because of possible velopharyngeal incompetence. Šurina and Jágr (1969) have made similar recommendations for cleft palate patients.

Pharynx

In view of the relative inaccessibility of the striated muscles of the pharynx, electromyography of the sphincters has not been widely attempted. Nonethe-less, a considerable number of publications that make reference to muscular action have appeared on the process of swallowing. In most of this material, no actual objective recordings of the sequence of events in the involved muscles have been reported. The recent resurgence of practical interest in the mechanism of swallowing and in the reparative surgery of the pharynx has made a bold approach to pharyngeal electromyography a necessity. So, Doty and Bosma (1956) performed emg studies of reflex swallowing in the areas of the mouth, pharynx and larynx of anesthetized monkeys, cats and dogs. Broadbent and Swinyard (1959) then made similar studies during operations on anesthetized human patients with cleft palate. Car (1970) and Miller (1972a, b) are other workers who have made real contributions to the field of swallowing reflexes, using indwelling electrodes in individual oral and pharyngeal muscles of experimental animals.

Shipp *et al.* (1970) in a study of patients who had had laryngectomies, found two types of swallowing patterns: the first, called "Type I," was observed when emg sampling was conducted early in surgery before any major structures were altered or divided. This pattern consisted of a single large burst of activity from the inferior constrictor with a coincident inhibition of the preswallowing low level activity from the cricopharyngeus. The second pattern, labeled "Type II," was found in emg sampling later in surgery and in all subsequent emg procedures on laryngectomized subjects up to 3 years postoperatively.

The typical Type II muscle activity pattern showed both muscles simultaneously initiating activity during the first burst and the quiet segment, whereas for the second burst the cricopharyngeus followed the inferior constrictor by 60 to 180 msec. The second burst appears to be analogous to the pattern found in Type I or "normal swallow."

Type II swallowing patterns were similar in the awake and anesthetized conditions although duration measures differed.

From the start of our own studies of the pharynx, we felt that we must not only examine reflex swallowing under anesthesia, but also swallowing under conditions that are as normal as possible. Therefore, we devised various procedures towards that end, and we also performed extensive dissections to clarify the anatomy of the region. Our first report (Dutta and Basmajian, 1960) dealt with anatomical studies in rabbits. It was followed by a report on the EMG of the rabbit's pharyngeal and palatal muscles (Basmajian and Dutta, 1961a). Our third report dealt with emg studies on conscious, normal human beings (Basmajian and Dutta, 1961b).

In about half our series of rabbits, direct recordings with concentric needle electrodes were made of the electrical potentials of the soft palate and individual constrictors during reflex swallowing under anesthesia.

In the other half of the series, operative exposure was followed immediately by implanting into the three constrictors special indwelling, flexible, wire electrodes to be used for postoperative electromyography. Electromyographic testing was usually done after recovery from the operation—a delay of several days being the rule.

In some of the rabbits in which implants were used, simultaneous records were made also from the fleshy part of the soft palate *via* a concentric needle electrode. Active voluntary swallowing was induced by running some water from an "eye-dropper" onto the tongue of the rabbit.

In our studies of conscious normal volunteers, the electrodes were passed under direct vision through the open mouth, using a laryngeal mirror and clinical headset, when necessary. A special type of bipolar electrode was designed for use in these experiments. It consists of two fine surgical stainless-steel wires about 10 cm long, glued together, yet insulated from each other, by lacquer. This type of electrode has the advantage of strength, lightness and flexibility (figs. 17.3, 17.4).

Except during obvious swallowing, in all the experiments on human beings, conscious rabbits and anesthetized rabbits, there is little or no activity in any of the constrictors. That is, there is no resting tonus.

During swallowing each constrictor of the rabbit contracts for about ¼ second, whether it is part of reflex or of conscious swallowing (fig. 17.5). The contraction of the superior constrictor begins simultaneously with that in the soft palate; that in the middle constrictor is delayed by about 25 msec; and that in the inferior constrictor, by about 75 msec. The entire duration of activity in the constrictors lasts about ⅓ second. These figures must be

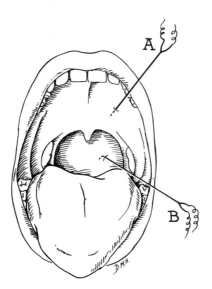

FIG. 17.3. Diagram of special bipolar electrodes, *A*, in soft palate and *B*, in pharyngeal wall.

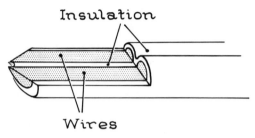

FIG. 17.4. Greatly enlarged cut-away diagram of tip of special electrodes (cf. fig. 17.3).

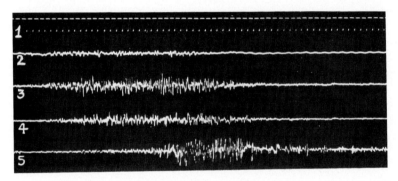

FIG. 17.5. Simultaneous EMGs during swallowing in the rabbit. *1*, calibration: 10-msec intervals; amplitude, ca. 300 μv; *2*, soft palate; *3*, superior constrictor; *4*, middle constrictor; and *5*, inferior constrictor. (Tracings slightly retouched to improve engraving.)

accepted as broad generalizations of our detailed results (Basmajian and Dutta, 1961a).

The amplitude of activity in the three constrictors of the rabbit gradually increases and becomes maximal just beyond the middle of its total period of activity. Then the amplitude falls to the base-line due to rapid relaxation. It is important to note that the base-line of *nil* activity is maintained except during swallowing, i.e., to repeat, there is no tonic activity of the palate and pharyngeal constrictor muscles. They act in an "all or none" fashion.

In the human volunteers, any activity picked up from these muscles following the insertion of the electrode apparently is a reflex "tightening up"; it can be abolished by the subject's voluntary relaxation. When the subject is relaxed and resting between swallows, the pharyngeal constrictors are inactive, as is the levator palati. During the sucking of water through a straw, all three constrictors remain silent, while—as might be expected—the levator palati is active. With each swallow, the duration of activity is very close to a ½ second in each of the pharyngeal constrictors (fig. 17.6).

Because we did not make recordings in the human muscles simultaneously (as we did in our study on rabbits) we cannot say what the exact sequence of activity is in man. However, there is no reason to suppose that it is different from that in rabbits.

Doty and Bosma (1956) described the emg pattern of activity during swallowing in the muscles of the mouth, pharynx and larynx of monkeys, cats and dogs. They found a definite pattern or sequence of activity, i.e., with minor variations, the "schedule of excitation and inhibition among the participating muscles was highly constant." No difference was found in reflex swallows evoked by various means including stimulation of the superior laryngeal nerve, stimulation with a cotton swab or the rapid squirting of water into the pharynx. A leading "complex," consisting of the superior constrictor, palatopharyngeus, posterior intrinsic muscles of the tongue and various muscles attached to the hyoid bone, becomes active for ¼ to ½ second to initiate the act. The middle constrictor, inhibited at first (as in our experiments), follows. The inferior constrictor is the last to become active, being deferred until the leading complex is nearly over (fig. 17.7).

Rather similar findings in dogs have been reported by Kawasaki *et al.* (1964) whose studies of air pressure gradients are particularly valuable.

Cricopharyngeus muscle has been the special concern of Levitt, Dedo and Ogura (1965) of St. Louis, who have recently made the greatest contribution to its understanding. In dogs they found that these sphincteric fibres at the junction of pharynx and esophagus relax much of the time, contrary to the widely held view that they are tonically contracting. Bursts of activity do occur but usually in response to external factors such as breath-holding, straining or stimulation of the hypopharynx. Murakami *et al.* (1972) found emg evidence of double-innervation (somatic and vagal) to the cricopharyngeus.

The striated muscle of the upper esophagus has been investigated by Arlazoroff *et al.* (1972). There was steady "high grade" EMG in all subjects (recorded from surface EMGs attached to a small inflated balloon); the activity increased further during the Valsalva manoeuvre.

Larynx and Vocal Cord

As Professor Georges Portmann of Bordeaux and Paris pointed out in his Semon Lecture (published in 1957), the larynx has two functions. The first

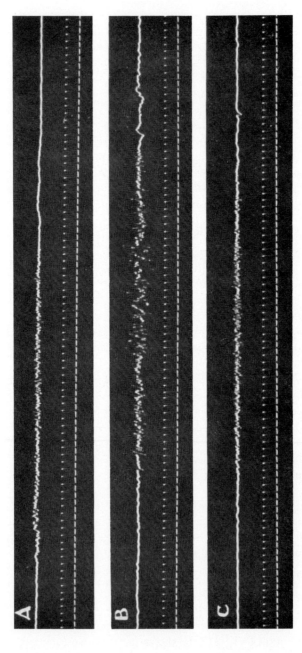

Fig. 17.6. EMGs of human pharyngeal constrictors: *A*, superior; *B*, middle; and *C*, inferior

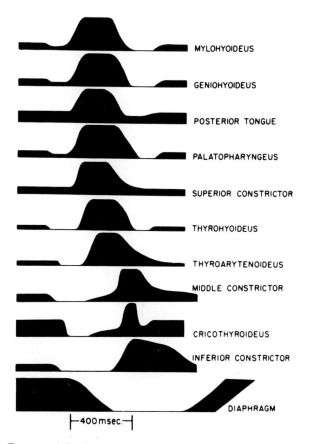

MYLOHYOIDEUS

GENIOHYOIDEUS

POSTERIOR TONGUE

PALATOPHARYNGEUS

SUPERIOR CONSTRICTOR

THYROHYOIDEUS

THYROARYTENOIDEUS

MIDDLE CONSTRICTOR

CRICOTHYROIDEUS

INFERIOR CONSTRICTOR

DIAPHRAGM

⊢400 msec.⊣

FIG. 17.7. Doty and Bosma's (1956) schematic summary of emg activity during swallowing in dogs.

comprises respiration and the protection of the pulmonary apparatus, these being inseparable. The other is phonation or the production of sound. Surprisingly, to the present time only provocative results have been obtained by electromyography for this vital function. Serra (1964) has given a thorough review of the various neuromuscular studies of this area.

Portmann and his colleagues first made some important contributions through their experiments on patients who had had operations that left the glottis exposed postoperatively. These investigators were able to insert needle electrodes directly into different parts of the vocal folds and the thyroarytenoid muscles.

During each expiration, the thyroarytenoid becomes very active, especially in the middle of expiration. At the beginning of inspiration, activity stops and does not reappear until expiration begins. Spontaneous and involuntary activity during expiration is quite independent of every effort of phonation.

Portmann's belief that during phonation the electromyographic oscillations have the same frequency as the sound emitted—strong support for the theory advanced by Husson (1950)—has been rejected widely (Rubin, 1960; Spoor and Van Dishoeck, 1960; Kirikae *et al.*, 1962; Milojevic and Hast, 1964; Dedo and Ogura, 1965; and still others). According to Floyd, Negus and Neil (1957) closure of the glottis is caused by tonic coordinated action of the sphincteric muscles in which thyroarytenoid plays its part. They have not succeeded in reproducing in dogs a higher frequency of vibration of the cords by stimulation of the laryngeal nerves with graded electrical frequencies. On the other hand, there is evidence being produced by independent workers which shows that complete tetanic fusion does not occur in laryngeal muscles until stimulation frequencies as high as 400 Hz are used (Mårtensson and Skogland, 1964). Also Louis-Sylvestre and MacLeod (1968) have advanced the provocative idea that the human vocal muscle mechanism may be similar to that of asynchronous insect muscle. Thus, an electrodynamic servo-system, linked to vibration frequency, allows instantaneous modifications of the mechanical impedance sensed by the muscle, which is kept in its natural environment. These investigators believe vibration frequency follows without latency the changes in impedance parameters; they find the sign and amplitude of variations to be identical in both insect flight muscle and human vocal muscle.

Green and Neil (1955) noted that impulses in the recurrent laryngeal nerve of cats coincide with inspiration and emg activity in the posterior cricoarytenoid. They found emg activity in the abductor muscles of the cords during inspiration alternating with activity in the adductors during expiration.

Under the general direction of Fritz Buchthal of the Institute of Neurophysiology at Copenhagen, Faaborg-Andersen completed a monumental study on the normal and abnormal muscular control of the vocal cords and published it as a monograph in 1957. In a long series of normal persons, he inserted needle electrodes under direct vision into the cricothyroid, vocalis, arytenoideus, thyroarytenoid and posterior cricoarytenoid muscles, two at a time, and recorded the subject's phonation simultaneously on a tape recorder.

At rest, there was a slight activity in all the adductors even if the breath was held quietly. This then is "postural activity," according to Faaborg-Andersen. It increased from the resting level during inspiration but was unchanged during expiration. At "rest," the abductors were very active by contrast; but during the inspiratory phase this activity decreased somewhat, although it was uninfluenced by expiration.

With phonation, an increase in electrical activity was found in all the adductor muscles. The change began 0.35 to 0.55 seconds before the audible sound. Unlike Portmann, Faaborg-Andersen found that single unit potentials fell below the range of 20 to 30 per second in basic frequency. This is much like ordinary skeletal muscles.

During phonation with increasing pitch there was no corresponding increase in the electrical activity of the adductor muscles. However, with increasing pitch the increase in the total electrical activity was marked provided that the increase in pitch was in the same register, i.e., within the same octave. If the increase in pitch was accompanied by a shift in register, the change was only slight.

Thus Faaborg-Andersen does not support the theory that the frequency of vibration of the vocal cords during phonation of a tone changes directly with the frequency of nerve impulses and contractions of the muscle fibers. Faaborg-Andersen upholds the orthodox view that the vibrations of the vocal cord are passive and independent of the frequency of motor unit contractions.

With phonation of different vowels there was no apparent electromyographic change in Faaborg-Andersen's series (fig. 17.8). With whispered voice or "silent speech," there was activity in the adductor muscles but it was far less than with ordinary voice.

Buchthal and Faaborg-Andersen (1964) found that the average time between the onset of the increase in electrical activity in cricothyroid muscle and the onset of phonation is between 1/10 and 1/5 of a second. The interval may be considerably shorter for some sounds.

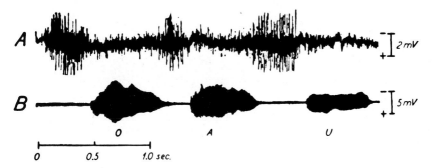

Fig. 17.8. EMG in thyroarytenoid (*tracing A*) and microphone recording (*tracing B*) during phonation of vowels o—a—u at frequency of 200 cps. (From Faaborg-Andersen, 1957.)

During a cough and during swallowing there was a considerable increase in the electromyographic potentials in all the adductors just before the onset of audible sound; conversely, the abductors relaxed during the cough.

The earlier morphological and emg studies have been summarized in a paper by Gomez Oliveros (1969). In recent years laryngeal EMG has gained a much wider following, partly because of Shipp's introduction of improved techniques employing our fine-wire electrode methodologies (Shipp, 1968; Shipp *et al.*, 1968; Shipp *et al.*, 1970). Figure 17.9 illustrates the technique of inserting electrodes into the posterior cricoarytenoideus. For other muscles, the direct anterior route through the skin also may be used in conscious human subjects (Hirose and Gay, 1972, 1973; Kotby, 1975; Dejonckere, 1975). Dedo and Hall (1969), in experiments with dogs, have used wires and needle electrodes, finding them both reliable.

Hiroto *et al.* (1967) of Kurume, Japan found little or no activity in posterior cricoarytenoideus during phonation while the cricothyroid, thyroarytenoid, lateral cricoarytenoid and interarytenoid muscles were very active. The

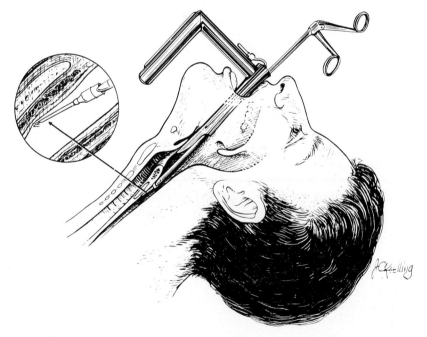

FIG. 17.9. A technique for introducing bipolar fine-wire electrodes into the posterior cricoarytenoideus muscle through a laryngoscope. (From Shipp *et al.*, 1970.)

change in electrical activity begins before the onset of speech sounds (with occasional exceptions for voiceless consonant syllables). The time interval from the onset of emg changes to the onset of sound was least in cricothyroid and for consonant syllables.

Hirose and Gay (1972, 1973) of the Haskins Laboratories in New Haven found that the posterior cricoarytenoideus participates in the production of voiceless consonants but not voiced consonants. A reciprocal pattern occurs in the interarytenoideus. They found three types of vocal attack that are characterized by the coordinated actions of the abductor and adductor muscles which act reciprocally for each type of attack.

The group associated with Thomas Shipp have shown that with normal adults there is greater airflow and greater activity in cricothyroid and interarytenoid muscles in modal phonation compared with those in vocal fry—the non-pathologic phonational register encompassing a range of frequencies below those of the modal register (McGlone and Shipp, 1971). Frequency through the vocal range is changed chiefly by the activity in cricothyroid and thyroarytenoid muscles, and their activity is highly correlated with measures of subglottal pressure. However, no systematic correlation occurs between (1) activity in posterior cricoarytenoid and interarytenoid muscles and air flow measures, and (2) frequency changes (Shipp and McGlone, 1971).

A series of papers from Oslo in 1970 have added to our understanding of laryngeal function (Kotby and Haugen, 1970a, b, c, d). They described a basic postural ('resting') activity in the posterior cricoarytenoid which fluctuates with breathing, rising most with deep inspiration. In phonation, it becomes very active as it does during sphincteric actions. Kotby and Haugen suggest that the concept of antagonists—abductors *vs* adductors—is false.

The elegant study of Gay, Strome, Hirose and Sawashima, who represent the Universities of Connecticut, Harvard and Tokyo, is especially noteworthy (Gay *et al.*, 1972). Both peroral and percutaneous insertion of our fine-wire electrodes was done in five subjects with apparently meticulous technique and computer analysis. They, too, showed that increases in fundamental frequency on phonation were accompanied progressively by increases of emg activity in cricothyroid muscle—but they were able to show this *in vocalis also*. Moreover, to a lesser extent the increase occurs in *all* the other intrinsic muscles too; but the posterior cricoarytenoid increase comes at the highest pitch level.

Gay *et al.*, disagreeing with Faaborg-Andersen (cited above), find that the posterior cricoarytenoid can also act as a tensor of the vocal folds, following the curve of activity of the cricothyroid. They warn, however, that there are individual differences between subjects—which also may account for earlier confusions in laryngeal emg research. They were unable to demonstrate a strict linear relationship of the activity in cricothyroid and vocalis to fundamental frequency, and they discounted adductor muscle actions as "probably secondary."

Changes in intensity seem to be related more to expiratory muscle force than laryngeal factors; there may be "a trading relationship between laryngeal and subglottal forces depending on whether a specific frequency level is aimed for or not" (Gay *et al.*, 1972).

Some Clinical Applications of Laryngeal EMG

Knutsson *et al.* (1969) and Sutton *et al.* (1972) demonstrated the possibility of studying the characteristics of individual motor units in various laryngeal muscles by means of percutaneous and peroral electrodes. Kotby and Haugen (1970) and Dedo (1970, 1972) took up the problem of diagnosing various vocal fold mobility disorders and paralysis. Ueda *et al.* (1971, 1972), Ohyama *et al.* (1972) and Lauerma *et al.* (1972) have investigated experimental artificial voice production, while Shipp (1970) has studied the function of the pharyngeal inferior constrictor and cricopharyngeus during alaryngeal voice production in laryngectomized patients.

Stuttering Larynx

Combining fiberoptics and EMG, Freeman and Ushijima (1974) studied stuttering in three patients. Recordings were made with percutaneous fine-wire electrodes in four of the five intrinsic laryngeal muscles. The so-called Wingate hypothesis was justified: thus laryngeal activity patterns differ between the stuttering moments and when the patients could be induced to speak fluently. The abnormal patterns at the stuttering moments reveal (1) disruption of the normal reciprocity of the abductors and adductors, (2) disruption of the normal synchrony between adductors and (3) generally higher activity in the intrinsic muscles.

Laryngeal Reflexes

A number of experimental studies employing EMG have been reported. Thus, Kurozumi *et al.* (1971) have shown the existence of bilateral motor innervation from the medulla oblongata to the laryngeal muscles. Murakami and Kirchner (1971, 1972) have investigated the effects of different frequencies of electrical stimulation on the various muscles and they have described the emg activity of intrinsic muscles during different reflexes and laryngeal closure. Abo-El-Enein and Wyke (1966, 1969) and Wyke (1968) have described the effects of anesthesia and nerve-sectioning on the myotatic reflexes. The laryngeal initiation of swallowing via sensory endings in the mucosa of the larynx has been described by Storey (1968a, b) of Toronto.

Laryngeal Muscle Training

Hardyck *et al.* (1966) first reported that emg feedback could be used (by our motor-unit training technique) in laryngeal muscles for treating persons whose silent reading was hampered by subvocalization. These matters are discussed in Chapters 5 and 6.

Chapter 18

Muscles of mastication, face and neck

Muscles of Mastication

T HE first concerted effort to apply electromyography to problems of orthodontics and normal temporomandibular physiology was made by a dentist, Robert Moyers, while working as a graduate student at Iowa. Upon his becoming Professor of Orthodontics at the University of Toronto in 1949, I had the privilege of being associated with him and of seeing his work first-hand. Following his lead, other workers have published a number of good studies although some of them have disagreed with certain details of Moyer's earliest work (1949 *et seq.*). The following account is a review of the published reports in the field. I have also drawn upon a number of special sources including the research of graduate students.

Also valuable are the newer reports of work with inserted electrodes by Ahlgren (1966, 1967) and Møller (1975, 1976) in Sweden, Munro (1974) and his colleagues in Sydney, and Vitti (1969a, b, 1970, 1971 *et seq.* with others) in Brazil. The privilege of having Munro and later Vitti as visiting colleagues with my group through 1972 to 1974 greatly enhances our understanding of mandibular and facial mechanisms. Another area that has emerged with vigour is that of mandibular reflexes. A discussion of the various reflexes now makes up a later section of this chapter.

Mandibular Rest Position

While the biomechanics and orthodontics of the mandibular rest position must remain the province of dental clinical-research specialists, a few words on its EMG are appropriate here. Møller (1975, 1976) correctly ascribed the confusion in the literature to the wide differences in emg equipment and technique and he pleads for better quantification, as did Frame *et al.* (1973).

Is the rest position controlled actively or determined passively? There now appears to be no doubt that in deliberately relaxed but fully upright postures, even with the lips together (but the teeth not together), there is virtually no emg activity (Vitti and Basmajian, 1975) except with occlusal interference (Funakoshi *et al.*, 1976). However, where the force of gravity acts on the jaw of an unrelaxed person, low-level myoelectric activity is the rule in the anterior part of temporalis; greater alertness and functional disorders both raise the level substantially (Møller, 1976). During sleep, intermittent increases and decreases of activity occurred throughout the night in the study of Fuchs (1975), as might be expected. However, Fuchs was able to elicit the patterns of change and his paper should be consulted for details.

Temporalis

RESTING TONUS. As noted above, fine-wire electrodes and the best surface-electrode techniques have clearly shown that at normal and deliberate rest (with the lips but not the teeth occluded) there is no activity in temporalis. However, the controversy began when Moyers (1949, 1950) reported for normal subjects a "remarkably even state of *tonus*" in all three parts of the muscle when it is at rest, stating also that the normal maintenance of mandibular posture is shared by all the parts. Carlsöö (1952), although he agreed that temporalis is the main postural muscle in the habitual rest position, insisted that the posterior part of temporalis was the more important part in this position. MacDougall and Andrew (1953) also agreed that resting postural tonus was obtainable, but they were less precise.

Latif (1957), while working under my direction, made a definitive emg study of both temporalis muscles in 25 normal teenage children. In the physiological resting position of the mandible in the upright subject, both the anterior and posterior fibres of temporalis were continuously active in almost all the subjects. However, this activity was much greater in the posterior fibres (fig. 18.1), as hinted at by Carlsöö's earlier findings and in contradiction to those of Moyers. In the same year, Latif's finding was duplicated independently by Kawamura and Fujimoto (1957) of Osaka. However, with coaxial needle electrodes in the anterior, middle and posterior part of the temporalis of 57 patients, Vitti (1969a) found no resting postural tonus in the majority and later Vitti and Basmajian (1975, 1977) thoroughly documented this finding in children and in adults. Vitti found some relationship between the state of dentition and the presence of activity in the posterior fibres. His similar study (Vitti, 1969b) of retraction of the angle of the mouth showed activity in the posterior fibres in most subjects and the reverse for the

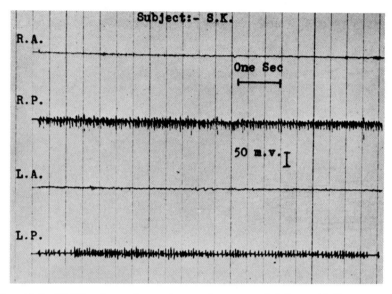

Fig. 18.1. EMGs of anterior and posterior parts of right and left temporalis muscles at rest show much greater postural activity in posterior parts (R.P. and L.P.) compared with anterior (R.A. and L.A.).

middle and anterior fibres. Christensen *et al.* (1969) of Copenhagen and Osaka indicate that actually there may be four distinct functional areas of the temporalis in cats.

END-TO-END OCCLUSION (INCISOR BITE). Latif found all parts of temporalis were active, the greater activity being somewhat more frequently in the anterior fibres (40% of muscles), but in many (22%) the posterior fibres predominated, while in a third of the muscles the activity was equal throughout (fig. 18.2). In several muscles, in contrast, the temporalis was inactive with incisor bite, even though this condition was considered to be the norm (quite erroneously) by Keith (1920). These results of Latif confirmed the findings of MacDougall and Andrew (1953). Vitti (1970) reported that the anterior and middle fibres were active with normal dentition and incomplete dentition with molar support; in edentulous patients, all three parts were active during incisive bite.

MOLAR OCCLUSION. All the fibres of the temporalis showed marked activity in all subjects (Latif, 1957; Ahlgren, 1967; Vitti, 1970; Vitti and Basmajian, 1975, 1977), as would be expected (fig. 18.3). This is the chief function of the temporalis.

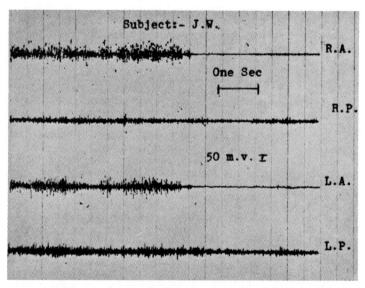

FIG. 18.2. EMGs of temporalis muscles during end-to-end occlusion from right anterior (R.A.), right posterior (R.P.), left anterior (L.A.) and left posterior (L.P.) parts. (From Latif, 1957.)

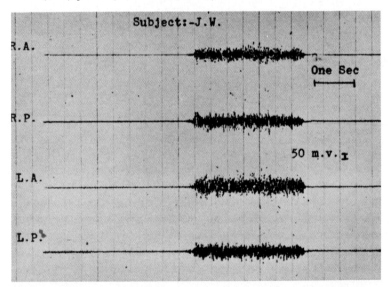

FIG. 18.3. EMGs of temporalis muscles during molar occlusion from right anterior (R.A.), right posterior (R.P.), left anterior (L.A.) and left posterior (L.P.) parts. (From Latif, 1957.)

RETRACTION OF THE JAW. A universal finding was a marked activity in the posterior fibres of temporalis with lesser activity in the anterior fibres during the drawing back of the jaw from the protruded (protracted) position. This is in keeping with the accepted teaching (Latif, 1957; Vitti, 1971, Vitti and Basmajian, 1975, 1977).

PROTRACTION. Latif's findings were not in agreement with the opinion of McCollum (1943) and the findings of Moyers that the anterior fibres are active during protraction. He found *no* activity, as did Carlsöö, MacDougall and Andrew, Woelfel *et al.* (1960) and Vitti (1971). Indeed the activity even dropped from the resting tonus (fig. 18.4). Apparently the temporalis shifts the burden of supporting the jaw to the muscles that protrude it, chiefly the lateral pterygoids.

LATERAL MOVEMENTS. Moyers' findings that the temporalis abducts the mandible was clearly confirmed by the observation that this action is almost universal (fig. 18.5). In repetitive side-to-side movements, first one temporalis and then the other acts, each pulling the mandible only to its own side (Latif, 1957; Vitti, 1971; Vitti and Basmajian, 1975, 1977).

DEPRESSION. During ordinary opening of the mouth, temporalis is inactive (MacDougall and Andrew, 1953; Latif, 1957; Vitti, 1971). When the man-

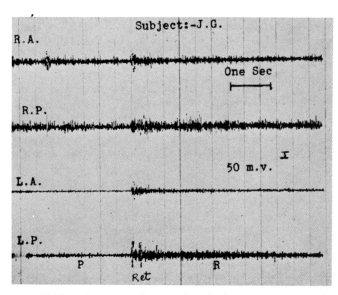

FIG. 18.4. EMGs of temporalis muscles during protraction of mandible (P.) and at rest (R.). Ret. = retraction. R.A. and R.P., from right anterior and posterior, L.A. and L.P., from left. (From Latif, 1957.)

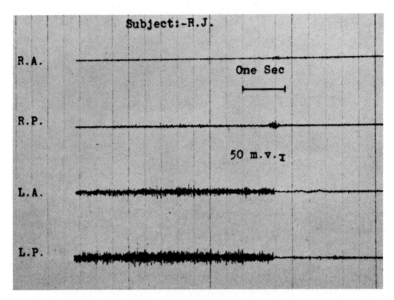

FIG. 18.5. EMGs of temporalis muscles during "left lateral position" of mandible. Note greatest activity in left posterior (L.P.) fibres, and least in right anterior (R.A.). (From Latif, 1957.)

dible is forcibly depressed (maximal opening of the mouth) some irregular muscle potentials do appear, suggesting that temporalis acts then as a protector against dislocation of the jaw.

The following paragraph summarizes the functions of temporalis. It maintains mandibular posture in the physiological resting position, the posterior fibres taking a more active part than the anterior. Its chief function is molar occlusion, and it is an ipsilateral abductor (and therefore a contralateral adductor) of the mandible. During maximal opening of the mouth, the temporalis acts as a preventative to dislocation but it plays no rôle either in ordinary opening of the mouth or in protraction of the jaw. In end-to-end occlusion (incisor bite) the anterior fibres are more active. The temporalis retracts the protruded jaw, the posterior fibres being especially active (Vitti and Basmajian, 1977).

Finally, Biggs, Pool and Blanton (1968) showed that the emg activity of monozygotic twins was similar in the duration of activity in their temporalis (and masseter) muscles.

Masseter

As would be expected, during forceful centric occlusion the masseter muscle is very active (Pruzansky, 1952; Moyers, 1950; Ahlgren, 1967; Vitti and Basmajian, 1975, 1977). During chewing movements the maximal activity occurs in the masseter about the time the jaw reaches the temporary position of centric occlusion which accounts for about a fifth of the chewing cycle (Gibbs, 1975). Masseter is not an important muscle in the habitual resting position (Carlsöö, 1952; Vitti and Basmajian, 1975, 1977), though it does show some activity in its superficial part during protrusion (Carlsöö, 1952; Vitti and Basmajian, 1977) and with increasing weights on the chin (Carlsöö, 1952) (fig. 18.6). It also acts in certain types of ipsilateral and contralateral movements of the mandible (Vitti and Basmajian, 1977). MacDougall and Andrew found that its deep fibres are occasionally active during retraction.

Hannam (1976) used on-line computer sampling and cross-correlation analysis of the bilateral myoelectric activities of the masseter muscles. Cross-correlation showed no significant patterns of difference from bite to bite when the 10 subjects were taken as a whole. However, individuals showed some slight differences. In almost half his sample, the muscle contralateral to the side chewed upon started to contract first, while the rest acted in miscellaneous ways.

In awake rabbits, Schärer and Pfyffer (1970) were able to stimulate cortical centers while recording from the masseter through fine-wire electrodes. The results were identical to normal chewing in masseters.

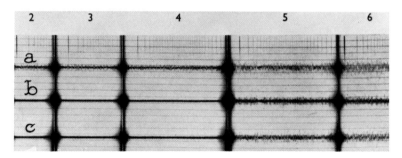

Fig. 18.6. EMGs from *a*, posterior part of temporalis; *b*, superficial part of masseter and *c*, medial pterygoid, with progressively increasing loads starting at "2." (From Carlsöö, 1952.)

Medial Pterygoid

During simple protraction, Moyers always found strong activity with needle electrodes in medial pterygoid. We agree (Vitti and Basmajian, 1977). This decreases slightly if the mandible was depressed while being protracted. Unilateral contractions of the muscle accompanied (caused ?) contralateral abduction of the chin and was particularly impressive when there was an added element of protraction (Moyers, 1950; Carlsöö, 1952). In the rabbit and rat, O'Dell *et al.* (1969) found with wire electrodes that medial pterygoid opposes forced lateral mandibular movement.

Lateral Pterygoid and Digastric

During mandibular depression, Moyer (1950) found that the first large potentials to appear are in the lateral pterygoid and the activity reaches its peak even before the other muscles become active (fig. 18.7). Furthermore, it continues throughout the movement. Woelfel *et al.* found the lateral pterygoid very active in contralateral excursions, uncontrolled openings and protrusions of the mandible. However, it was inactive during hinge openings of up to 1 cm. Apparently its function is to draw forward the articular disc in the temporomandibular joint along with the head of the mandible. In rhesus monkeys (which have the same morphological arrangement as man in this muscle) McNamara (1973) found that the inferior head of the lateral pterygoid is active only during opening and protrusive movements. The superior head was active only during such closing movements as chewing, swallowing and clenching of the teeth. Supporting McNamara, Grant (1973) insists that the muscle is two muscles, both functionally and structurally.

During mandibular depression, the digastric muscle comes into action after the lateral pterygoid and is not as important. However, its action is essential for the maximum depression of forced or complete opening of the mouth. The right and left digastric muscles do not function as individuals. They contract bilaterally with greatest activity during uncontrolled openings and retrusions of the mandible (Woelfel *et al.*, 1960). Munro (1974) agreed but added a new dimension: when tooth contact occurs during elevation of the mandible in chewing, the posterior belly contracts without activity in the anterior belly. Apparently this is related to positioning of the hyoid bone, not the jaw.

Other work on the digastric muscles has dealt with their general functions—e.g., coughing, breathing and swallowing (Lesoine and Paulsen, 1968; Hrycyshyn and Basmajian, 1972) and mandibular reflexes (Munro and

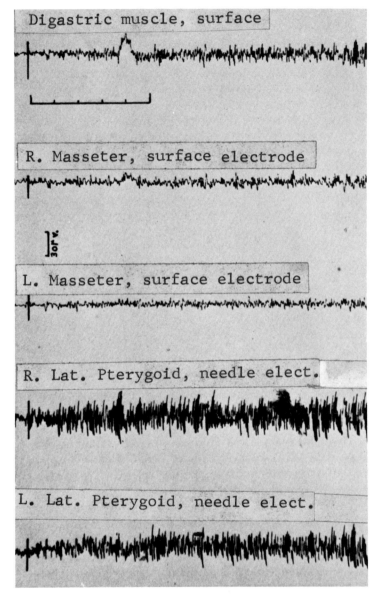

F IG. 18.7. EMGs of mandibular depression. (From Moyers, 1950.)

Basmajian, 1971). Coughing and swallowing strongly recruit the activity of the digastric muscles.

Coordination during Chewing

Using subcutaneous wire electrodes, Ahlgren (1966) studied the muscular coordination and emg activity during chewing of peanuts and of chewing-gum. Considerable variation of patterns from subject to subject was a salient finding. The commonest pattern during gum-chewing (in 44%) was one in which the ipsilateral temporalis contracts first, then the contralateral temporalis and both masseters simultaneously. Vitti and I (1977) also found marked activity in the medial pterygoid muscles bilaterally. The peak of activity and the ending of activity occurs simultaneously most of the time. With peanut-chewing, the commonest pattern was both masseters and temporalis contracting simultaneously.

Comparing the activity of masseter and temporalis muscles while chewing hard and soft foods, Steiner *et al.* (1974) found marked variations between normal subjects. Masseter always led the action. With hard foods (carrots) the delay before recruitment of temporalis was shorter and the total activity was considerably greater.

In a series of experiments with dogs, Vitti (1965) found a constant activity of temporalis, masseter and medial pterygoid during mandibular elevation in chewing towards either side; there was a decrease with "incisive chewing," in which the greatest activity was in the anterior and middle fibres of temporalis, middle and deep part of masseter, and medial pterygoid. Unlike Vitti's finding in dogs, bats (*myotis lucifugus*) have clear alternations of activity between the muscles of the two sides (Kalen and Gans, 1970).

Summary of Mandibular Movements

The various movements of the jaw are produced by cooperative activity of several muscles bilaterally or unilaterally. Mandibular elevation is performed by the temporalis, masseter and medial pterygoids and depression by the lateral pterygoids and digastrics. The digastrics show their greatest activity in forceful opening of the mouth at the limit of depression of the mandible. Lateral movements are performed by the ipsilateral temporalis and masseter and the contralateral medial pterygoid (and, to a lesser extent, the lateral pterygoid). Protraction is performed by the medial and lateral pterygoids while retraction is by the temporalis, chiefly its posterior fibres, and perhaps the deep part of masseter.

Mandibular Reflexes

Electromyographic investigations by several observers have shown that following tooth contact in man, there is an inhibition of the mandibular elevator muscles (Ahlgren, 1969; Beaudrea, Daugherty and Masland, 1969; Griffin and Munro, 1969; Kloprogge, 1975; Nagasawa *et al.*, 1976; Lund, 1976). It is believed that this inhibition is induced reflexly by stimulation of periodontal end organs. Griffin and Munro (1969) found that, at the time of inhibition of the mandibular elevators, the lateral pterygoid muscle was inactive, whereas localised activity of the digastric muscle occurred, mainly towards the end of the inhibitory phase of the mandibular elevators. They also found localised activity of the digastric muscle with inhibition of the masseter muscle after the tapping of a single tooth; these effects were abolished by local anaesthesia of the tooth. Beaudrea *et al.* (1969) also found a motor pause in the temporalis and masseter muscles following the hitting of teeth, but the digastric muscle in their series also exhibited a motor pause at the time of inhibition of the mandibular elevators.

Munro and I (1971) investigated the six mandibular muscles with fine-wire electrodes in normal persons performing the open-close-open cycle. A microphone applied to the skin over the zygoma was used to determine the time of initial tooth contact.

A period of inhibition of 12 to 20 msec mean duration was found in each mandibular elevator, commencing after a mean interval of 10 to 14 msec from initial tooth contact. In each cycle the inhibitory periods of the six mandibular elevators were found to be almost synchronous. The mean duration of the overlap of inhibition of the mandibular elevators was 11.9 msec, commencing after a mean interval of 14.6 msec from initial tooth contact.

In a further five subjects the activity of the anterior temporalis muscles was recorded with that of the anterior bellies of the digastric muscles. The activity of the digastric muscles at the time of inhibition of the mandibular elevators was analyzed. No activity was seen in 13% of digastric traces at this time; a further 25% of traces showed activity continuing through part or all of the period of inhibition of the mandibular elevators. In the remaining 62% of cycles there was a localised burst of digastric muscle activity at this time, usually either throughout or towards the latter end of the period of inhibition of the mandibular elevators. It has previously been shown that the lateral pterygoid muscle is inactive at this time.

This pattern of muscular activity would be expected to result in jaw opening (or at least in a reduction in occlusal pressure) in the interval from 15 to 27 msec after initial tooth contact.

A lateral jaw movement reflex has been described by Lund *et al.* (1971) and central coordination of rhythmical mastication by Dellow and Lund (1971). My former colleagues Anderson and Mahan (1971) at Emory University have described the interaction of receptors in tooth pulp and periodontal ligaments in the jaw-depression reflex. Klineberg, Greenfield and Wyke (1970) emphasize the role of mechanoreceptors in the temporomandibular joint capsule in the reflex control of chewing, while Bratzlavky (1972) explains the reflex pauses in the elevators entirely on spindle afferent discharges. Further work has been done by many investigators on the reflex activities of the mandibular elevators to elucidate postural and muscle-spindle neuromotor functions (see for example: Clark and Wyke, 1965; Lamarre and Lund, 1975; Övall and Elmqvist, 1975; Godaux and Desmedt, 1975a, b; Amano and Funakoshi, 1976; Ongerboer de Visser and Goor, 1976; Hagbarth *et al.*, 1976; Widmalm and Hedegård, 1976). A most elegant monograph by Widmalm (1976), his thesis at the University of Gothenburg, Sweden, offers an excellent review and illustrations (fig. 18.8).

Muscles of Facial Expression

Systematic electromyography of the muscles of expression was badly neglected until a few years ago, although clinical electromyographers are constantly concerned with facial palsies. (Orbicularis oculi had been investigated with the eye muscles; see p. 410.)

Mentalis

Mentalis muscle, which wrinkles the chin, was the subject of an investigation primarily concerned with its reactions during sleep. Hishikawa, Sumitsuji, Matsumoto and Kaneko (1965) of Osaka, showed a fairly continuous "tonic" activity of mentalis in man except during deep sleep when it relaxes completely for short periods. More concerned with dental problems, Schlossberg and Harris (1956) described considerable activity in mentalis muscle during swallowing, as did Jacob *et al.* (1971). We found only slight activity in mentalis during gentle check puffing and with various gentle grimacings with the lips. When these were more forceful, the activity rose; during forceful pulling of the corners of the mouth downward and pursing the lips, activity rose to 'marked' (Isley and Basmajian, 1973).

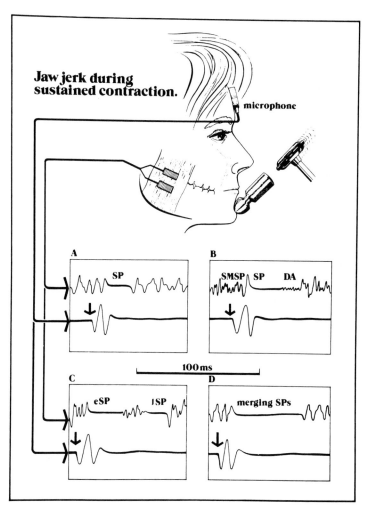

FIG. 18.8. Widmalm's method of eliciting jaw jerks during sustained contraction of masseter.

Buccinator

Although, in some respects, this muscle lining the cheek is more related to chewing, it has many other functions too. The first major paper on buccinator was by de Sousa and Vitti (1965) at São Paulo. They reported that, with needle electrodes, the muscle consistently showed activity in retracting the mouth laterally and in various complex functions (which has been described with the *Mouth,* (p. 362). However, distension of the cheeks left buccinator silent. Our work with fine-wire electrodes (Isley and Basmajian, 1973; Basmajian and Newton, 1973) convincingly shows its complexity of function. Buccinator undoubtedly is recruited in individualized patterns and even exhibits variations within the same subject depending on the amount of force and on slight shifts of emphasis in lip movement. Gentle puffing fails to recruit buccinator activity; more forceful effort increases it to varying degrees with various subjects but not to a marked level. Pulling the corners of the mouth back—or any movements involving such a component—recruits variable (but rarely marked) activity even when done forcefully. A wide forced smile is most effective, however, resulting in marked activity in almost half of the subjects (Isley and Basmajian, 1973). Blanton, Biggs and Perkins (1970) found rather similar results although their subjects showed more activity in blowing. They also demonstrated activity during mandibular movements, which they ascribed to sympathetic behaviour and not prime action. Sucking movements were also quite effective, they said, in recruiting buccinator; however, our results with thumb-sucking showed very little activity in buccinator (Vitti, Basmajian *et al.,* 1975).

In a related study, we showed the marvelous control man has over various parts of buccinator (Basmajian and Newton, 1973). With minimal feedback training, a series of subjects rapidly learned to consciously control contractions in different parts of the muscle while maintaining silence in the remainder—and they could do this while performing on the clarinet!

Lip Muscles

Two muscles are included entirely within the confines of the lips—orbicularis oris superior and inferior (OOS and OOI); they really are two muscles and not just a continuous sphincter, as textbooks imply. Others radiate outward to bone and fascia to move the corners of the mouth: upward—levator anguli oris (LAO); upward and laterally—two zygomaticus muscles, major and minor; laterally—risorius and buccinator (see above); and downward—depressor anguli oris (DAO). The upper lip has an "elevator," the levator labii superioris (LLS) and the lower lip a "depressor," the depressor labii inferioris (DLI). There are also various wisps here and there which vary with the individual. All facial muscles (including others higher on the face) are supplied by branches of the same nerve—the facial or 7th cranial nerve.

As noted before, the complexity of the arrangement and the superficial nature of their disposal has retarded emg progress of these muscles until lately. Speech pathologists and musicians could not wait for anatomists and physiologists and have taken the initiative for obvious practical reasons. Here, let us consider first the general kinesiologic studies, second, the applied speech studies, third, the musician-related work and finally, clinical (surgical) applications.

LIP KINESIOLOGY. A number of investigators have now emphasized that OOS and OOI are separate muscles (Jacob *et al.*, 1971; Basmajian and White, 1973; Isley and Basmajian, 1973; White and Basmajian, 1973). They act in complex fashions during a multitude of lip positionings. Thus, forceful puffing, pulling the corners of the mouth in different directions, pursing the lips and curling the lips over the teeth, all recruit varying levels of activity, and there again they vary from subject to subject (Isley and Basmajian, 1973). DAO has been investigated separately by Vitti *et al.* (1972) who find similar rational results.

Zygomaticus major acts intermittently, and only a wide forced smile recruits maximal activity. The corner-moving muscles are aptly named—they do what their names proclaim. The same is true of the LLS. All these muscles are most active during broad smiling. Puffing the cheeks out brings out the least response in this group.

While only slight myoelectric activity can be picked up in orbicularis oris during the rest position, we found that during aberrant oral activity, such as thumb-sucking, there is marked activity in the lips (Vitti, Basmajian *et al.*, 1975). Kelman and Gatehouse (1975) found that diffuse surface-electrode pickup of the lip muscles remained surprisingly consistent in individuals repeating the same utterances. Garrity (1975) reported a high correlation between surface-electrode pickups from the lips and from the chin and she then made the wrong conclusion (that one was as good as the other to reveal lip functions). All she was doing was recording cross-talk between pickup sites, no serious problem in the general type of speech research she represents. See also the section below on *Musical Performance,* which must be much more precise.

LEVATOR LABII SUPERIORIS ALAEQUE NASI. Vitti *et al.* (1974) confirmed the assumption that this muscle raises the upper lip and dilates the nostril, but he found it remains silent when the face is resting and composed. In various facial grimaces it springs into activity, but in other facial expressions it remains relaxed, e.g., raising the eyebrows, closing the eyes, opening the mouth and blowing without distension of the cheeks.

Procerus and Frontalis Muscles

Recently we studied these two muscles with surface electrodes and fine-wire electrodes simultaneously (Vitti and Basmajian, 1976). This was done because these two muscles are being often used by psychologists as a reference area for muscular tension studies. In the resting state, as with other facial muscles, we found no activity, contrary to the report of Sumitsuji *et al.* (1967) that there are high-amplitude emg discharges in the frontalis. Akamatsu (1960) studied the EMG of frontalis in schizophrenic patients; he reported a relationship between symptom pattern and patient's posture on the one hand and emg activity on the other. Earlier, Tokizane (1954) reported that except for frontalis and levator labii superioris, facial muscles showed electrical silence in the resting state. Although they found no significant change in frontalis due to postural variation, Sumitsuji *et al.* found that frontalis activity decreased when their subjects lie down. Not only frontalis but also mentalis and other facial muscles decreased their emg amplitude in the supine posture. The resting silence from sensitive surface and wire electrodes in our study of normal subjects agrees with wide clinical emg experience that striated muscle at rest shows complete relaxation. The reports to the contrary, mentioned above, apparently arise from varying levels of tension in the face over a period of some time. To reiterate: at rest, frontalis and procerus are silent and become active on demand or in response to specific (perhaps uncontrolled) emotional states or expressions.

Elevation of the eyebrows, frowning with the forehead, closing the eyes normally and whistling, all did not recruit procerus muscle. The same silence resulted from the frontalis muscle in the movements of pulling the eyebrows downward, closing the eyes either normally or strongly, in the expressions of sadness, gladness and revulsion, transverse wrinkling over the bridge of the nose, whistling and winking either the right or the left eye.

Our data showed that the frontalis has a very marked activity only in elevating the eyebrows and frowning with the forehead. It showed its antagonism to the procerus, which is inactive in these movements, and confirmed the electrophysiologic experiments of Duchenne, who showed the procerus is a direct antagonist of the frontalis.

Confirming the classical descriptions given by some anatomy textbooks and by Martone (1962), the procerus muscle shows a very marked activity during the production of transverse wrinkles over the bridge of the nose and moderate activity in pulling the eyebrows downward. Closing the eyes strongly recruits a marked activity in the muscle. However, in this movement our subjects always provoked transverse wrinkles over the bridge of the nose, which, as already noted, always makes the muscle active. In winking the right and left eyes, the procerus was active with a little less intensity. This was also true during facial expressions of revulsion and aggression. In all these movements, subjects always showed transverse wrinkles over the bridge of the nose.

Finally, in the facial expression of sadness and gladness, some subjects showed negligible activity in procerus muscle. These expressions having been performed by the subjects voluntarily and not "naturally" could lead to some confusion. Further, as our subjects performed called-for expressions, these sometimes did not correspond to our own concept of spontaneous or involuntary expressions. However, our videotape record of those facial expressions and movements always showed that whenever the subjects provoked transverse wrinkles over the bridge of the nose, the procerus was active while the frontalis was not.

Speech Research

The lip muscles have been the subject of research by a growing number of investigators. Only the group at Haskins Laboratory and the University of Connecticut has used fine-wire electrodes in a limited fashion for speech research on the lips (Gay and Hirose, 1973; Gay *et al.,* 1974; Gay, 1975), a natural extension of their laryngeal research technique. Most speech research has avoided intramuscular electrodes. At the University of North Carolina, Lubker, with Parris (1970) has extended his speech studies of the palate (p. 365) to lip muscles. Some progress has emerged in improving electrode techniques when fine-wire inserted electrodes are unacceptable to the investigator; thus, Allen, Lubker and Turner (1973) have combined our fine-wire techniques with paint-on techniques (using the NASA spray-on electrode method for EKGs) with promising early results. Concentrating on orbicularis oris (with surface electrodes) are a group at Essex University in Colchester, England (Tatham and Morton, 1968a, b; Tatham, 1971) and Fibiger (1971) in Stockholm. These investigators are concerned with articulated speech and with disturbances such as stuttering. Fibiger has spread his interest to other muscles of the lip (1972) as has still another group in Stockholm (Persson *et al.,* 1969; Leanderson *et al.,* 1970, 1971; Leanderson, 1972).

The last named group has shown among other applied findings that linguistic context considerably influences the recruitment of muscles used to produce any given phoneme; i.e., the emg activity pattern associated with a phoneme is not constant. When, for instance, *i* was preceded by a rounded vowel *y* there was a considerable activity in depressor labii inferioris, while this activity was much reduced when *i* was preceded by ε. A difference in the amount of motor unit activity was observed in one and the same muscle when *p* and *b* were spoken in the same context. The activity was more vigorous at the closure of the *p* than at that of the *b*. These differences between voiceless and voiced consonants can probably be explained on the basis of a difference in the time course of the intra-oral pressure during the closure, which has been established by previous measurements. Orbicularis oris has been studied effectively during articulated speech with tiny paint-on electrodes connected to fine wires to conduct the potentials (McAllister *et al.*, 1974). At the Haskins Laboratories our fine-wire hook electrodes have been used quite effectively also (Gay *et al.*, 1974).

At the Speech Research Laboratory of the University of Wisconsin, Ronald Netsell and his colleagues are also exploring the use of EMG in lip function during normal and distorted speech (Netsell and Cleeland, 1973). Daniel and Guitar (1973) of that group have studied the effects of emg feedback on facial and speech gestures during recovery following neural anastomosis. In Seattle, Folkins and Abbs (1975) found that the muscles of the lips and jaw are capable of on-line compensatory motor re-organization in response to resistive loading of the jaw during speech.

Musical Performance

My musician colleagues, who joined me from Appalachian State University from time to time, helped pioneer the use of fine-wire electrodes in the lips and cheeks for elucidating the part played by these muscles in the *embouchure* of wind instrument performance. The details are reported elsewhere in a series of papers (Basmajian and White, 1973; White and Basmajian, 1973; Isley and Basmajian, 1973; Basmajian and Newton, 1973). Here we will touch upon the main results very briefly.

Musical performance, like facial expressions, recruits muscles in patterns that differ from person to person. No absolute pattern is demonstrably more efficient than another to produce desired expressions or notes. However, we found significant correlation between certain patterns of lip muscle contraction and proved proficiency in a long series of performers. Also, we found that performers can deliberately change their patterns during a performance if they are given artificial emg feedback *with no apparent alteration* of the sounds produced.

Surgical Application

DePalma, Leavitt and Hardy (1958) made a brief electromyographic study of the lip musculature in full thickness flaps that had been rotated by plastic surgery from the upper to the lower lip to fill a wedge-shaped defect. In four patients, there was regeneration and reinnervation of the motor nerve to the flap with return of apparently normal function. A similar study has been reported in Sweden by Isaksson *et al.* (1962).

Neck Muscles

Platysma

The first and probably the only careful study of this broad sheet of subcutaneous muscle was done by de Sousa (1964) who investigated 20 men using needle electrodes. The obvious actions of this rather obvious muscle produced the greatest activity (pulling the thoracic skin up and the angle of the mouth down).

During abrupt inspiration (but not expiration) there was activity that de Sousa interprets as helping to reduce the constricting effect of the skin on the subcutaneous veins of the neck. Widening the opening of the mouth ("buccal rima") elicited marked activity, but the natural opening of mouth and jaws did not. Active and resisted movements (flexion, extension and rotation) of the neck did not recruit platysma, nor did swallowing.

Sternomastoid, Scalenes and Longus Colli

The longus colli muscle is unique: it is the only spinal muscle that both lies in front of the spinal column and has attachments confined to the vertebrae. Therefore, its effects must be directly upon and restricted to the cervical spine—unlike other anterior muscles of the head and neck which exert their effects more or less indirectly. Other than the single case study by Fountain *et al.* (1966), no electromyographic reports were available on the role of this interesting but deeply located muscle until the 1970's.

In quick succession, two studies were completed on the sterno(cleido) mastoid. In São Paulo, de Sousa, Furlani and Vitti (1973) and then Vitti *et al.* (1973) reported results that overlap and agree; these studies probably will remain definitive. The latter study also included the longus colli muscle. Using bipolar fine-wire electrodes, we examined the right and left longus colli (LC) and sternocleidomastoideus (StM) muscles electromyographically in 10 healthy young adults (Vitti *et al.*, 1973). Action potentials were recorded on FM magnetic tape and each experiment was also videotaped. The head-neck motions were recorded using a special neck goniometer. The muscles were studied in sitting, supine, prone and lateral positions both during free movements and against resistance.

We found complete inactivity in both sternomastoid and longus colli muscles in relaxed sitting, normal breathing, deep expiration and wet and dry swallowing. There was very marked synchronous emg activity of the LC and StM muscles during resisted forward flexion, marked activity during neck flexion against head weight in the supine position, and during resisted right and left side-bending. Variable activity was found in both muscles during deep breathing, coughing, forceful blowing, loading on top of the head, resisted backward extension, neck holding against head weight in the prone position, and in twisting movements downwards and upwards. During free flexion-extension movements, LC and StM act synchronously. During free lateral bending they work homolaterally; but during free rotation to the right, the right LC works with the left StM and *vice versa*.

Longissimus Cervicis

Only Fountain *et al.* (1966) have examined this deep muscle in one human subject. It was a strong extensor of the cervical spine but had no lateral-bending effect. In relaxed sitting, it was silent but acted in synchrony with longus colli during rotation.

When the human subject is sitting or standing in a relaxed position, both longus and longissimus cervicis are almost completely inactive. This is in keeping with most postural responses in the human erect trunk. The same is not true in the dog unless the head is supported externally.

Longus colli is a strong flexor of the cervical spine, acting reciprocally with the longissimus, a strong extensor. Both muscles are synchronously active during rotation. Influences on lateral bending of the neck appear to be minimal, contrary to widespread belief. Most surprising and interesting is a pronounced increase in activity in longus colli during talking, coughing and swallowing. Apparently this represents a reflex stabilizing of the neck during pharyngeal contractions.

Semispinalis Capitis and Splenius Capitis

Although Tourney and Paillard (1952) gave a general account of the splenius, probably the only emg study with modern techniques and intramuscular electrodes of the specific functions of these two huge, but obscure neck muscles was reported by us (Takebe, Vitti and Basmajian, 1974). Fifteen normal adult subjects had bipolar fine-wire electrodes placed in the middle of the bellies of these muscles (determined by repeated dissecting room experiments). Myopotentials were integrated and summated by computer analysis.

Semispinalis is almost limited to one main function: extension of the head on the neck. No doubt it also has antigravity functions when one leans forward; but in well-balanced, erect postures it falls economically silent, contrary to widespread opinion. It is not a rotator.

Splenius capitis is equal to semispinalis in its exuberance during extensor activity; but it is also just as active during rotation of the head to its own side. It is, then, an important extensor-rotator. Probably it is as important in neck rotation as the better known sternomastoid.

NECK MOVEMENTS. The slightest attempt of a supine subject to raise his head from the couch is accompanied by marked activity in the sternomastoid and scalenes. This was the finding of Campbell (1955) and I have also confirmed this in scattered observations.

No other systematic reports are available on the activity of these muscles as demonstrated electromyographically. However, Hellebrandt *et al.* (1956) published an account of their research on the tonic neck reflexes in "exercises of stress" in human subjects. This group had noted previously that spontaneous ipsilateral head-turning occurred regularly during unilateral wrist-extension in hypertonic neurological conditions. From their study they concluded that exercise against resistance calls upon synergists which may be far removed from the part exercised. The pattern of concurrent action in the neck that developed as a result of stress in exercising the upper limb is sufficient to modify the position of the head. Following this discovery, the use of strong voluntary head-movement in the direction of the spontaneous positioning was tested; it appeared to augment the output of work. Therefore, this reflex positioning of the head does, in fact, appear to help the normal work of the upper limb.

Other Neck Muscles

Trapezius is discussed with the upper limb in Chapter 10 (p. 189), the oral diaphragm (floor of the mouth) in Chapter 17 (p. 361), the laryngeal muscles in Chapter 17 (p. 369) and digastricus earlier in the present chapter with the muscles of mastication (p. 379). No systematic studies have been reported on the infrahyoid muscles except for cricothyroid, which was studied (along with the anterior belly of the digastric) by Lesoine and Paulsen (1968). The greatest intensity of contraction in cricothyroid occurs in swallowing. The next level of intensity occurs during phonation; then, in reducing order: during strong expiration and coughing (equal); strong inspiration; and finally quiet breathing (very slight).

Chapter *19*

Extraocular muscles and muscles of middle ear

Extraocular Muscles

ALTHOUGH Scandinavian, Japanese and Italian workers have also published in this field, most of the ocular electromyography that has received wide attention has been done by several small groups of investigators in the U.S.A. The two most prolific groups are located in San Francisco and New York City. The former includes Marg, Tamler, Jampolsky and other occasional associates; the latter is centred on one investigator, Goodwin M. Breinin. Of necessity, this chapter will be entirely a review and synthesis of the work of these and other investigators, for I have had no extensive experience with the extraocular muscles.

Before proceeding further, it must be clear that *electro-oculography,* which is sometimes confused with electromyography, is distinctly a different technique and therefore is not the subject of our present chapter. Electro-oculography is recorded through superficial skin electrodes in the region of the orbit and consists of potentials arising in the retina, extraocular muscles and other sources in the orbital cavity. With a venerable but rather dull history dating back about a century, it has continued to show more promise than results. For an excellent early summary of electro-oculography, see Marg (1951). More recent papers with more than passing interest are: a brief up-to-date review by Peters (1971) and a description of the basic aspects by Bicas (1972). These should serve as a jumping-off place for readers interested in exploring electro-oculography. The widest use of the technique is in sleep studies—where the emphasis is on the presence or absence of rapid eye movements—REM (see p. 404).

Electronystagmography is related to electro-oculography. It depends on the existence of a corneo-retinal electrical potential that varies with the eye movements during nystagmus. It, too, is beyond the scope of this chapter; for a good review, see Milojevic (1965).

Electromyography of the muscles that move the eyeball is a comparatively recent development and uses direct recording of motor unit potentials from these muscles. Jampolsky *et al.* (1959) warn, with proper alarm, against attempts to measure relative strength of extraocular muscles by using only the amplitude of potentials as the criterion, because there are many sources of emg artifact. Variations in the structure of the muscles being tested and in the positioning of the electrodes can also lead to errors. Even in the last few years, Jampolsky (1970) continues to advise caution, flatly stating that EMG " . . . is of no practical usefulness in management of the usual strabismus patient." He concedes that it is " . . . a very valuable laboratory and investigative tool" to study ocular rotations.

Normal Position and Movement of the Eyeball

Several groups have shown that the recti exhibit fairly strong, persistent activity to maintain the position of the eye during waking hours. Changing visual inputs constantly affect the level of motor unit activity, apparently without imbalance resulting (Serra, 1970). We may conclude, then, that positioning results from a balance of activity among these muscles. Björk and Kugelberg (1953) demonstrated a persistence of this activity even in darkness.

These last-named workers also showed that each change of ocular fixation is accompanied by a gradual increase of activity in the prime mover or agonist with a reciprocal decrease in the antagonist whether the movements were slow or fast. With extreme positions of gaze, e.g., far to one or other side, the antagonist is usually completely inhibited. Kadefors, Petersén and Tengroth (1974) found that the rectified average EMG of the lateral rectus tends to be an exponential rather than a linear function of the degree of ocular deviation. Muscle length is not a major function.

During fast movements, there is a complete inhibition of the antagonist accompanying the sharp short burst of activity in the agonist (fig. 19.1). During a saccadic eye movement, Tamler, Marg, Jampolsky and Nawratzki (1959) found a heightened burst of activity in the agonist, inhibition of the antagonist, and co-activity of the auxiliary muscles. A saccadic eye movement is a type that occurs in changing the gaze from one point in space to another and includes small terminal oscillations. Tamler *et al.* confidently assert that rapid movements of this sort are not ballistic in nature (see Chapter 8, p. 155). Miller's (1958) independent findings would seem to agree (fig. 19.2), although he states that for large movements one or more bursts of activity, in addition to the initial burst, may be found. Nonetheless, the antagonist remains completely inhibited. Yamazaki (1968) found essentially the same (but limited to 13 msec duration) in the medial and lateral recti as part of tiny horizontal flick movements when monocular fixation is performed.

Robinson (1965) has shown that both smooth pursuit and saccadic movements may occur with completely temporal independence. Smooth movements may occur just before, after or with saccadic ones.

There seems to be no doubt, then, that the muscles of the eye are in more-or-less continuous activity except during sleep (see below). Indeed, their rôle may be compared best to dynamic guy-wires.

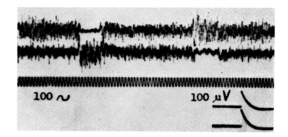

FIG. 19.1. EMGs from lateral rectus (*upper tracing*) and medial rectus of one eye during quick changes of fixation from 10° outward to centre of scale and back to 10° outward. There is abrupt inhibition in the antagonist simultaneous with increase in agonist activity. (From Björk and Kugelberg, 1953.)

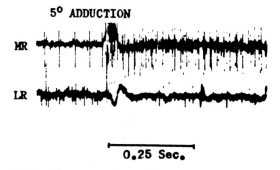

FIG. 19.2. EMGs from medial and lateral rectus during a 5° saccadic movement. (From Miller, 1958.)

Position of Ocular Rest

Among oculists, the position of the eyes during sleep and anesthesia has been the subject of much interest, speculation and investigation. Reported observations have been conflicting and sometimes confusing. During anesthesia, for example, different degrees of vergence have been reported, as well as versions during the induction stage. In surgical anesthesia, the movements cease and during the asphyxial stage of anesthesia, a convergent depressed position may be assumed. A similar behaviour may be noted during fainting (Wójtowicz, 1966) and sleep (Breinin, 1957 *et seq.*).

According to Breinin (1957b), the disappearance of esotropia during anesthesia is a frequent observation, with straight eyes or divergence taking the place of esotropia. While investigating the electromyographic changes during anesthesia and sleep, Breinin found that the "innervation" to all horizontal recti rapidly decreased during the induction of general anesthesia with intravenous Pentothal in a patient with intermittent exotropia. When the surgical plane was reached, all nerve impulses were repressed and the eyes occupied a moderately divergent position. As the level of anesthesia lightened, individual motor units appeared. When the plane of anesthesia was again deepened, the motor units disappeared promptly. Therefore, the position of the eyeball during surgical anesthesia suggests that the motor nerve impulses to the extraocular muscles cease completely.

The position of the eyeball during anesthesia is apparently dictated by anatomical-mechanical factors—bony, fascial and muscular—and represents the anatomical position of rest. It is, then, the *basic* position which is modified by the normal factors of motor innervation during consciousness. If this basic position is one of divergence, then a basic stress toward divergence must underlie whatever position the eyes assume in consciousness, according to Breinin. "It opposes esodeviation and facilitates exodeviation." The frequency of divergence under surgical anesthesia supports the concept of a basic anatomical divergence, to which, in exotropia, may be added innervational divergence.

Eye Movements during Sleep

During the light dozing phase preceding ordinary sleep, very irregular discharges and bursts of potentials appeared in Breinin's investigations. He reports that a run of double and single motor units occurred at intervals, as well as rhythmical trains of potentials (fig. 19.3). During what appeared to be lightening of sleep induced by particularly cacophonous snores, a single burst of potentials occurred, accompanied by an obvious movement of the eyeball.

The commoner method of sleep researchers is the much less precise DC electro-oculography. Initially eyelid closure induces upward movement of the globes; there they remain for 55% to 85% of the time spent in stages 2, 3 and 4. With the onset of the stage now known as the rapid eye movement (REM) sleep, they move down to the central position and begin to flick rapidly—usually in oblique directions but sometimes in the horizontal or vertical directions (Jacobs *et al.*, 1971). Eye movements do not occur in infants during quiet sleep but appear as both slow rolling movements and REMs (in varying amounts) during irregular sleep (Prechtal and Lenard, 1967).

Asymmetric Convergence

Asymmetric convergence may be defined as convergence of the eyes occurring in any direction other than along the median plane. For example, this would occur if an object approached one eye directly from the front. This would require a much greater convergent movement of the other eye to fixate on the approaching object. Breinin (1955) has observed that whenever an eye turned in either a version or a vergence movement there was an

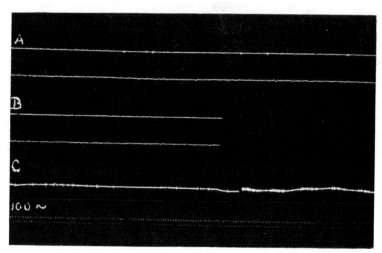

Fig. 19.3. *A* and *B*, EMGs (*upper tracing in each case*) with simultaneous electro-oculograms of medial rectus during sleep. Tracing *C*: shows resumption of normal waking pattern (toward its right end). (From Breinin, 1957b.)

increase in the electrical activity of the agonist accompanied by a reciprocal inhibition of the antagonist. When he noted no change in activity in either horizontal rectus of the stationary eye during asymmetric convergence, he concluded that the antagonistic innervations for a version and vergence movement cancelled each other in the brain and never reached the muscles. If the motor impulse to that eye were to change, it would have to move. However, fixation precludes such movement. Miller (1959) seems to agree. He found that sudden asymmetric convergence is accompanied by a burst in the yoke muscles (i.e., pairs of muscles which are agonists in performing versions) followed by a convergent pattern in both medial recti. No increase in coactivity of horizontal antagonists was found during slow asymmetric convergence on an approaching target until the near point of convergence was approached.

An electromyographic study by Blodi and Van Allen (1957) also agrees with that of Breinin. They found that there was no change in the electrical potentials from the horizontal muscles of the apparently stationary eye during asymmetric convergence. They also suggested that the sum total of innervation of the two horizontal muscles has to remain the same regardless of the position of the eye.

On the other hand, Tamler, Jampolsky and Marg (1958) have presented electromyographic data which support Hering's concept of peripheral receipt and adjustment of opposing stimuli to the apparently stationary eye during asymmetric convergence. According to them, the reason the eye remains stationary is due to cocontraction of opposing horizontal recti, that is, there is a simultaneous increase in innervation of the lateral and medial rectus muscles. One possible reason why other investigators have not found this by electromyography was, they suggest, a failure to induce a sufficient amount of angular convergence in the moving eye in order to register observable changes on the electromyogram of the stationary eye.

Hering's law appears to be supported by both of two test methods of asymmetric convergence. The first is that of smooth, binocular convergence along the axis of one eye; and the second is that of uncovering one eye to force a fusional convergent movement while the other eye continues to fixate.

In "breaking fusion" Tamler *et al.* often found simultaneous decrease in innervation of the horizontal recti of the stationary eye. They explain that if it did not occur, then continued refusion movements with repeated covering and uncovering of an eye would cause greater and greater build-up of electrical activity in the horizontal muscles of the stationary eye.

Blodi and Van Allen (1960) finally laid to rest the theory that vergence movements by the medial recti were somehow controlled by the sympathetic nervous system. After anesthetizing the cervical sympathetic trunk in human subjects, they found no electromyographic change and no directly observable change from the normal response of convergence on command.

Cocontraction

Cocontraction as applied to extraocular muscles (as with synergists in other skeletal muscles) is defined as the simultaneous increased contraction of extraocular muscles which are normally antagonistic in their primary field of action. One suggestion has been that cocontraction occurs when the eye moves from its primary position to any secondary or tertiary position. According to this hypothesis, as the lateral rectus abducts an eye, the vertical recti as well as the superior and inferior obliques cocontract to steady the eye in its horizontal path. At the same time, the cocontraction is believed to prevent undue torsion of the globe and perhaps helps to maintain the abduction because the lateral rectus loses its mechanical advantage as the movement progresses. The same reasoning would apply to adduction, supraduction and infraduction (Tamler, Marg, Jampolsky and Nawratski, 1959).

Tamler, Marg and Jampolsky (1959) studied the electrical activity of four extraocular muscles simultaneously in many subjects. They concluded that there is little or no increased coactivity of auxiliary extraocular muscles in adduction, abduction, supraduction and infraduction during slow, vertical and horizontal following movements in planes through the primary position (fig. 19.4). This does not necessarily mean that they do not contribute to the movement. Apparently, the effect of muscle activity on eye movement depends upon the position of the eye at that particular moment. Hence, they suggest that the "primary position innervational tonus" of the auxiliary

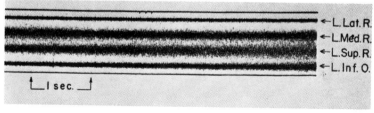

Fig. 19.4. EMGs from various muscles of "up gaze" in sagittal plane of left eye. No obvious change in lateral and medial recti. (From Tamler *et al.*, 1959.)

muscles, which does not appear to change during these movements, may indeed contribute to the movement after the eye has gone a certain distance. Also, the primary position tonus may be all that is necessary to steady the eye and prevent torsion during the movement. Tamler and his group conclude that though one can deduce muscle contraction with certainty from the electrical activity of a muscle, one cannot firmly state the function of the muscle at that moment.

Furthermore, they insist that normal muscle activity, as found in the primary position, is present in the auxiliary muscles during movement, but that systematic changes in coactivity do not generally occur. Occasionally, they did find barely perceptible changes in the recordings from those muscles which are auxiliary during movements. These changes, when elicited, had no systematic pattern and therefore are not consistent with a cocontraction hypothesis that requires systematic changes with a given direction of movement. Therefore, these authors, whom I have quoted so extensively, are in basic agreement with Breinin's conclusion regarding the same problem.

Innocenti (1971) studied the emg activity of extraocular muscles in unanesthetized rabbits during oculocompensatory positions. During inclination around a longitudinal axis, units belonging to the superior and medial rectus and to the superior oblique were seen to be activated by ipsolateral and inhibited by contralateral inclination; the units belonging to the inferior rectus, inferior oblique and lateral rectus, showed an opposite pattern.

During upward inclination around a bitemporal axis, the units of medial and inferior rectus and superior oblique of both eyes were activated while the units belonging to the other extraocular muscles were inhibited and often unaffected. An opposite pattern was exhibited by the same units during downwards inclination. A linear ratio was found between the degree of inclination and the increase in discharge rate of extraocular units.

Stretch and Proprioceptive Effects

Sears, Teasdall and Stone (1959) were concerned about the proprioceptive function of the tension in the extraocular muscles. Previously, muscle spindles had been observed in the muscles of the eye by Cooper and Daniel (1949).

Sears *et al.* investigated by electromyography the possible existence of stretch mechanisms in human extraocular muscles using patients prior to surgical enucleation of the eye. Under topical anesthesia they inserted a concentric needle electrode into one of the horizontal recti muscles. Recordings were made with the eyes both in primary and in horizontal gaze, the latter about 30° beyond the midline. A suture was placed through the tendon of one horizontal rectus muscle which was then cut away from its insertion distal to this ligature. The effects of disinsertion and of gentle manual pulling on the suture were recorded from the agonist muscle, while the muscle was acting as either agonist, antagonist or yoke muscle. With the eyes in primary gaze, they found that the action potentials were unchanged by disinsertion or manual pull. During agonist contraction, however, stretch of the agonist itself, of the antagonist or of the yoke muscle produced a decrease in the frequency and amplitude of the discharges. This inhibitory effect was not observed in a patient with oculomotor palsy when stretch was applied to the muscles of an eye which had a previous retrobulbar alcohol injection.

We can only agree with Sears and his colleagues that probably the extraocular muscles of man do have receptors which can be stimulated by passive stretch. Breinin (1957a) reached the same conclusion, stating: " . . . it appears that a proprioceptive mechanism must be postulated for the extraocular muscles. It is not implied that such a mechanism provides muscle position sense or awareness. Failure to make this distinction has resulted in confusion in the past."

Bearing on the problem of control mechanisms is the possibility of innervational plasticity of the oculomotor system. This was investigated experimentally by Metz and Scott (1970). Studies in 14 Rhesus monkeys utilizing superior and lateral rectus muscle cross-transplantation resulted in return of conjugate eye movements within 10 days. Electromyographic recordings from transplanted and normal extraocular muscles indicated no change in innervational patterns. Central sixth nerve stimulation and postmortem orbital examination demonstrated that observed ocular movement can be explained mechanically and need not result from innervational plasticity.

Eye-Head Coordination

Eye-head coordination was investigated by recording from the neck and eye muscles in monkeys by Bizzi *et al.* (1971). The results show that (1) during eye-head turning, neural activity reaches the neck muscles before the eye muscles; and (2) all agonist neck muscles are activated simultaneously regardless of the initial head position. Since overt movement of the eyes precedes that of the head, it was concluded that the central neural command initiates the eye-head sequence but does not specify its serial order. Furthermore, it was determined that the compensatory eye movement is not initiated centrally but instead is dependent upon reflex activation arising from movement of the head.

Eyelids

Björk and Kugelberg (1953) performed the first and definitive studies on the levator palpebrae superioris and orbicularis oculi. They found the levator to be continually active during waking hours except when looking sharply downwards or when the eyes are closed. Lowering of the upper lid is accompanied by progressive decrease in levator activity but with little or no activity in the orbicularis. In other words, the lid is lowered passively by the relaxing of the levator. The lowering of the lid in the act of blinking, however, is caused by activity in orbicularis with immediate cessation of activity in the levator (fig. 19.5). This is followed immediately by the reverse (inhibition of orbicularis and activity in levator).

Blinking was also studied by Van Allen and Blodi (1962). They too found a quick, well-controlled, reciprocal relationship between the orbicularis oculi and levator palpebrae superioris. This occurred as expected in normal people, but it also was preserved in the presence of gross neurological conditions affecting the region.

Though not an appropriate topic here, the blink reflex may interest some readers. Since the early work of Kugelberg (1952) scattered papers have appeared (e.g., Wójtowicz and Gwóźdź, 1968; Young and Shahani, 1969; Ferrari and Messina, 1972).

In view of the indecision surrounding the innervation of the levator

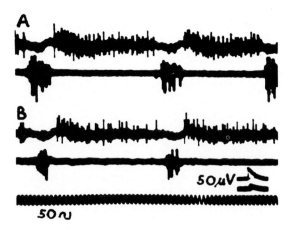

Fig. 19.5. *A*, EMGs of blinking from levator palpebrae superioris (*upper trace*) and orbicularis oculi. *B*, spontaneous blinking. (From Björk and Kugelberg, 1953.)

palpebrae and arising from the drooping or ptosis of the lid that is part of Horner's syndrome, Blodi and Van Allen (1960) investigated this muscle further. Duplicating a Horner's syndrome by producing a block in the cervical sympathetic trunk with local anesthetic, they found no significant effect on the levator. Even with marked ptosis of the upper lid, the activity in levator was not reduced. Blodi and Van Allen also found normal EMGs in the levators of patients with long-standing Horner's syndrome. Therefore, taking all this together, we must conclude that the sympathetic nervous system does not innervate the striated levator palpebrae. The drooping of the lids in Horner's syndrome is not due to any direct paralytic effect on the striated fibres but appears to be due to paralysis of smooth muscle.

Muscles of the Middle Ear

As the reader would suspect, the tiny stapedius and tensor tympani muscles have not been studied extensively. However, Berlendis and De Caro (1955) in Italy, and more recently Wersäll (1958) in Stockholm, Kirikae (1960) of Tokyo, Carmel and Starr (1963) of Bethesda, U.S.A., Dewson *et al.* (1965) at Stanford University, Henson (1965) at Yale and Teig (1973) in Oslo have performed EMGs within the middle ear of experimental animals. In Copenhagen, Salomon and Starr (1963) have made direct recordings from two human patients. But most studies of tensor tympani and stapedius in man have been made by an ingenious *indirect* method now gaining fairly wide acceptance. This depends on measurements of changes in acoustic impedance in the external ear canal. Interested readers should consult Møller (1961) for a technical description.

Wersäll's beautiful experiments, which were performed on rabbits and cats and described in his monograph, established several important points. He showed that there is continuous activity at rest, i.e., resting tonus, in these middle ear muscles. When sound reaches the eardrum there is a small delay (less than 0.1 second) before a reflex contraction of the muscles becomes evident. This delay is constant regardless of the intensity of the sound. There is however a threshold above which the sound must rise to elicit the reflex. It is generally higher in the tensor tympani than the stapedius. Nonetheless, contrary to widely held opinion, protective action against sudden loud noises is no more developed than against a moderate noise.

Wersäll showed a characteristic difference between the two muscles in the dependence of reflex tension on sound intensity. Stapedius, but not tensor tympani, reaches a plateau before the sound is so intense as to become injurious. Under the stimulation of steady sound the reflex contractions of both muscles often showed undulatory fluctuations. During strong sustained stimulation, the tension drops to a steady state in both muscles. Wersäll ascribed this fall in tension to fatigue.

Kirikae (1960) measured the latency of the acoustic reflex more precisely than Wersäll. For stapedius it was 3.4 msec and is shorter than it is for tensor. Activating time, which requires muscle contraction, is 2 to 3 msec. Maximal contraction takes a further several msecs. Fatigue or adaption phenomenon was also prominent in his studies.

In unrestrained cats, Carmel and Starr (1963) added a number of new observations or confirmed Wersäll's findings. As in his investigation, the middle-ear muscles were found to relax gradually during prolonged sound stimulation. Both muscles contract during certain bodily movements and vocalization. Reflexes are modified by previous exposure to loud noises. Application of electric shock concurrently with test signals also modifies the responses of the muscles. Thus they concluded that a wide variety of dynamic processes influence the activity of middle-ear muscles, not just the physical characteristics of sound impinging on the ear.

The limited studies on man reported by Salomon and Starr (1963) revealed the middle-ear muscles to be active with general motor events, such as closing the eyes, movements of the face and head, vocalization, yawning, swallowing, coughing and laughing. Thus they proposed the theory that central controls for these functions are closely integrated. Dewson *et al.* (1965) found (in cats) a close similarity or relationship in the emg reaction of the middle-ear muscles and the stage of rapid eye movements during sleep.

The activity and function of middle-ear muscles in echo-locating bats forms the subject of an intriguing paper by O. W. Henson (1965). He found that stapedius begins to contract just before vocalization and reaches its maximum as each pulse begins, apparently to dampen the ossicles during the sound-emission stage. Relaxation starts immediately and continues during the echo phase, apparently to allow its maximum perception. Tensor tympani seems to play no part in this system. Jen and Suga (1976) have provided elegant expansion to Henson's earlier work.

Finally, mention should be made of a related technique for the study of reflex contractions of stapedius and tensor tympani by the indirect method of acoustic impedance at the tympanic membrane. The Zwislocki acoustic impedance bridge in the external ear canal is used in both experimental animals and man (e.g., Borg, 1972; Deutsch, 1972).

Chapter 20

The future of normal electromyography

E
LECTROMYOGRAPHY fulfils a real need in diagnostic or clinical examinations and it will thrive in the large and medium sized medical
centres around the world. My concern in this final chapter is not
with clinical electromyography; rather, it is with basic scientific electromyography of the type which has occupied our attention in the previous chapters.
What directions will such investigations take—or, indeed, should they take?

First of all, we should abandon any idea that this is a technique that needs
to go looking for a job. We must agree that it is a technique standing ready
to be called upon whenever problems present themselves. Do such problems
still exist or are they all solved?

To answer these questions we should review the various parts of the
musculature and draw attention to functions which are at present inade-
quately explained, particularly those which are the subject of great controversy, and, finally, those which are the richest in dogma. In all three categories
there are problems that cry out for solution by electromyography. Almost
every page of this book illustrates how many problems have been solved while
still others have been left unsolved. Any anatomist, physiologist, kinesiologist
or orthopedic surgeon can point out many such unsolved problems.

TECHNIQUE. As suggested in Chapter 2, a general improvement in technique
is needed. Reliance on the exclusive use of surface electrodes must be
abandoned by electromyographers who hope to add to the future of electromyography. Adequate apparatus and training are now quite readily available. Therefore, research in electromyography should be improved in quality
and reliability. Improved techniques for synchronized photographic videotape recording of movements are now available and should be used more
widely. The obvious usefulness of computer analysis should greatly enhance
all phases of EMG.

FATIGUE AND NEUROMUSCULAR STUDIES. Though electromyography has
produced some useful results there is enough conflict—as revealed in Chapter

4—to make a final verdict on the mechanisms of fatigue impossible. An attack on a broader front is required; and when it is made electromyography will be the main factor in its success. The recent awakening of physical-training specialists to the possibilities of electromyographic evaluation raises the question of the direction their interest will take. If they do not avoid the psuedo-scientific approach which often is the curse of studies in athletics, they will get so few valid results from electromyography that they will only succeed in casting unwarranted suspicion upon the technique.

TRAINING OF MOTOR UNITS AND BIOFEEDBACK. Being deeply involved in our laboratory in this special study, my colleagues and I are convinced of its fundamental importance. Widespread employment of our techniques can only lead to many useful applications.

MUSCLE MECHANICS. Anatomists and physiologists in particular are faced with the solution of problems, both large and small, many of which eventually have practical implications. These problems very often will have arisen from purely clinical problems. Therefore, in view of this mutual exchange, coop-eration between research staffs and clinicians is especially needed in this sphere.

POSTURE. The disagreements described in Chapter 9 about the rôles of various postural muscles is not a reflection on the value of electromyography. It is, however, an indication of the lack of enough confirmation. All too often, opposite opinions are based on short unconfirmed series. Obviously, then, we must expect—indeed, demand—further unbiased confirmation by independ-ent groups of all the opinions now available. Certainly, the genuine scientist welcomes a critical repetition of his work. Although he does not like to see it capriciously questioned or slighted, he respects an honest and careful evalu-ation.

UPPER AND LOWER LIMBS. Looking back over Chapters 10 to 12, I am impressed with how almost all of the emg studies on the limbs are scattered and unrelated. Many have been done with very special problems in mind, but only a few have been systematic. Fewer still have had adequate confir-mation. Finally, the hand and foot have been unaccountably neglected. The future needs in the limb regions, therefore, are self-evident.

THE BACK. Although there has been an excellent cross-confirmation of the basic principles governing the action of back muscles, the finer function of the intrinsic muscles has been ignored. Clinical problems, such as the etiology of scoliosis, require solutions that electromyography can help provide.

GAIT. As pointed out in Chapter 14 on Locomotion this is a fruitful field of study, but it requires unusually complex apparatus. Fortunately a number of laboratories are embarked on full-scale studies of this practical type of work with modern on-line computer analysis.

ABDOMINAL WALL AND PERINEUM. Almost all the emg work done on the anterior abdominal wall and the sphincters of the urethra and anus require confirmation. The levator ani remains completely ignored though it is of great importance and not entirely inaccessible.

RESPIRATORY MUSCLES. Although in recent years much excellent work has been reported on the intercostal muscles, these studies have been isolated and are often conflicting. Much of this work requires confirmation with inserted electrodes before complete acceptance can be expected. Work on the diaphragm of real practical value is now started but will require considerable new effort.

PHARYNX AND MOUTH. Although some electromyography of the swallowing mechanisms has been done recently, much additional work is needed with many novel approaches and a variety of techniques. Clinicians working closely as partners with experienced electromyographers are the best qualified to broaden our knowledge. The tongue continues to be almost a complete no-man's-land as far as electromyography is concerned, and almost any investigator with an interest in the tongue is confronted with a minimum of technical problems if fine-wire electrodes are used.

LARYNX. Anyone reviewing the literature of this region will be satisfied that the early heated controversy has shed limited light on the true functions of the muscles of the vocal cord. Inserted fine-wire electrodes have revolutionized this area of study. Here, too, several different approaches are needed and the best results will come again from teams of experienced clinicians working imaginatively with experienced electromyographers. Neither group of specialists can operate successfully without the other.

MUSCLES OF MASTICATION. In recent years there has been a new wave of interest with new publications on the electromyography of the mandibular muscles. A glance at Chapter 18 will reveal that a particular divergent opinion is supported only by the experiments of the author of that opinion. Though clinical dental electromyography is still enjoying the dregs of several vintage years, it can only provide valid results if the basic physiology is well grounded. Indeed, practising dentists are likely to turn away from electromyography completely unless a solid basis of normal electromyography is made available. Again, the new fine-wire electrode technique can revolutionize EMG in the study of mandibular function.

FACE AND NECK. The muscles of expression, the infrahyoid muscles and the suprahyoid muscles of the floor of the mouth remain almost virgin territories for electromyographic exploration. Their neglect, except for the scattering of new studies with fine-wire electrodes, is particularly puzzling in view of their relative accessibility. The very vital rôle of the scalenes in respiration is of such great importance that it is rather amazing how little attention has been paid to them except by a handful of investigators.

MUSCLES OF THE EYE AND MIDDLE EAR. These regions must be investigated by specialists. We are fortunate that a considerable number of ophthalmologists have been involved in electromyography. However, the vigorous controversies that have resulted appear to have alienated many other ophthalmologists. Undoubtedly, with the expenditure of time and effort, truth will emerge. There can be little doubt that electromyography provides great promise in the practical problems of eye movements. The muscles of the middle ear in the human being remain uninvestigated. Here again, judicious team-work between a clinical specialist and an electromyographer would provide fascinating and rewarding results.

ENVOY. These, then, survive as the problems of the future in normal or "non-clinical" electromyography. Patience and persistence combined with flair and imagination will be needed. Little of real value comes easily in science. In this field, where one must apply the techniques of one science to the solution of problems in others, the need for all these ingredients is especially acute. In the final analysis, the future of electromyography is exciting and real especially for those who are adequately equipped and trained.

Some useful commercial equipment

This appendix is only a simple guide for a minority of readers. As such, it is not a complete account of all the apparatus available; that would fill a book. However, inquiries often arise about the appropriateness and availability of various electronic systems. The following paragraphs are a summary of the sort of preliminary advice one gives. Finally, under the heading of "a basic set-up" (p. 419) a brief account is given of how to use inexpensive apparatus (less than $1,000) to make a start in EMG.

Commercial EMGs

In the previous editions of *Muscles Alive,* this section included substantial descriptions of specific commercial EMGs. Instead, and perhaps more useful, a partial alphabetical list of manufacturers and suppliers is provided below for those readers who wish to obtain information from commercial sources. Reference to standard catalogues can also be recommended. Unsophisticated readers should seek the advice of medical electronics experts before investing in *any* equipment. Some of the list below are American suppliers of equipment made in other countries where direct contacts also can made with the manufacturers.

Beckman Instruments, Inc., 2500 Harbor Blvd., Fullerton, California
Bethesda Instrument, 4925 Fairmont Avenue, Bethesda, Maryland
Biocom, 9522 W. Jefferson Boulevard, Culver City, California
Biomedical Electronics, Inc., 653 Lofstrand Lane, Rockville, Maryland
Bionic Instruments, Inc., 221 Rock Hill Road, Bala Cynwyd, Pennsylvania
Disa Elektronik A/S, Herlev, Denmark
Grass Instrument, 101 Old Colony Avenue, Quincy, Massachusetts
Hewlett Packard, 1501 Page Mill Road, Palo Alto, California
Honeywell, Box 5227, Denver, Colorado
Lehigh Valley Electronics, Inc., Box 125, Fagelsville, Pennsylvania
London Co., 811 Sharon Drive, Westlake, Ohio
Marietta Apparatus, 118 Maple Street, Marietta, Ohio
Medelec Co., Ltd., London, England
Medical Instrument Co., 1128 N. Las Palmas Avenue, Los Angeles, California

Newport Labs Inc., 630 Young Street, Santa Anna, California
Nihon Kohden Kogya Co., Ltd., Tokyo 161, Japan
Northern Precision Labs., Inc., 202 Fairfield Road, Fairfield, New Jersey
San-ei Instrument Co., Ltd., Shinjuku-ku, Tokyo, Japan
Siemans, 186 Wood Avenue, Iselin, New Jersey
Stoelting, 424 N. Homan Avenue, Chicago, Illinois
Teca, 220 Ferris Avenue, White Plains, New York

Direct Recorders

Several companies are producing direct read-out recorders that use an ultra-violet light source and sensitive paper. Miniature galvanometers having

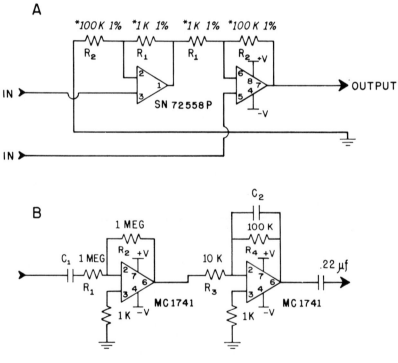

Fig. 1. *A*, circuit diagram of the miniature source-attached, differential amplifier; stars indicate that the 1% resistors are required to give high common mode rejection. *B*, a suitable second amplifier stage at the stationary part of equipment.

a frequency response to 4800 Hz are used to deflect the light beams onto the moving paper. Among others, a galvanometer with a response of 1000 Hz is available; this is good for most emg applications.

A Basic Set-Up

Even without the standard emg units described in the earlier paragraphs, good results can be obtained with readily available and inexpensive commercial equipment designed for other general uses. For example, in certain projects one might need only a minimal set-up. We often rely on a cheap single-channel assembly consisting of a Tektronix Oscilloscope (Type 502), a Tektronix Low Level Preamplifier (Type 122) and a Heathkit Audio-amplifier (Type A-9C) with an ordinary loudspeaker. Equivalent units of other manufacturers can be and are substituted. For example, the preamplifier might be an Argonaut Type LRA-042 or a Sanborn Type 350, etc. Many inexpensive general-purpose oscilloscopes could be substituted for the Tektronix 502 for visual display. One should aim for a minimum sensitivity of about 25 μv per inch.

Small audio or hi-fi amplifiers (using either transistors or tubes) are readily available. They can be satisfactory if they have a power output of about 3 watts (this is not critical) and provision for magnetic phono input. They can be connected to the output of the preamplifier for audio monitoring in conjunction with a standard loudspeaker. A start can be made in electromyography without spending huge sums of money. Further, the simple assemblies will still be useful as adjuncts when more elaborate and comprehensive apparatus has been obtained.

Reducing the length of wire connections between the amplifier and indwelling or surface electrodes is an excellent manoeuver for greatly improving signal-to-noise ratio. Miniaturization and attachment of the primary amplifiers to the skin with direct connection of the electrodes is the solution (Basmajian and Hudson, 1974). Our device is very cheap (less than ten dollars), easy to build, practical and effective (fig. 1). Built in a few minutes, it can be enclosed in a case or package 20 x 25 x 10 mm. It can replace the much more expensive stationary commercial amplifiers.

References

AARONS, L. (1968) Diurnal variations of muscle action potentials and word associations related to psychological orientation. Psychophysiology, 5: 77–91.

ABO-EL-ENEIN, M. A. AND WYKE, B. (1966) Laryngeal myotatic reflexes. Nature, 209: 682–686.

ADRIAN, E. D. AND BRONK, D. W. (1928) The discharge of impulses in motor nerve fibres. Part I. Impulses in single fibres of the phrenic nerve. J. Physiol., 66: 81–101.

ADRIAN, E. D. AND BRONK, D. W. (1929) The discharge of impulses in motor nerve fibres. Part II. The frequency of discharge in reflex and voluntary contractions. J. Physiol., 67: 119–151.

AGOSTONI, E. (1962) Diaphragm contraction as a limiting factor to maximum expiration. J. Appl. Physiol., 17: 427–428.

AGOSTONI, E. (1963) Diaphragm activity during breath holding: factors related to onset. J. Appl. Physiol., 18: 30–36.

AGOSTONI, E., SANT'AMBROGIO, G. AND CARRASCO, H. DEL P. (1960) Elettromiografia del diaframma e pressione transdiaframmatica durante la tosse, lo sternuto ed il riso. Atti Accad. Nazion. Lincei, 28: 493–496.

AGOSTONI, E., SANT'AMBROGIO, G. AND GARRASCO, H. DEL P. (1960) Electromyography of the diaphragm in man and transdiaphragmatic pressure. J. Appl. Physiol., 15: 1093–1097.

AHLGREN, J. (1966) Mechanism of mastication: a quantitative cinematographic and electromyographic study of masticatory movements in children, with special reference to occlusion of the teeth. Acta Odont. Scandinav., 24: suppl. 44, 109 pp.

AHLGREN, J. (1967) Kinesiology of the mandible: an EMG study. Acta Odont. Scandinav., 25: 593–611.

AHLGREN, J. (1969) The silent period in the EMG of the jaw muscles during mastication and its relationship to tooth contact. Acta Odont. Scandinav., 27, 3: 219–227.

AHLGREN, J. AND LIPKE, D. P. (1977) Electromyographic activity in digastric muscles and opening force of mandible during static and dynamic conditions. Scand. J. Dent. Res., 85: 152–154.

AKAMATSU, K., (1960) Electromyographic studies on the expressionless faces of schizophrenics. Hiroshima Med. J., 8: 2249.

ÅKERBLOM, B. (1948) Standing and Sitting Posture (tr. by Ann Synge). A.-B. Nordiska Bokhandeln., Stockholm.

ÅKERBLOM, B. (1969) Anatomische und Physiologische Grundlagen zur Gestaltung von Sitzen. Ergonomics, 12: 120–131.

ALEXANDER, A. B. (1975) An experimental test of assumptions relating to the use of electromyographic biofeedback as a general relaxation training technique. Psychophysiology, 12: 656–662.

ALLEN, C. E. L. (1948) Muscle action potentials used in the study of dynamic anatomy. Brit. J. Phys. Med., 11: 66–73.

ALLEN, G. D., LUBKER, J. F. AND TURNER, D. T. (1973) Adhesion to mucous membrane for electromyography. J. Dent. Res., 52: 394.

ALLERT, M. L. (1969) Anal reflex in the electromyogram of bladder and intestinal sphincter muscles. Wien. Z. Nervenh., 27: 281–287.

AMANO, N. AND FUNAKOSHI, M. (1976) Tonic periodontal-masseteric reflex in man. J. Gifu Dent. Soc., 3: 182–183.

AMATO, A., HERMSMEYER, C. A. AND KLEINMAN, K. M. (1973) Use of electromyographic feedback to increase inhibitory control of spastic muscles.

Phys. Ther., *53:* 1063–1066.

ANDERSON, K. V. AND MAHAN, P. E. (1971) Interaction of tooth pulp and periodontal ligament receptors in a jaw-depression reflex. Exper. Neurol., *23:* 295–302.

ANDERSSON, J. G., JONSSON, B. AND ÖRTENGREN, R. (1974) Myoelectric activity in individual lumbar erector spinae muscles in sitting: a study with surface and wire electrodes. Scand. J. Rehab. Med. *3:* suppl., 91–108.

ANDERSSON, J. G., ÖRTENGREN, R., NACHEMSON, A. L., ELFSTROM, G. AND BROMAN, H. (1975) The sitting posture: an electromyographic and discometric study. Orthop. Clin. N.A., *6:* 105–120.

ANDERSSON, S. AND STENER, B. (1959) Experimental evaluation of the hypothesis of ligamento-muscular protective reflexes. II. A study in cat using the medial collateral ligament of the knee joint. Acta Physiol. Scandinav., *48:* suppl. 166, 27–49.

ANGEL, R. W. (1975) Electromyographic patterns during ballistic movement of normal and spastic limbs. Brain Res., *99:* 387–392.

ANGGARD, L. AND OTTOSON, D. (1963) Observations on the functional development of the neuromuscular apparatus in fetal sheep. Exper. Neurol., *7:* 294–304.

ARANDA, L. AND JARA, O. (1969) Estudio electromiografico del reflejo bulbocavernoso en condicion normal y patologica. Neurocirurgia (Santiago) *27:* 248–250.

ARCHIBALD, K. C. AND GOLDSMITH, E. I. (1967) Sphincteric electromyography. Arch. Phys. Med., *48:* 387–392.

ARIENTI, A. (1948a) Estudos electromiográficos da locomoção humana (Portuguese text). Resenha Clin.-Cient. Instit. Lorenz., *17:* 175–178.

ARIENTI, A. (1948b) Analyse oscillographique de la marche de l'homme (French text). Acta physiotherap. Rheumat. Belg., *3:* 190–192.

ARLAZOROFF, A., RAPOPORT, Y., SHANON, E. AND STREIFLER, M. (1972) Observations on the electromyogram of the oesophagus at rest and during Valsalva's manoeuvre. Electroenceph. Clin. Neurophysiol., *33:* 110–113.

ASHWORTH, B., GRIMBY, L. AND KUGELBERG, E. (1967) Comparison of voluntary and reflex activation of motor units: functional organization of motor neurones. J. Neurol. Neurosurg. Psychiat., *30:* 91–98.

ASMUSSEN, E. (1960) The weight-carrying function of the human spine. Acta Orthop. Scandinav., *29:* 276–290.

ASMUSSEN, E. AND KLAUSEN, K. (1962) Form and function of the erect human spine. Clin. Orthop., No. 25, 55–63.

AVNI, J. AND CHACO, J. (1972) Objective measurements of postural changes in depression. Paper presented to Annual Meeting of Am. Psych. Assoc., Boston, Mass.

AVON, G. AND SCHMITT, L. (1975) Electromyographie du trapèze dans diverses positions de travail à la machine à écrire. Ergonomics, *18:* 619–626.

BÄCKDAHL, M. AND CARLSÖÖ, S. (1961) Distribution of activity in muscles acting on the wrist (an electromyographic study). Acta Morph. Neer.-Scandinav., *4:* 136–144.

BACKHOUSE, K. M. (1968) The mechanics of normal digital control in the hand and analysis of the ulnar drift of rheumatoid arthritis. Ann. Roy. Col. Surg. Engl., *43:* 154–173.

BACKHOUSE, K. M. AND CATTON, W. T. (1954) An experimental study of the functions of the lumbrical muscles in the human hand. J. Anat., *88:* 133–141.

BAHORIC, A. AND CHERNICK, V. (1975) Electrical activity of phrenic nerve and diaphragm in utero. J. Appl. Physiol., *39:* 513–518.

BAILEY, J. A., POWERS, J. J. AND WAYLONIS, G. W. (1970) A clinical evaluation of electromyography of the anal sphincter. Arch. Phys. Med., *51:*

403–409.

BAJD, T., STANIČ, U., TOMŠIČ M. AND KRAJNIK, J. (1974). Symposium and Seminars, "Informatica 74," Bled, Jugoslavia, 7–12 October, 1974.

BAKER M., REGENOS, E., WOLF, S.L. AND BASMAJIAN, J. V. (1977) Developing strategies for biofeedback: applications in neurologically handicapped patients. Phys. Ther., *57:* 402–408.

BAKER, P. S. (1972) A method for obtaining muscle action potentials from a locust during free turning in flight. Proceedings of Physiol. Soc., November, 1972, 2 P–3 P.

BALLESTEROS, M. L. F. See Fernandez-Ballesteros.

BALLINTIJN, C. M. (1969) Muscle coordination of the respiratory pump of the carp (cyprinus carpio L.) J. Exper. Biol., *50:* 569–591.

BALLINTIJN, C. M. AND HUGHES, G. M. (1965a) The muscular basis of the respiratory pumps in the trout. J. Exper. Biol., *43:* 349–362.

BALLINTIJN, C. M. AND HUGHES, G. M. (1965b) The muscular basis of the respiratory pumps in the dogfish (*scyliorhinus canicula*). J. Exper. Biol., *43:* 363–383.

BALSHAN, I. D. (GOLDSTEIN). (1962) Muscle tension and personality in women. Arch. Gen. Psychiat., *7:* 436–448.

BANKOV, S. AND JØRGENSEN, K. (1969) Maximum strength of elbow flexors with pronated and supinated forearm. Comm. Dan. Nat. Ass. Infant. Paral., No. 29.

BARANOVA-KRYLOV, I. N. (1969) Testing of the excitability of the spinal centres of a pair of antagonistic muscles (Russian text). Zh. Vyssh. Nerv. Diat., Pavlov, *16:* 889–891.

BARNETT, C. H. AND HARDING, D. (1955) The activity of antagonist muscles during voluntary movement. Ann. Phys. Med., *2:* 290–293.

BARNETT, C. H. AND RICHARDSON, A. (1953) The postural function of the popliteus muscle. Ann. Phys. Med., *1:* 177–179.

BARTOSHUK, A. K. AND KASWICH, J. A. (1966) Electromyographic gradients as a function of tracking cues. Psychon. Sci., *6:* 43–44.

BASMAJIAN, J. V. (1952) Electromyography. Univ. Toronto M. J., *30:* 10–18.

BASMAJIAN, J. V. (1955a) *Primary Anatomy,* ed. 3. The Williams & Wilkins Co., Baltimore.

BASMAJIAN, J. V. (1955b) in Letters to the Editor, Lancet, *2:* Nov. 26, 1140.

BASMAJIAN, J. V. (1957a) Electromyography of two-joint muscles. Anat. Rec., *129:* 371–380.

BASMAJIAN, J. V. (1957b) New views on muscular tone and relaxation. Canad. M. A. J., *77:* 203–205.

BASMAJIAN, J. V. (1958a) A new six-channel electromyograph for studies on muscle. I.R.E. Transactions Med. Electronics, *PGME-11:* 45–47.

BASMAJIAN, J. V. (1958b) Electromyography of iliopsoas. Anat. Rec., *132:* 127–132.

BASMAJIAN, J. V. (1959) "Spurt" and "shunt" muscles: an electromyographic confirmation. J. Anat., *93:* 551–553.

BASMAJIAN, J. V. (1961) Weight-bearing by ligaments and muscles. Canad. J. Surg., *4:* 166–170.

BASMAJIAN, J. V. (1963a) Control and training of individual motor units. Science, *141:* 440–441.

BASMAJIAN, J. V. (1963b) Conscious control of single nerve cells. New Scientist, *20:* 662–664.

BASMAJIAN, J. V. (1970) *Primary Anatomy,* ed. 6. The Williams & Wilkins Co., Baltimore.

BASMAJIAN, J. V. (1972) Electromyography comes of age. Science, *176:* 603–609.

BASMAJIAN, J. V. (1975) A simple method to improve performance of fine-wire emg electrodes, Am. J. Phys. Med., *53:* 269–270.

BASMAJIAN, J. V. (1976a) Facts *vs.* myths in emg biofeedback. Biofeedback and Self-Reg., *1:* 396–371.

BASMAJIAN, J. V. (1976b) The human bicycle: an ultimate biological convenience. Orthop. Clin. N.A., *7:* 1027–1029.

BASMAJIAN, J. V. (1977) Motor learning and control: a working hypothesis. Arch. Phys. Med. Rehab. *58:* 38–41.

BASMAJIAN, J. V., BAEZA, M. AND FABRIGAR, C. (1965) Conscious control and training of individual spinal motor neurons in normal human subjects. J. New Drugs, *5:* 78–85.

BASMAJIAN, J. V. AND BAZANT, F. J. (1959) Factors preventing downward dislocation of the adducted shoulder joint: an electromyographic and morphological study. J. Bone & Joint Surg., *41-A:* 1182–1186.

BASMAJIAN, J. V. AND BENTZON, J. W. (1954) An electromyographic study of certain muscles of the leg and foot in the standing position. Surg. Gynec. & Obst., *98:* 662–666.

BASMAJIAN, J. V. AND BOYD, W. H. (1960) (Motion picture film) Electromyography of the Diaphragm: Direct Recording Technique in the Rabbit. Exhibited at Am. Assoc. of Anatomists Annual Meeting, Chicago, March 21–23 1961; abstract in Anatomical Record, 1961, *139:* 337.

BASMAJIAN, J. V. AND CROSS, G. L. (1971) Duration of motor unit potentials from fine-wire electrodes. Am. J. Phys. Med., *50:* 144–148.

BASMAJIAN, J. V. AND DUTTA, C. R. (1961a) Electromyography of the pharyngeal constrictors and soft palate in rabbits. Anat. Rec., *139:* 443–449.

BASMAJIAN, J. V. AND DUTTA, C. R. (1961b) Electromyography of pharyngeal constrictors and levator palati in man. Anat. Rec., *139:* 561–563.

BASMAJIAN, J. V., FORREST, W. J. AND SHINE, G. (1966) A simple connector for fine-wire electrodes. J. Appl. Physiol., *21:* 1680.

BASMAJIAN, J. V. AND GREENLAW, R. K. (1968) Electromyography of iliacus and psoas with inserted fine-wire electrodes (abstract). Anat. Rec., *160:* 310.

BASMAJIAN, J. V. AND GRIFFIN, W. R. (1972) Function of anconeus muscle: an electromyographic study. J. Bone & Joint Surg., *54-A:* 1712–1714.

BASMAJIAN, J. V. AND HUDSON, J. E. (1974) Miniature source-attached differential amplifier for electromyography, *53:* 234–236

BASMAJIAN, J. V., KUKULKA, C. G., NARAYAN, M. G. AND TAKEBE, K. (1975) Biofeedback treatment of foot-drop after stroke compared with standard rehabilitation technique: effects on voluntary control and strength. Arch. Phys. Med. Rehab., *56:* 231–236.

BASMAJIAN, J. V. AND LATIF, A. (1957) Integrated actions and functions of the chief flexors of the elbow: a detailed electromyographic analysis. J. Bone & Joint Surg., *39-A:* 1106–1118.

BASMAJIAN, J. V. AND LOVEJOY, J. F., JR. (1971) Functions of the popliteus muscle in man: a multifactorial electromyographic study. J. Bone & Joint Surg., *53-A:* 557–562.

BASMAJIAN, J. V. AND NEWTON, W. J. (1973) Feedback training of parts of buccinator muscle in man. Psychophysiology, 92.

BASMAJIAN, J. V. AND SAMSON, J. (1973) Standardization of methods in single motor unit training. Am. J. Phys. Med., *52:* 250–256.

BASMAJIAN, J. V. AND SIMARD, T. G. (1967) Effects of distracting movements on the control of trained motor units. Am. J. Phys. Med., *46:* 1427–1449.

BASMAJIAN, J. V. AND SPRING, W. B. (1955) Electromyography of the male (voluntary) sphincter urethrae. Anat. Rec., *121:* 388.

BASMAJIAN, J. V. AND STECKO, G. A. (1962) A new bipolar indwelling electrode for electromyography. J. Appl. Physiol., *17:* 849.

BASMAJIAN, J. V. AND STECKO, G. (1963) Role of muscles in arch support of the foot. J. Bone & Joint Surg., *45-A:* 1184–1190.

BASMAJIAN, J. V. AND SUPER, G. (1973) Dantrolene sodium in the treatment of spasticity. Arch. Phys. Med., *54:* 60–64.

BASMAJIAN, J. V. AND SZATMARI, A. (1955a) Effect of largactil (chlorpromazine) on human spasticity and electromyogram. A.M.A. Arch. Neurol., *73:* 224–231.

BASMAJIAN, J. V. AND SZATMARI, A. (1955b) Chlorpromazine and human spasticity: an electromyographic study. Neurology, *5:* 856–860.

BASMAJIAN, J. V. AND TRAVILL, A. (1961) Electromyography of the pronator muscles in the forearm. Anat. Rec., *139:* 45–49.

BASMAJIAN, J. V. AND WHITE, E. R. (1973) Neuromuscular control of trumpeters' lips. Nature, 241, 70.

BAŠTECKÝ, J., VINAŘ, O. AND ROTH, Z. (1968) Delayed auditory feedback and EMG of mimic muscles in schizophrenia. Activ. Nerv. Supp. (Praha), *10:* 212–214.

BAŠTECKÝ, J., VINAŘ, O. AND ROTH, Z. (1969) Lee effect and EMG of the mimic muscles in schizophrenia (Czech. text; English summary) Bratislavské Lekaraké Listy, *49:* 618–622.

BASTIAN, J. (1972) Neuro-muscular mechanisms controlling a flight maneuver in the honeybee. J. Comp. Physiol., *77:* 126–140.

BASTRON, J. A. AND LAMBERT, E. H. (1960) The clinical value of electromyography and electrical stimulation of nerves. Med. Clin. N. A., *44:* 1025–1036.

BATTYE, C. K. AND JOSEPH, J. (1966) An investigation by telemetering of the activity of some muscles in walking. Med.

& Biol. Engng., *4:* 125–135.

BAUMANN, F. (1969) Zur Frage der muskulären Verspannung bei der Koxarthrose auf Grund elektromyographischer Untersuchungen. Zeit. Orthop. Grenz., *106:* 500–508.

BAUMANN, F. (1971) Die Elektromyographie in der Orthopädie. Zeit. Orth. Grenz., *109:* 326–338.

BAUMANN, F. (1972) Der muskuläre Entspannungseffekt bei Hüftoperationen, in *Bücherei des Orthopäden*, vol. 9, pp. 43–103, P. Otte and K.-F. Schlegel, eds. Ferdinand Enke Verlag, Stuttgart.

BAUMANN, F. AND BEHR, O. (1969) Elektromyographische Untersuchungen der Hüftmuskulatur nach Arthrodese. Arch. Orthop Unfall-Chir., *66:* 1–17.

BAUWENS, P. (1948) Electromyography. Brit. J. Phys. Med., *11:* 130–136.

BAUWENS, P. (1950) The analysis of action potentials in electromyography. Proc. Inst. Electr. Engin., *97:* No. 107, Pt. I, 217–222.

BAUWENS, P. AND RICHARDSON, A. T. (1951) Electrodiagnosis, in *British Encyclopaedia of Medical Practice*, ed. 2, vol. 5, pp. 22–31. Butterworth & Co., London.

BAYKUSHEV, S. (1973) Two variants of feedback artificially created in central motor disturbances on the base of EMG. Electromyogr. Clin. Neurophysiol., *13:* 477–483.

BEARN, J. G. (1961a) The significance of the activity of the abdominal muscles in weight lifting. Acta Anat., *45:* 83–89.

BEARN, J. G. (1961b) An electromyographic study of the trapezius, deltoid, pectoralis major, biceps and triceps muscles, during static loading of the upper limb. Anat. Rec. *140:* 103–108.

BEAUDREAU, D. E., DAUGHERTY, W. F., JR. AND MASLAND, W. S. (1969) Two types of motor pause in masticatory muscles. Am. J. Physiol., *216:* 16–21.

BECK, A. (1930) Elecktromyographische

Untersuchungen am Sphincter ani (German text). Pflüger's Arch. Ges. Physiol., *224:* 278–292.

BECKER, R. O. (1960) The electrical response of human skeletal muscle to passive stretch. J. Bone & Joint Surg., *42-A:* 1091–1103.

BECKER, R. O. AND CHAMBERLIN, J. T. (1960) A modified coaxial electrode for electromyography. Arch. Phys. Med., *41:* 149–151.

BEEVOR, C. E. (1903) Croonian lectures on muscular movements and their representation in the central nervous system. Lancet, *1:* 1715–1724.

BEEVOR, C. E. (1904) *Croonian Lectures on Muscular Movements and Their Representation in the Central Nervous System.* Adlard and Son, London.

BEKOFF, A. (1976) Ontogeny of leg motor output on the chick embryo. Brain Res., *106:* 271–291.

BELL-BERTI, F. (1974) Control of pharyngeal cavity size for English voiced and voiceless stops. J. Acoust. Soc. Am., *57:* 456–461.

BELL-BERTI, F. (1976) An electromyographic study of velopharyngeal function in speech. J. Speech Hearing Res., *19:* 225–240.

BELL-BERTI, F. AND HIROSE, H. (1973) Patterns of palatoglossal activity and their implications for speech organization. Status Report on Speech Research. Report SR-34 Haskins Laboratories, New Haven, Connecticut.

BENTLEY, D. R. AND HOY, R. R. (1970) Postembryonic development of adult motor patterns in crickets: a neural analysis. Science, *170:* 1409–1411.

BERÁNEK, R. AND NOVOTNÝ, I. (1959) Spontaneous electrical activity and excitability of the denervated insect muscle. Physiol. Bohemoslovenica, *8:* 87–92.

BERGER, S. M. AND HADLEY, S. W. (1975) Some effects of a model's performance on an observer's electromyographic activity. Am. J. Psychol., *88:* 263–276.

BERGSTRÖM, R. M. (1959) The relation between the number of impulses and the integrated electric activity in electromyogram. Acta Physiol. Scandinav., *45:* 97–101.

BERGSTRÖM, R. M. (1962) The relation between the integrated kinetic energy and the number of action potentials in the electromyogram during voluntary muscle contraction. Ann. Acad. Scient. Fennicae, Series A; V. Medica 93 Suomalainen Tiedeakatemia, Helsinki, Finland.

BERGSTRÖM, R. M. AND KERTTULA, Y. (1961) On the neural control of breathing as studied by electromyography of the intercostal muscles of the rat. Ann. Acad. Scient. Fennicae, Series A, V. Medica, 79.

BERLENDIS, P. A. AND DeCARO, L. G. (1955) L'unitá motoria del muscolo stapedio (Italian text). Boll. Soc. Med.-Chir., Pavia, *69:* 33–36.

BERNSHTEIN, V. M. (1967) Statistical parameters of the electric signal of a muscle model. Biophysics, *12:* 693–703.

BERNSTEIN, N. A. (1940) *Investigations on the Biodynamics of Walking, Running, and Jumping. Part II.* Central Scientific Institute of Physical Culture, Moscow.

BERNSTEIN, N. A., SALZBERGER, O., PAVELNKO, P., POPOVA, T., SADCHIKOV, N. AND OSIPOV, L. (1935) Untersuchungen über die Biodynamik der Locomotion. I Teil. Verlag des Instituts für Experimentelle Medizin der Soviet-Union, Moscow.

BERTOZ, A. AND METRAL, S. (1970) Behavior of a muscular group subjected to a sinusoidal and trapezoidal variation of force. J. Appl. Physiol., *29:* 378–384.

BICAS, H. E. A. (1972) Electro-oculography in the investigation of ocular imbalance—I. Basic aspects. Vision Res., *12:* 993–1010.

BIERMAN, W. AND RALSTON, H. J. (1965) Electromyographic study during passive and active flexion and extension of the knee of the normal human subject. Arch. Phys. Med., *46:* 71–75.

BIERMAN, W. AND YAMSHON, L. J. (1948) Electromyography in kinesiologic evaluations. Arch. Phys. Med., *29:* 206–211.

BIGGS, N. L., POOL, H. E. AND BLANTON, P. L. (1968) An electromyographic study of the temporalis and masseter muscles of monozygotic twins. Trans. of Int. Soc. EMG Kinesiol., Montreal, Electromyography, *8:* suppl. 163–171.

BIGLAND, B. AND LIPPOLD, O. C. J. (1954a) The relation between force, velocity and integrated electrical activity in human muscles. J. Physiol., *123:* 214–224.

BIGLAND, B. AND LIPPOLD, O. C. J. (1954b) Motor unit activity in the voluntary contraction of human muscle. J. Physiol., *125:* 322–335.

BIGLAND, B., LIPPOLD, O. C. J. AND WRENCH, A. (1953) The electrical activity in isotonic contractions of human calf muscle. Proc. of the Physiol. Society, 20–21 March, J. Physiol., *120:* 40P–41P.

BIRO, G. AND PARTRIDGE, L. D. (1971) Analysis of multiunit spike records. J. Appl. Physiol., *30:* 521–526.

BISHOP, B. (1964) Reflex control of abdominal muscles during positive-pressure breathing. J. Appl. Physiol., *19:* 224–232.

BIZZI, E. KALIL, R. E. AND TAGLIASCO, V. (1971) Eye-Head coordination in monkeys: evidence for centrally patterned organization. Science, *173:* 452–454.

BJÖRK, Å. AND KUGELBERG, E. (1953) The electrical activity of the muscles of the eye and eyelids in various positions and during movement. Electroencephalog. & Clin. Neurophysiol., *5:* 595–602.

BJÖRK, Å. AND WÅHLIN, Å. (1960) The effect of succinylcholine on the cat diaphragm: an electromyographic study. Acta Anaesth. Scandinav., *4:* 13–20.

BLANTON, P. L., BIGGS, N. L. AND PERKINS, R. C. (1970) Electromyographic analysis of the buccinator muscle. J. Dent. Res., *49:* 389–394.

BLODI, F. C. AND VAN ALLEN, M. W. (1957) Electromyography of extraocular muscles in fusional movements: I. Electric phenomena at the breakpoint of fusion. Am. J. Ophth., *44:* (No. 4, special Part II (not I) of the October issue), 136–142.

BLODI, F. C. AND VAN ALLEN, M. W. (1960) The effect of paralysis of the cervical sympathetic system on the electromyogram of extraocular muscles in the human. Am. J. Ophth. *49:* 679–683.

BOEËTHIUS, J. AND KNUTSSON E. (1970) Resting membrane potential in chick muscle cells during ontogeny. J. Exper. Zool., *174:* 281–286.

BOLE, C. T. (1965) Electromyographic kinesiology of the genioglossus muscles in man. M.S. Thesis, Ohio State University, Columbus, Ohio.

BOLZANI, L. (1955) La curva intensitá-durata (i/t) ela velocitá di conduzione del nervo nell'uomo. Arch. Suisses de Neurol. et Psychiat., *74:* 148–154.

BOOKER, H. E., RUBOW, R. T. AND COLEMAN, P. J. (1969) Simplified feedback in neuromuscular retraining: an automated approach using electromyographic signals. Arch. Phys. Med., *50:* 621–625.

BORDEN, G. J. AND HARRIS, K. S. (1973) Oral feedback. II, An electromyographic study of speech under nerve-block anesthesia. J. Phonet., *1:* 297–308.

BORG, E. (1972) Excitability of the acoustic m. stapedius and m. tensor tympani reflexes in the nonanesthetized rabbit. Acta Physiol. Scand., *85:* 374–389.

BORS, E. (1926) Ueber das Zahlenverhältnis zwischen Nerven- und Muskelfasern. Anat. Anz., *60:* 14–416.

BORS, E. J. AND BLINN, K. A. (1957) Spinal reflex activity from the vesical mucosa in paraplegic patients. A.M.A. Arch. Neurol., *78:* 339–354.

Bors, E. and Blinn, D. A. (1959) Bulbo-cavernosus reflex. J. Urol., *82:* 128–130.

Bors, E. and Blinn, K. A. (1965) Abdominal electromyography during micturition. California Med., *102:* 17–22.

Botelho, S. Y. and Steinberg, S. A. (1965) Electromyography in canine fetus (abstract). Bull. Am. Assoc. EMG and Electrodiag., *12:* 17.

Bouisset, S,, Denimal, J. and Soula, C. (1963) Relation entre l'accélération d'un raccourcissement musculaire et l'activité electromyographique intégrée. J. Physiol., Paris, *55:* 203.

Bouisset, S. and Goubel, F. (1967) Relation entre l'activité électromyographique intégrée et la vitesse d'execution de mouvements monoarticulaires simples. J. Physiol., *59:* 359.

Bouisset, S. and Maton, B. (1972) Quantitative relationship between surface EMG and intramuscular electromyographic activity in voluntary movement. Am. J. Phys. Med., *51:* 285–295.

Bouisset, S. and Maton (1973) Comparison between surface and intramuscular EMG during voluntary movement. In *New Developments in EMG and Clinical Neurophysiology,* edited by J. E. Desmedt, Vol. 1, 533–539, Karger, Basel.

Boyd, I. A. and Roberts, T. D. M. (1953) Proprioceptive discharges from stretch-receptors in the knee-joint of the cat. J. Physiol., *122:* 38–58.

Boyd, W. H. (1968) The electrical activity of intercostal muscles during quiet breathing in rabbits. Can. J. Physiol. Pharm., *46:* 749–755.

Boyd, W. H. (1969) Coordinated electrical activity of diaphragm and intercostal muscles in rabbits during quiet breathing. Arch. Phys. Med., *50:* 127–132.

Boyd, W. H. and Basmajian, J. V. (1963) Electromyography of the diaphragm in rabbits. Am. J. Physiol., *204:* 943–948.

Boyd, W. H., Blincoe, H. and Hayner, J. C. (1965) Sequence of action of the diaphragm and quadratus lumborum during quiet breathing. Anat. Rec., *151:* 579–582.

Braithwaite, F., Channell, G. D., Moore, F. T. and Whillis, J. (1948) The applied anatomy of the lumbrical and interosseous muscles of the hand. Guy's Hosp. Rep., *97:* 185–195.

Brandell, B. R. (1970) An electromyographic-cinematographic study of the muscles of the index finger. Arch. Phys. Med., *51:* 278–285.

Brandell, B. R. (1977) Functional roles of the calf and vastus muscles in locomotion. Am. J. Phys. Med., *56:* 59–74.

Brandell, B. R., Huff, G. J. and Spark, G. J. (1968) An electromyographic-cinematographic study of the thigh muscles using M.E.R.D. (muscle electronic recording device) I. Trans. Int. Soc. EMG Kines., Montreal, 1968, 67–76.

Brandstater, M. E. and Lambert, E.H. (1973) Motor unit anatomy. In *New Devleopments in Electromyography and Clinical Neurophysiology,* edited by J. E. Desmedt, *1:* 14–22, Karger, Basel.

Brandt, K. and Fenz, W. D. (1969) Specificity in verbal and physiological indicants of anxiety. Percept. Motor Skills, *29:* 663–675.

Brantner, J. N. and Basmajian, J. V. (1975) Effects of training on endurance in hanging by the hands. J. Motor Behav., *7:* 131–134.

Bratanova, Ts. Kh. (1966) On bioelectric activity of muscles-antagonists in the course of elaboration of a motor habit (Russian text). Zh. Vyssh. Nerv. Diat., Pavlov, *16:* 411–416.

Bratzlavsky, M. (1972) Pauses in activity of human jaw closing muscle. Exper. Neurol., *36:* 160–165.

Braune, C. W. and Fischer, O. (1895) Der Gang des Menschen. I Teil. Versuche unbelasten und belasten Menschen. Abhandl. d. Math.-Phys. Cl. d. k. Sächs. Gesellsch. Wissensch.,

21: 153–322.

BREININ, G. M. (1955) The nature of vergence revealed by electromyography. A.M.A. Arch. Ophth., *54:* 407–409.

BREININ, G. M. (1957a) Electromyographic evidence for ocular muscle proprioception in man. A.M.A. Arch. Ophth., *57:* 176–180.

BREININ, G. M. (1957b) The position of rest during anesthesia and sleep: electromyographic observations. A.M.A. Arch. Ophth., *57:* 323–326.

BREININ, G. M. (1958) Analytic studies of the electromyogram of human extraocular muscle. Am. J. Ophth., *46:* 123–142.

BRENNAN, J. B. (1959) Clinical method of assessing tonus and voluntary movement in hemiplegia. Brit. Med. J., *1:* March 21, 767–768.

BRIGGS, D. I., LANDAU, B., AKERT, K. AND YOUMANS, W. B. (1960) Kinesiology of the abdominal compression reaction (abstract). Physiologist, *3:* 30, August, 1960.

BRISCOE, G. (1920) The muscular mechanism of the diaphragm. J. Physiol. (Lond.), *54:* 46–53.

BRISCOE, J. C. (1920) The mechanism of post-operative massive collapse of the lungs. Quart. J. Med., *13:* 293–334.

BROADBENT, T. R. AND SWINYARD, C. A. (1959) The dynamic pharyngeal flap, its selective use and electromyographic evaluation. Plast. & Reconstruct. Surg., *23:* 301–312.

BRODY, G., BALASUBRAMANIAN, R., SCOTT, R.N. (1974) A model for myo-electric signal generation. Med. Biol. Eng., *12:* 29–41.

BROOKS, J. E. (1968) Intracellular electromyography: resting and action potentials in normal human muscle. Arch. Neurol., *18:* 291–300.

BROOKS, J. E. (1970) Disuse atrophy of muscle: intracellular electromyography. Arch. Neurol., *22:* 27–30.

BROOME, H. L. AND BASMAJIAN, J. V. (1971) The function of the teres major muscle: an electromyographic study. Anat. Rec., *170:* 309–311.

BROOME, H. L. AND BASMAJIAN, J. V. (1971) Survival of iliopsoas muscle after Sharrard procedure. Am. J. Phys. Med., *50:* 301–302.

BROWN, D. M. AND BASMAJIAN, F. V. (1977) A bioconverter for upper extremity rehabilitation. Am. J. Phys. Med., in press.

BROWN, M. E., LONG C. AND WEISS, G. (1960) Electromyographic kinesiology of the hand. Part 1. Method. Phys. Ther. Rev., *40:* 453–458.

BROWN, W. E. K., HARLAND, R. AND BASMAJIAN, J. V. (1957) Electromyography of quadriceps femoris. (Unpublished work at the University of Toronto.)

BRUDNY, J., KOREIN, J., GRYNBAUM, B. B., FRIEDMAN, F. W., WEINSTEIN, S., SACHS-FRANKEL, G. AND BELANDRES, P. V. (1976) EMG feedback therapy: review of treatment of 114 patients. Arch. Phys. Med. Rehab., *57:* 55–61.

BRUNO, L. J. J., DAVIDOWITZ, J. AND HEFFERLINE, R. F. (1970) EMG waveform duration: a validation method for the surface electromyogram. Behav. Res. Meth. & Instru., *2:* 211–219.

BUCHTHAL, F. (1959) The functional organization of the motor unit: a summary of results. Am. J. Phys. Med., *38:* 125–128.

BUCHTHAL, F. AND CLEMMESEN, S. (1941) On differentiation of muscle atrophy by electromyography. Acta Psychiat. Neurol., *16:* 143–181.

BUCHTHAL, F. AND ENGBAEK, L. (1963) Refractory period and conduction velocity of the striated muscle fibre. Acta Physiol. Scandinav., *59:* 199–219.

BUCHTHAL, F. ERMINIO, F. AND ROSENFALCK, P. (1959) Motor unit territory in different human muscles. Acta Physiol. Scandinav., *45:* 72–87.

BUCHTHAL, F. AND FAABORG-ANDERSEN, K. (1964) Electromyography of laryngeal and respiratory muscles; correlation with phonation and respiration. Ann. Otol. Rhin. Laryng., *73:* 118–124.

BUCHTHAL, F. AND FERNANDEZ-BALLESTEROS, M. L. (1965) Electromyographic study of the muscles of the upper arm and shoulder during walking in patients with Parkinson's disease. Brain, *88:* 875–896.

BUCHTHAL, F., GULD, C. AND ROSENFALCK, P. (1954) Action potential parameters in normal human muscle and their dependence on physical variables. Acta Physiol. Scandinav., *32:* 200–218.

BUCHTHAL, F., GULD, C. AND ROSENFALCK, P. (1957) Multielectrode study of the territory of a motor unit. Acta Physiol. Scandinav., *39:* 83–103.

BUCHTHAL, F., GULD, C. AND ROSENFALCK, P. (1957b) Volume conduction of the spike of the motor unit potential investigated with a new type of multielectrode. Acta Physiol Scandinav., *38:* 331–354.

BUCHTHAL, F. AND MADSEN, E. E. (1950) Synchronous activity in normal and atrophic muscle. Electroenceph. Clin. Neurophysiol., *2:* 425–444.

BUCHTHAL, F., PINELLI, P. AND ROSENFALCK, P. (1954) Action potential parameters in normal human muscle and their physiological determinants. Acta. Physiol. Scandinav., *32:* 219–229.

BUCHTHAL, F. AND ROSENFALCK, P. (1973) On the structure of motor units. In *New Developments in EMG and Clinical Neurophysiology*, edited by J. E. Desmedt, Vol. 1, pp. 71–85, Karger, Basel.

BUCHTHAL, F. AND SCHMALBRUCH, H. (1969) Contraction time and fibre types of normal and diseased human muscle, in *Muscle Disease* (proceedings of an International Congress, Milan, May, 1969). Excerpta Medica International Congress Series No. 199.

BUDZYNSKI, T. H. AND STOYVA, J. M. (1969) An instrument for producing deep muscle relaxation by means of analog information feedback. J. Appl. Behav. Anal., *2:* 231–237.

BUDZYNSKI, T., STOYVA, J. AND ADLER, C. (1970) Feedback-induced muscle relaxation: application to tension headache. J. Behav. Exp. Psychiat., *1:* 205–211.

BULLER, A. J., ECCLES, J. C. AND ECCLES, R. M. (1960) Interactions between motoneurones and muscles in respect of the characteristic speeds of their responses. J. Physiol., *150:* 417–439.

BURKE, R. E., RUDOMIN, P. AND ZAJAC, III, F. E. (1976) The effect of activation history on the tension production by individual muscle units. Brain Res. *109:* 515–529.

BURKE, R. E. AND TSAIRIS, P. (1973) Anatomy and innervation ratios in motor units of cat gastrocnemius. J. Physiol. (London), *234:* 749–765.

CAIN, W. C. (1973) Nature of perceived effort and fatigue: roles in strength and blood flow in muscle contractions. J. Motor Behav., *5:* 33–47.

CALDWELL, C. W. AND RESWICK, J. B. (1975) A percutaneous wire electrode for chronic research use. IEEE Trans. Biomed. Eng. (Sept.) 429–432.

CAMPBELL, E. J. M. (1952) An electromyographic study of the role of the abdominal muscles in breathing. J. Physiol., *117:* 222–233,

CAMPBELL, E. J. M. (1954) The muscular control of breathing in man, Ph.D. Thesis, University of London; quoted in his 1958 monograph (see reference below).

CAMPBELL, E. J. M. (1955a) An electromyographic examination of the role of the intercostal muscles in breathing in man. J. Physiol., *129:* 12–26.

CAMPBELL, E. J. M. (1955b) The role of the scalene and sternomastoid muscles in breathing in normal subjects. An electromyographic study. J. Anat., *89:* 378–386.

CAMPBELL, E. J. M. (1955c) The functions of the abdominal muscles in relation to the intra-abdominal pressure and the respiration. Arch. Middlesex Hosp., *5:* 87–94.

CAMPBELL, E. J. M. (1958) *The Respiratory Muscles and the Mechanics of Breathing*, Lloyd-Luke (Medical Books) Ltd., London.

CAMPBELL, E. J. M. AND GREEN, J. H. (1953a) The expiratory function of the abdominal muscles in man. an electromyographic study. J. Physiol., *120:* 409–418.

CAMPBELL, E. J. M. AND GREEN, J. H. (1953b) The variations in intra-abdominal pressure and the activity of the abdominal muscles during breathing; a study in man. J. Physiol., *122:* 282–290.

CAMPBELL, E. J. M. AND GREEN, J. H. (1955) The behaviour of the abdominal muscles and the intra-abdominal pressure during quiet breathing and increased pulmonary ventilation: a study in man. J. Physiol, *127:* 423–426.

CAMPBELL, K. M., BIGGS, N. L., BLANTON, P. L. AND LEHER, R. P. (1973) Electromyographic investigation of the relative activity among four components of the triceps surae. Am. J. Phys. Med., *52:* 30–41.

CANTER, A., KONDO, C. Y. AND KNOTT, J. R. (1975) A comparison of emg feedback and progressive muscle relaxation training in anxiety neurosis. Brit. J. Psychiat., *127:* 470–477.

CAPPOZZO, A., LEO, T. AND PEDOTTI, A. (1975) A general computing method for the analysis of human locomotion. J. Biomech., *8:* 307–320.

CAR, A. (1970) La commande corticale du centre déglutiteur bulbaire. J. Physiol. (Paris), *62:* 361–386.

CARDUS, D., QUESADA, E. M. AND SCOTT, F. B. (1963) Studies on the dynamics of the bladder. J. Urol., *90:* 425–433.

CARLSON, K. E., ALSTON W. AND FELDMAN, D. J. (1964) Electromyographic study of aging in skeletal muscle. Am.

J. Phys. Med., *43:* 141–145.

CARLSÖÖ, S. (1952) Nervous coordination and mechanical function of the mandibular elevators: an electromyographic study of the activity, and an anatomic analysis of the mechanics of the muscles. Acta Odont. Scandinav., *10:* suppl. 11, 1–132.

CARLSÖÖ, S. (1961) The static muscle load in different work positions: an electromyographic study. Ergonomics, *4:* 193–211.

CARLSÖÖ, S. (1962) A method of studying walking on different surfaces. Ergonomics, *5:* 271–274.

CARLSÖÖ, S. (1964) Influence of frontal and dorsal loads on muscle activity and on the weight distribution in the feet. Acta. Orthop. Scandinav., *34:* 299–309.

CARLSÖÖ, S. (1966) The initiation of walking. Acta Anat., *65:* 1–9.

CARLSÖÖ, S. (1973) *How Man Moves.* Translated from Swedish edition by W. P. Michael, Heinemann, London and Crane Russak, New York.

CARLSÖÖ, S. AND EDFELDT, A. W. (1963) Attempts at muscle control with visual and auditory impulses as auxiliary stimuli. Scand. J. Psychol., *4:* 231–235.

CARLSÖÖ, S. AND FOHLIN, L. (1969) The mechanics of the two-joint muscles rectus femoris, sartorius and tensor fasciae latae in relation to their activity. Scand. J. Rehab. Med., *1:* 107–111.

CARLSÖÖ, S. AND GUHARAY, A. R. (1967) A study of the muscle activity and blood flow in the muscles of the upper arm in supporting heavy loads. Acta Physiol. Scandinav., *72:* 366–369.

CARLSÖÖ, S. AND JOHANSSON, O. (1962) Stabilization of and load on the elbow joint in some protective movements: an experimental study. Acta Anat., *48:* 224–231.

CARLSÖÖ, S. AND MAYR. J. (1974) A study of the load on joints and muscles in work with a pneumatic hammer and a bolt gun. Work-environm.-hlth., *11:* 32–38.

CARLSÖÖ, S. AND MOLBECH, S. (1966) The function of certain two-joint muscles in a closed muscular chain. Acta Morph. Neer.-Scand., *4:* 377–386.

CARLSÖÖ, S. AND NORSTRAND, A. (1968) The coordination of the knee muscles in some voluntary movements and in the gait in cases with and without knee joint injuries. Acta Chir. Scand., *134:* 423–426.

CARLSÖÖ, S. AND SKOGLUND, G. (1969) En rörelseananatomisk studie av fall med gångrubbningar (Swedish text). Lakartidningen, *66:* 3966–3972.

CARLSSON, S. G., GALE, E. N., AND ÖHMAN, A. (1975) Treatment of temporomandibular joint syndrome with biofeedback training. J. Am. Dent. Assoc., *91:* 602–605.

CARMAN, D. J., BLANTON, P. L. AND BIGGS, N.L. (1971) Electromyography of the anterolateral abdominal musculature (abstract). Anat. Rec., *169:* 474.

CARMEL, P. W. AND STARR, A. (1963) Acoustic and nonacoustic factors modifying middle-ear muscle activity in waking cats. J. Neurophysiol., *26:* 598–616.

CARVALHO, C. A. F., GARCIA, O. S., VITTI, M. AND BERZIN, F. (1972) Electromyographic study of the m. tensor fasciae latae and m. sartorius. Electromyogr. Clin. Neurophysiol., *12:* 387–400.

CARVALHO, C. A. F., KÖNIG, B., JR. AND VITTI, M. (1967) Electromyographic study of the muscles "extensor digitorum brevis" and "extensor hallucis brevis." Rev. Hosp. Fac. Med. S. Paulo, *22:* 65–72.

CARVALHO, C. A. F. AND VITTI, M. (1965) Estudo Eletromiográfico do músculo extensor longo do halux. Folio Clin. Biol., *34:* 77–91.

CASARIN, G., PURICELLI, R. AND VILLA, A. (1974) Una tecnica telemetrica via radio nello studio elettromiografico della

deambulazione. Riabilitazione, *7:* 177–189.

CATTON, W. T. AND GRAY, J. E. (1951) Electromyographic study of the action of the serratus anterior muscle in respiration (abstract). J. Anat., *85:* 412.

CHACO, J. AND WOLF, E. (1971) Subluxation of the glenohumeral joint in hemiplegia. Am. J. Phys. Med., *50:* 139–143.

CHACO, J. AND YULES, R. B. (1969) Velopharyngeal incompetence post tonsilloadenoidectomy: an electromyographic study. Acta Otolaryng., *68:* 276–278.

CHANTRAINE, A. (1974) Exploration électrique des sphincters striés urétral et anal. J. Urol. Nephr., *80:* 207–213.

CHAPMAN, A. J. (1974) An electromyographic study of social facilitation: a test of the 'mere presence' hypothesis. Brit. J. Psychol., *65:* 123–128.

CHENG, I.-S. KOOZEKANANI, S. H. AND FATCHI, M. T. (1975) A simple computer-television interface system for gait analysis. IEEE Trans. Biomed. Eng., May, 1975, 259–260.

CHRISTENSEN, E. H. (1962) Muscular work and fatigue, a paper delivered at "Muscle as a Tissue" International Conference, Lankenau Hospital, Philadelphia in 1960. Pp. 176–189 in *Muscle as a Tissue,* K. Rodahl and S. M. Horvath, editors, McGraw-Hill, New York, Toronto, London.

CHRISTENSEN, J. AND HAUSER, R. L. (1971) Circumferential coupling of electrical slow waves in circular muscle of cat colon. Am. J. Physiol., *221:* 1033–1037.

CHRISTENSEN, J., TAKATA, M. AND KAWAMURA, Y. (1969) Electrophysiologic analysis of innervation of temporalis muscle in the cat. J. Dent. Res., *48:* 327.

CHRISTIE, B. G. B. AND COOMES, E. N. (1960) Normal variation of nerve conduction in three peripheral nerves. Ann. Phys. Med., *5:* 303–309.

CLAMANN, H. P. (1968) A quantitative

analysis of the firing pattern of single motor units of a skeletal muscle of man and their utilization in isometric contractions. Ph.D. Thesis, Johns Hopkins University.

CLAMANN, H. P. (1969) Statistical analysis of motor unit firing patterns in a human skeletal muscle. Biophys. J. *9:* 1233–1251.

CLAMANN, H. P. (1970) Activity of single motor units during isometric tension. Neurology, *20:* 254–260.

CLARK, R. F. K. AND WYKE, B. D. (1975) Temporomandibular articular reflex control of the mandibular musculature. Intern. Dent. J., *25:* 289–296.

CLARKE, A. M. (1967) Relationship between the electromyogram and isometric reflex response in pre-strain conditions in normal humans. Nature, *214:* 1114–1115.

CLEELAND, C. S. (1973) Behavioral technics in the modification of spasmodic torticollis. Neurology, *23:* 1241–1247.

CLEMMESEN, S. (1951) Some studies of muscle tone. Proc. Roy. Soc. Med., *44:* 637–646.

CLENDENIN, M. A. AND SZUMSKI, A. J. (1971) Influence of cutaneous ice application on single motor units in humans. J. Am. Phys. Ther. Assoc., *51:* 166–175.

CLOSE, J. R. (1964) *Motor Function in the Lower Extremity: Analysis by Electronic Instrumentation.* Charles C Thomas, Springfield, Ill.

CLOSE, J. R., NICKLE, E. D. AND TODD, F. N. (1960) Motor-unit action-potential counts: their significance in isometric and isotonic contractions. J. Bone & Joint Surg., *42–A:* 1207–1222.

CNOCKAERT, J. C. (1975) Comparaison electromyographique du travail en allongement et en raccourcissement au course de mouvement de va-et-vient. Electromyogr. Clin. Neurophysiol., *15;* 477–489.

COËRS, C. AND WOOLF, A. L. (1959) The

Innervation of Muscle, a Biopsy Study. Blackwell Scientific Publications, Ltd., Oxford, and Charles C Thomas, Springfield, Ill.

COGGSHALL, J. C., BEKEY, G. A. (1970) A stochastic model of skeletal muscle based on motor unit properties. Math. Biosc., *7:* 405–419.

COHEN, A. H. AND GANS, C. (1975) Muscle activity in rat locomotion: movement analysis and electromyography of the flexors and extensors of the elbow. J. Morphol., *146:* 177–196.

COHEN, L. (1970) Interaction between limbs during bimanual voluntary activity. Brain, *93:* 259–272.

COHEN, M. J. (1973) The relation between heart rate and electromyographic activity in a discriminated escape-avoidance paradigm. Psychophysiology, *10:* 8–20.

COHEN, M. J. AND JOHNSON, H. J. (1971) Relationship between heart rate and muscular activity within a classical conditioning paradigm. J. Exper. Psychol., *90:* 222–226.

COMTET, J. J. AND AUFFRAY, Y. (1970) Physiologie des muscles élévateurs de l' épaule. Rev. Chirurg. Orthop. (Paris), *56:* 105–117.

COOK, W. A. (1967) Antagonistic muscles in the production of clonus in man. Neurology, *17:* 779–782.

COOPER, R. R. (1972) Alterations during immobilization and regeneration of skeletal muscle in cats. J. Bone & Joint Surg., *54-A:* 919–953.

COOPER, S. AND DANIEL, P. M. (1949) Muscle spindles in human extrinsic eye muscles. Brain, *72:* 1–24.

COPACK, P. B., FELMAN, E., LIEBERMAN, J. S. AND GILMAN, S. (1975) Differences in proximal and distal conduction velocity of efferent nerve fibers to the medial gastrocnemius muscle. Brain Res., *91:* 147–150.

COTTON, F. J. (1921) Subluxation of the shoulder—downward. Boston M. &

Surg. J., *185:* 405–407.

COURSEY, R. S. (1975) Electromyograph feedback as a relaxation technique. J. Consult. Clin. Psychol., *43:* 825–834.

COX, D. J., FREUNDLICH, A. AND MEYER, R. G. (1975) Differential effectiveness of electromyographic feedback, verbal relaxation instructions, and medication placebo with tension headaches. J. Consult. Clin. Psychol., *43:* 892–899.

CUNNINGHAM, D. P. AND BASMAJIAN, J. V. (1969) Electromyography of genioglossus and geniohyoid muscles during deglutition. Anat. Rec., *165:* 401–410.

CURRIER, D. P. (1969) Measurement of muscle fatigue. J. Am. Phys. Th. Assoc., *49:* 724–730.

CURRIER, D. P. (1972) Maximal isometric tension of the elbow extensors at varied positions. Part 2. Assessment of extensor components by quantitative electromyography. Phys. Ther., *52:* 1265–1276.

CZÉH, G. AND SZÉKELY, G. (1971) Muscle activities recorded simultaneously from normal and supernumerary forelimbs in ambystoma. Acta Physiol. Acad. Sc. Hungar., *40:* 287–301.

CZIPOTT, Z. AND HERPAI, S. (1971) Elektromyographische Untersuchungen bie Meniskus-, Knie- und Bandverletzungen. Z. Orthop., *109:* 768–778.

DA HORA, B. (1959) O "musculus anconeus." Contribuição ao estudo da sua arquitetura e das suas funções (Portuguese text). Thesis, University of Recife, Recife, Brazil, 127 pp.

DANIEL, B. AND GUITAR, B. (1973) EMG feedback and recovery of facial and speech gestures following neural anastomosis. Unpublished Document (No. 7), Speech Research Laboratory, University of Wisconsin.

DANIEL, E. E. (1968) The electrical activity of the alimentary tract. Am. J. Digest. Dis., *13:* 297–319.

DANIELSSON, C.-O., FRANKSSON, C. AND PETERSÉN, I. (1956) Stress incontinence following the Manchester operation for prolapse. Acta Obst. Gynec. Scandinav., *35:* 335–344.

DAVIS, M. H., SAUNDERS, D. R., CREER, T. L. AND CHAI, H. (1973) Relaxation training facilitated by biofeedback apparatus as a supplemental treatment in bronchial asthma. J. Psychosomat. Res., *17:* 121–128.

DAWSON, G. D. (1956) The relative excitability and conduction velocity of sensory and motor nerve fibers in man. J. Physiol., *131:* 436–451.

DAWSON, G. D. AND SCOTT, J. W. (1949) The recording of nerve action potentials through the skin in man. J. Neurol. Neurosurg. Psychiat., *12:* 259–267.

DEANDRADE, J. R., GRANT, C. AND DIXON, A. ST. J. (1965) Joint distension and reflex muscle inhibition in the knee. J. Bone & Joint Surg., *47-A:* 313–322.

DEDO, H. H. (1970) The paralyzed larynx: an electromyographic study in dogs and humans. Laryngoscope, *80:* 1455–1517.

DEDO, H. H. (1971) Electromyographic and visual evaluation of recurrent laryngeal nerve anastomosis in dogs. Ann. Otol. Rhinol. Laryng., *80:* 664–669.

DEDO, H. AND HALL, W. N. (1969) Electrodes in laryngeal electromyography: reliability comparison. Otol. Rhinol. Laryngol., *78:* 172–181.

DEDO, H. H. AND OGURA, J. H. (1965) Vocal cord electromyography in the dog. Laryngoscope, *75:* 201–311.

DE GIRARDI-QUIRION, C. (1976) La rétroaction biologique en physiothérapie: une expérience vécue en pédiatrie. Physiother. Can., *28:* 14–19.

DE JESUS, P. V., HAUSMANOWA-PETRUSEWICZ, I. AND BARCHI, R. L. (1973) The effect of cold on nerve conduction of human slow and fast nerve fibers. Neurology, *23:* 1182–1189.

DEJONCKERE, P. (1975) La technique externe d'électromyographie laryngée en clinique laryngologique et phonia-trique. Acta Oto-Laryng. Belg., 29: 4: 677–691.

DELHEZ, L. (1964) Evolution de l'activ-ité électrique intégrée du diaphragme durant l'hyperventilation. Compte Rend. Séances Soc. Biol., 158: 2496–2500.

DELHEZ, L., BOTTIN, R., DAMOISEAU, J. AND PETIT, J. M. (1964a) Examen com-paratif des électromyogrammes des pi-liers du diaphragme dérivés au moyen de trois modèles de sondes-électrodes. Electromyography, 4: 5–14.

DELHEZ, L., DAMOISEAU, J. AND DE-ROANNE, R. (1964b) Comportement elec-trique du diaphragme et des muscles abdominaux durant la respiration sous pression positive intermittente chez des sujets normaux et emphysémateux. Rev. Électrodiag.-Thérap., 1: 197–209.

DELHEZ, L., BOTTIN-THONON, A. AND PE-TIT, J. M. (1968) Influence de l'en-traînement sur la force maximum des muscles respiratoires. Trav. Soc. Méd. Belge d'Educ. Phys. Sports, 20: 52–63.

DELHEZ, L. AND PETIT, J. M. (1966) Don-nées actuelles de l'électromyographie respiratoire chez l'homme normal. Electromyography, 6: 101–146.

DELHEZ, L., PETIT, J. M., DEROANNE, R., PIRNAY, F. AND SNEPPE, R. (1969) Ev-olution de l'activité électrique integrée de quatre muscles locomoteurs durant la marche et la course sur tapis roulant. Electromyography, 9: 417–431.

DELHEZ, L., TROQUET, J., DAMOISEAU, J. AND PETIT, J. M. (1963) Activité antag-oniste du diaphragme à la fin de l'expiration forcée. J. Physiol., Paris, 55: 241–242.

DELHEZ, L., TROQUET, J., DAMOISEAU, J. AND PETIT, J. M. (1963) Influence des modalités d'activité électrique des mus-cles abdominaux et du diaphragme sur les diagrammes volume/pression de re-lâchement du thorax et des poumons. Arch. Internat. Physiol. Biochim., 71:

175–194.

DELHEZ, L., TROQUET, J., DAMOISEAU, J. AND PETIT, J. M. (1964c) Nécessité de l'électromyographie dans les mesures d'elastance thoraco-pulmonaire. Rev. Electrodiag.-Thérap., 1: 39–45.

DELHEZ, L., TROQUET, J., DAMOISEAU, J., PIRNAY, F., DEROANNE, R. AND PETIT, J. M. (1964d) Influence des modalités d'exécution des manoeuvres d'expira-tion forcée et d'hyperpression thoraco-abdominale sur l'activité électrique du diaphragme. Arch. Internat. Physiol. Biochim., 72: 76–94.

DELLOW, P. G. AND LUND, J. P. (1971) Evidence for central timing of rhythm-ical mastication. J. Physiol., 215: 1–13.

DE LUCA, C. J. (1968) Myoelectric anal-ysis of isometric contractions of the human biceps brachii. M. Sc. Thesis, University of New Brunswick, Fredric-ton, New Brunswick, Canada.

DE LUCA, C. J. (1975) A model for a motor unit train recorded during con-stant force isometric contractions. Biol. Cybernetics, 19: 159–167.

DELUCA, C. J. AND FORREST, W. J. (1972) An electrode for recording single motor unit activity during strong muscle con-tractions. IBEEE Trans. Biomed. Engng., BME-19: 367–372.

DE LUCA, C. J. AND FORREST, W. J. (1973a) Properties of motor unit action potential trains. Kybernetics, 12: 160–168.

DE LUCA, C. J. AND FORREST, W. J. (1973b) Force analysis of individual muscles acting simultaneously on the shoulder joint during isometric abduc-tion. J. Biomech., 6: 385–393.

DE LUCA, C. J. AND FORREST, W. J. (1973c) Probability distribution func-tion of the inter-pulse intervals of single motor unit action potentials during iso-metric contraction. In *New Developments in Electromyography and Clinical Neuro-physiology* edited by J. E. Desmedt, 1: 638–647, Karger, Basel.

De Luca, C. J. and van Dyk, E. J. (1975) Derivation of some parameters of myoelectric signals recorded during sustained constant force isometric contractions. Biophys. J. *15:* 1167–1180.

Dempster, W. T. and Finerty, J. C. (1947) Relative activity of wrist moving muscles in static support of the wrist joint: an electromyographic study. Am. J. Physiol., *150:* 596–606.

Denny-Brown, D. (1929) On the nature of postural reflexes. Proc. Roy. Soc., *104B:* 252–301.

Denny-Brown, D. and Foley, J. M. (1948) Myokymia and the benign fasciculation of muscular cramps. Tr. A. Am. Physicians, *61:* 88–96.

Denny-Brown, D. and Pennybacker, J. B. (1938) Fibrillation and fasciculation in voluntary muscle. Brain, *61:* 311–334.

Denslow, J. S. and Gutensohn, O. R. (1967) Neuromuscular reflexes in response to gravity. J. Appl. Physiol., *23:* 243–247.

DePalma, A. T., Leavitt, L. A. and Hardy, S. B. (1958) Electromyography in full thickness flaps rotated between upper and lower lips. Plast. & Reconstruct. Surg., *21:* 448–452.

de Sousa, O. M. (1958) Aspectos da arquitetura e da ação dos músculos estriados, baseada na eletromiografia (Portuguese text). Folia Clin. Biol., *28:* 12–42.

de Sousa, O. Machado (1964) Estudo eletromiográfico do m. platysma (Portuguese text with English summary). Folia Clin. Biol. (Brazil), *33:* 42–52.

de Sousa, O. M., Berzin, F. and Berardi, A. C. (1969) Electromyographic study of the pectoralis major and latissimus dorsi muscles during medial rotation of the arm. Electromyography, *9:* 407–416.

de Sousa, O. M., de Moraes, J. L. and de Morais Vieira, F. L. (1961) Electromyographic study of the brachiora-

dialis muscle. Anat. Rec., *139:* 125–131.

de Sousa, O. M., de Morais, W. R., and Ferraz, E. F. (1957) Observações anatômicas e eletromiográficas sôbre o "m. pronator quadratus" (Portuguese text). Folia Clin. Biol., *27:* 214–219.

de Sousa, O. M. de Morais, W. R. and Ferraz, E. C. de F. (1958) Estudo eletromiográfico de alguns músculos do antebraço durante a pronação (Portuguese text). Rev. Hosp. Clin., *13:* 346–354.

de Sousa, O. M. and Furlani, J. (1974) Electromyographic study of the m. rectus abdominis. Acta Anat., *88:* 281–298.

de Sousa, O. M., Furlani, J. and Garcia, O. S. (1975) Atividade de músculos antagonistas: estudo electromiográfico. Rev. Hosp. Clin. Fac. Med. S. Paulo, *30:* 471–473.

de Sousa, O. M., Furlani, J. and Vitti, M. (1973) Étude électromyographique du m. sternocleidomastoideus. Electromyogr. Clin. Neurophysiol., *13:* 93–106.

de Sousa, O. M. and Vitti, M. (1965) Estudo eletromiográfico do m. buccinator. O. Hospital (São Paulo), *68:* 105–117.

de Sousa, O. Machado and Vitti, M. (1966) Estudio electromiográfico de los músculos adductores largo y mayor (abstract). Arch. Mex. Anat., *7:* 52–53.

Deutsch, L. J. (1972) The threshold of the stapedius reflex for pure tone and noise stimuli. Acta Otolaryng., *74:* 248–251.

deVries, H. A. (1965) Muscle tonus in postural muscles. Am. J. Phys. Med., *44:* 275–291.

deVries, H. A. (1968) Method for evaluation of muscle fatigue and endurance from electromyographic fatigue curves. Am. J. Phys. Med., *47:* 125–135.

deVries, H. A., Burke, R. K., Hopper,

R. T. AND SLOAN, J. H. (1976) Relationship of resting EMG level to total body metabolism with reference to the origin of "tissue noise." Am. J. Phys. Med., *55:* 139–147.

DEVRIES, H., BURKE, R. K., HOPPER, R. T. AND SLOAN, J. H. (1977) Effect of EMG Biofeedback in Relaxation Training. Am. J. Phys. Med., *56:* 75–81.

DEWSON, J. H. III, DEMENT, W. C. AND SIMMONS, F. B. (1965) Middle ear muscle activity in cats during sleep. Exper. Neurol., *12:* 1–8.

DI BENEDETTO, A., SIEBENS, A. A., CINCOTTI, J. J., GRANT, A. R., AND GLASS, P. (1959) A study of diaphragmatic and crural innervation by the direct recording of action potentials in the dog. J. Thoracic & Cardiovas. Surg., *38:* 104–107.

DIETZ, V., BISCHOFBERGER, E., WITA, C. AND FREUND, H.-J. (1976) Correlation between the discharges of two simultaneously recorded motor units and physiological tremor. Electroenceph. Clin. Neurophysiol., *40:* 97–105.

DIXON, H. H. AND DICKEL, H. A. (1967) Tension headache. Northw. Med., *66:* 817–820.

DONISCH, E. W. AND BASMAJIAN, J. V. (1972) Electromyography of deep muscles in man. Am. J. Anat., *133:* 25–36.

DORAI RAJ, B. S. (1964) Diversity of crab muscle fibers innervated by a single motor axon. J. Cell. & Comp. Physiol., *64:* 41–54.

DOTY, R. W. AND BOSMA, J. F. (1956) An electromyographic analysis of reflex deglutition. J. Neurophysiol., *19:* 44–60.

DOWNIE, A. W. AND SCOTT, T. R. (1964) Radial nerve conduction studies. Neurology, *14:* 839–843.

DOYLE, A. M. AND MAYER, R. F. (1969) Studies of the motor unit in the cat. Bull. Sch. Med., Univ. Md., *54:* 11–17.

DRAPER, M. H., LADEFOGED, P. AND

WHITTERIDGE, D. (1957) Expiratory muscles involved in speech. J. Physiol., *138:* 17–25.

DUBO, H. I. C., PEAT, M., WINTER, D. A., QUANBURY, A. O., HOBSON, D. A., STEINKE, T. AND REIMER, G. (1976) Electromyographic temporal analysis of gait: normal human locomotion. Arch. Phys. Med. Rehab., *57:* 415–418.

DUCHENNE, G. B. A. (1867) *Physiologie des mouvements,* transl. by E. B. Kaplan (1949) (re-issued in 1959). W. B. Saunders Co., Philadelphia and London.

DUNN, H. G., BUCKLER, W. ST. J., MORRISON, G. C. E. AND EMERY, A. W. (1964) Conduction velocity of motor nerves in infants and children. Pediatrics, *34:* 708–727.

DUTHIE, H. L. AND WATTS, J. M. (1965) Contribution of the external anal sphincter to the pressure zone in the anal canal. Gut, *6:* 64–68.

DUTTA, C. R. AND BASMAJIAN, J. V. (1960) Gross and histological structure of the pharyngeal constrictors in the rabbit. Anat. Rec., *137:* 127–134.

DUVAL, A. (1975) Analyse functionelle des muscles de la paroi antero-laterale de l'abdomen par électromyocartographie. Bull. Assoc. Anat., *59:* 743–755.

EASON, R. G. (1960) Electromyographic study of local and generalized muscular impairment. J. Appl. Physiol., *15:* 479–482.

EASTMAN, M. C. AND KAMON (1976) Posture and subjective evaluation of flat and slanted desks. Human Factors, *18:* 15–26.

EBERHART, H. D., INMAN, V. T. AND BRESLER, B. (1954) The principal elements in human locomotion. In *Human Limbs and Their Substitutes.* ed. by P. E. Klopsteg and P. D. Wilson. McGraw-Hill Book Co., New York.

EBERHART, H. D., INMAN, V. T., SAUNDERS, J. B. deC. M., LEVENS, A. S., BRESLER, B. AND COWAN, T. D. (1947) Fundamental studies of human locomotion and other information related to the design of artificial limbs. A Report to the N.R.C. Committee on Artificial Limbs. University of California, Berkeley.

EBLE, J. N. (1961) Reflex relationships of paravertebral muscles. Am. J. Physiol., *200:* 939–943.

ECCLES, J. C. AND O'CONNOR, W. J. (1939) Responses which nerve impulses evoke in mammalian striated muscles. J. Physiol., *97:* 44–102.

EDELWEJN, Z. (1964) Effect of high temperature on muscle bioelectric activity. Acta Physiol. Polon., *15:* 433–439.

EDSTRÖM, L. AND KUGELBERG, E. (1968) Histological composition, distribution of fibres and fatiguability of single motor units. J. Neurol. Neurosurg. Psychiat., *31:* 424–433.

EDWARDS, R. G. AND LIPPOLD, O. C. (1956) The relation between force and integrated electrical activity in fatigued muscle. J. Physiol. *132:* 677–681.

EKSTEDT, J. (1964) Human single muscle fiber action potentials. Acta Physiol. Scandinav., *61:* suppl. 226, 96 pp.

EKSTEDT, J., LINDHOLM, B., LJUNGGREN, S. AND STÅLBERG, E. (1971) The jittermeter: a variability calculator for use in single fiber electromyography. Electroenceph. Clin. Neurophysiology, *30:* 154–158.

EKSTEDT, J., NILSSON, G. AND STÅLBERG, E. (1974). Calculation of the electromyographic jitter. J. Neurol. Neurosurg. Psychiat., *37:* 526–539.

ELBLE, R. J. AND RANDALL, J. E. (1976) Motor-unit activity responsible for 8- to 12-Hz component of human physiological finger tremor. J. Neurophysiol. *39:* 370–383.

ELFTMAN, H. (1939) The function of the arms in walking. Human Biol., *11:* 529–535.

ELFTMAN, H. (1966) Biomechanics of muscle: with particular application to

studies of gait. J. Bone & Joint Surg., *48-A:* 363–377.

ELKUS, R. AND BASMAJIAN, J. V. (1973) Endurance: why do people hanging by their hands let go? Am. J. Phys. Med., *52:* 124–127.

ELLIOTT, B. C. AND BLANKSBY, B. A. (1976) Reliability of averaged integrated electromyograms during running. J. Hum. Mov. Studies, *2:* 28–35.

EMANUEL, M. (1965) The pathophysiology of the urinary sphincter. Surg. Clin. N. A. *45:* 1467–1480.

ENG, G. D. (1976) Spontaneous potentials in premature and full-term infants. Arch. Phys. Med. Rehab. *57:* 120–121.

Engberg, I. (1964) Reflexes to foot muscles in the cat. Acta Physiol. Scandinav., *62:* suppl. 235, 64 pp.

ENGELHARDT, J. K., ISHIKAWA, K., LISBIN, S. J. AND MORI, J. (1976) Neurotrophic effects on passive electrical properties of cultured chick skeletal muscles. Brain Res., *110;* 170–174.

ERTEKIN, C. AND REEL, F. (1976) Bulbocavernosus reflex in normal men and in patients with neurogenic bladder and/or impotence. J. Neurol. Sc., *28:* 1–15.

FAABORG-ANDERSEN, K. L. (1957) Electromyographic investigation of intrinsic laryngeal muscles in humans: an investigation of subjects with normally movable vocal cords and patients with vocal cord paresis. Acta Physiol. Scandinav., *41:* suppl. 140, 1–148.

FAIRBANK, T. J. (1948) Fracture-subluxations of the shoulder. J. Bone & Joint Surg., *30-B:* 454–460.

FARRAR, W. B. (1976) Using electromyographic biofeedback in treating orofacial dyskinesia. J. Prosthet. Dent., *35:* 384–387.

FATT, P. (1959) Skeletal neuromuscular transmission, in *Handbook of Physiology, Section 1: Neurophysiology, Vol. 1.* American Physiological Society, Washington, 199–212.

FEINSTEIN, B., LINDEGÅRD, B., NYMAN, E. AND WOHLFART, G. (1955) Morphological studies of motor units in normal human muscles. Acta Anat., *23:* 127–142.

FELDKAMP, L., ABBINK, F. AND GÜTH, V. (1976) Elektromyographische studie des bewegungsverhaltens bei sänglingen mit zerebraler bewegungsstörung. Z. Orthop., *114:* 32–37.

Fényes, I., Gergely, Ch. and Tóth, Sz. (1960) Clinical and electromyographic studies of "Spinal Reflexes" in premature and full-term infants. J. Neurol. Neurosurg. & Psychiat., *23:* 63–68.

FERNANDEZ-BALLESTEROS, M. L., BUCHTHAL, F. AND ROSENFALCK, P. (1964) The pattern of muscular activity during the arm swing of natural walking. Acta Physiol. Scandinav., *63:* 296–310.

FERRARI, E. AND MESSINA, C. (1972) Blink reflexes during sleep and wakefulness in man. Electroenceph. Clin. Neurophysiol., *32:* 55–62.

FERRAZ, E. C. F., DE MORAES, J. L. AND PAROLARI, J. B. (1958) Atividade dos músculos fibulares longo e curto (Portuguese text). Folia Clin. Biol., *28:* 140–142.

FETZ, E. E. AND FINOCCHIO, D. V. (1971) Operant conditioning of neural and muscular activity. Science, *174:* 431–435.

FIBIGER, S. (1971) I. Communication disorders. A. Stuttering explained as a physiological tremor. Report from the Dept. of Speech Communication, Royal Instit. Technol. (KTH), Stockholm, privately printed.

FIBIGER, S. (1972) Further discussion on "stuttering explained as a physiological tremor." Report from the Dept. of Speech Communication, Royal Instit. Technol. (KTH) Stockholm, privately printed.

FICK, R. (1911) *Handbuch der Anatomie und Mechanik der Gelenke,* vol. 3. Gustav Fischer, Jena, Germany.

FINK, B. R. (1960) A method of monitoring muscular relaxation by the integrated abdominal electromyogram. Anesthesiology, *21:* 178–185.

FINK, B. R., HANKS, E. C., HOLADAY, D. A. AND NGAI, S. H. (1960) Monitoring of ventilation by integrated diaphragmatic electromyogram. J. A. M. A., *172:* 1367–1371.

FINLEY, F. R., CODY, K. A. AND FINIZIE, R. V. (1969) Locomotion pattern in elderly women. Arch. Phys. Med., *50:* 140–146.

FLECK, H. (1962) Action potentials from single motor units in human muscle. Arch. Phys. Med., *43:* 99–107.

FLINT, M. M. (1965a) Abdominal muscle involvement during performance of various forms of sit-up exercise. Am. J. Phys. Med., *44:* 224–234.

FLINT, M. M. (1965b) An electromyographic comparison of the function of the iliacus and the rectus abdominis muscles. A preliminary report. J. Am. Phys. Therap. Assoc., *45:* 248–253.

FLINT, M. M., DRINKWATER, B. L. AND MCKITTRICK, J. E. (1970) Shoulder dynamics subsequent to a radical mastectomy. Electromyography, *10:* 171–182.

FLINT, M. M. AND GUDGELL, J. (1965) Electromyographic study of abdominal muscular activity during exercise. Res. Quart., *36:* 29–37.

FLOYD, W. F., NEGUS, V. E. AND NEIL, E. (1957) Observations on the mechanism of phonation. Acta Oto-laryng., *48:* 16–25.

FLOYD, W. F. AND SILVER, P. H. S. (1950) Electromyographic study of patterns of activity of the anterior abdominal wall muscles in man. J. Anat., *84:* 132–145.

FLOYD, W. F. AND SILVER, P. H. S. (1951) Function of erector spinae in flexion of the trunk. Lancet, Jan. 20, 133–138.

FLOYD, W. F. AND SILVER, P. H. S. (1955) The function of the erectores spinae muscles in certain movements and postures in man. J. Physiol., *129:* 184–203.

FLOYD, W. F. AND WALLS, E. W. (1953) Electromyography of the sphincter ani externus in man. J. Physiol., *122:* 599–609.

FOLKINS, J. W. AND ABBS, J. H. (1975) Lip and jaw motor control during speech: responses to resistive loading of the jaw. J. Speech Hearing Res., *18:* 207–220.

FORBES, A. (1922) The interpretation of spinal reflexes in terms of present knowledge of nerve conduction. Physiol. Rev., *2:* 361–414.

FORREST, W. J. AND BASMAJIAN, J. V. (1965) Function of human thenar and hypothenar muscles: an electromyographic study of twenty-five hands. J. Bone & Joint Surg., *47-A:* 1585–1594.

FORREST, W. J. AND KHAN, M. A. (1968) Electromyography of the flexor pollicis brevis and adductor pollicis in twenty hands: a preliminary report. (Transactions of the Internat. Soc. EMG. Kines., Montreal 1968.) Electromyography *8:* suppl. 1, 49–53.

FOUNTAIN, F. P., MINEAR, W. L. AND ALLISON, R. D. (1966) Function of longus colli and longissimus cervicis muscles in man. Arch. Phys. Med., *47:* 665–669.

FRAME, J. W., ROTHEWELL, P. S. AND DUXBURY, A. J. (1973) The standardization of electromyography of the masseter muscle in man. Arch. Oral Biol., *18:* 1419–1423.

FRANKSSON, C. AND PETERSÉN, I. (1955) Electromyographic investigations of disturbances in the striated muscles of the urethral sphincter. Brit. J. Urol., *27:* 154–161.

FREEMAN, F. AND USHIJIMA, T. (1974) The stuttering larynx: an EMG, fiberoptic study of laryngeal activity accompanying the moment of stuttering. Status Report SR-41, Haskins Laboratories, New Haven, Connecticut.

FREEMAN, M. A. R. AND WYKE, B. (1966) Articular contributions to limb muscle reflexes: the effects of partial neurectomy of the knee-joint on postural reflexes. Brit. J. Surg., *53:* 61–69.

FREEMAN, M. A. R. AND WYKE, B. (1967) Articular reflexes at the ankle joint: an electromyographic study of normal and abnormal influences of ankle-joint mechanoreceptors upon reflex activity in the leg muscles. Brit. J. Surg., *54:* 990–1001.

FRENCKNER, B. AND EULER, C. V. (1975) Influence of pudendal block on the function of the anal sphincters. Gut, *16:* 482–489.

FRUEND, H.-J., BÜDINGEN AND DIETZ, V. (1975) Activity of single motor units from human forearm muscles during voluntary isometric contractions. J. Neurophysiol., *38:* 933–946.

FRIEDEBOLD, G. (1958) Die aktivität normaler Rückenstreckmuskulatur im Elektromyogramm unter verschiedenen Haltungsbedingungen; eine Studie zur Skelettmuskelmechanik (German text). Ztschr. Orthop., *90:* 1–18.

FRITZELL, B. (1963) An electromyographic study of the movements of the soft palate in speech. Folia Phoniat., *15:* 307–311.

FRITZELL, B. AND KOTBY, M. N. (1976) Observations on thyroarytenoid and palatal levator activation for speech. Folia Phonet., *28:* 1–7.

FRUHLING, M., BASMAJIAN, J. V. AND SIMARD, T. G. (1969) A note on the conscious controls of motor units by children under six. J. Motor Behav., *1:* 65–68.

FUCHS, P. (1975) The muscular activity of the chewing apparatus during night sleep. J. Oral Rehab., *2:* 35–48.

FUDEL-OSIPOVA, S. I., AND GRISHKO, F. E. (1962) Features specific to electromyograms taken during voluntary muscle contraction in old age. Biull. Eksp. Biol. Med., *3:* 9–14.

FUDEMA, J. J., FIZZELL, J. A. AND NELSON, E. M. (1961) Electromyography of experimentally immobilized skeletal muscles in cats. Am. J. Physiol., *200:* 963–967.

FUJIWARA, M. AND BASMAJIAN, J. V. (1975) Electromyographic study of two-joint muscles. Am. J. Phys. Med., *54:* 234–242.

FUNAKOSHI, M., FUJITA, N. AND TAKE- HANA, S. (1976) Relations between oc- clusal interference and jaw muscle ac- tivities in response to changes in head position. J. Dent. Res., *55:* 684–690.

FURLANI, J. (1976) Electromyographic study of the m. biceps brachii in move- ments of the glenohumeral joint. Acta Anat., *96:* 270–284.

FURLANI, J., BÉRZIN, F. AND VITTI, M. (1974) Electromyographic study of the gluteus maximus muscle. Electro- myogr. Clin. Neurophysiol., *14:* 377– 388.

FURLANI, J., VITTI, M. AND BÉRZIN, F. (1973) Estudo eletromiográfico do m. biceps femoral. Fol. Clin. Biol., *1:* 188–192.

FURLANI, J., VITTI, M. AND BÉRZIN, F. (1977) Musculus biceps femoris, long and short head: an electromyographic study. Electromyogr. Clin. Neurophys- iol., *17:* 13–19.

GAARDNER, K. (1971) Control of states of consciousness, Parts I and II. Arch. Gen. Psychiat., *25:* 429–435; 430–441.

GANS, C. AND HUGHES, G. M. (1967) The mechanism of lung ventilation in the tortoise *testudo graeca linné.* J. Exper. Biol., *47:* 1–20.

GARDNER, E., GRAY, D. J. AND O'RAHILLY, R. (1960) *Anatomy: a Re- gional Study of Human Structure.* W. B. Saunders Co., Philadelphia and Lon- don.

GARRITY, L. I. (1975) Measurement of subvocal speech: correlations between two muscle leads and between the recording methods. Percept. Motor Skills, *40:* 327:330.

GASSEL, M. M. (1964) Sources of error in motor nerve conduction studies. Neu-

rology, *14:* 825–835.

GASSEL, M. M. AND DIAMANTOPOULOS, E. (1964) Patterns of conduction times in the distribution of the radial nerve: a clinical and electrophysiological study. Neurology, *14:* 222–231.

GASSELL, M. M. AND TROJABORG, W. (1964) Clinical and electrophysiologi- cal study of the pattern of conduction times in the distribution of the sciatic nerve. J. Neurol. Neurosurg. & Psy- chiat., *27:* 351–357.

GASSER, H. S. AND NEWCOMER, H. S. (1921) Physiological action currents in the phrenic nerve. An application of the thermionic vacuum tube to nerve physiology. Am. J. Physiol., *57:* 1–26.

GATEV, V. (1967) Studies of the electrical activity of the antagonistic muscles of the arm in normal children aged be- tween 1 and 5 months. Compte Rend. l'Acad. Bulg. Sc., *20:* 743–747.

GATH, I. (1974) Analysis of point process signals applied to motor unit firing patterns, I. superposition of independ- ent spike trains. Math. Biosci., *22:* 211–222.

GAY, T. (1974) Some electromyographic measures of coarticulation in VCV ut- terances. Proc. Fifth Essex Phonetics Symposium, 1975, pp. 15–28.

GAY, T. AND HIROSE, H. (1973). Effect of speaking rate on labial consonant production: a combined electromy- ographic/high-speed motion picture study. Phonetica, *22:* 44–56.

GAY, T., STROME, M., HIROSE, H. AND SAWASHIMA, M. (1972) Electromyog- raphy of the intrinsic laryngeal mus- cles during phonation. Ann. Otol. Rhinol. Laryngol., *81:* 401–410.

GAY, T., USHIJIMA, T., HIROSE, H. AND COOPER, F. S. (1974) Effect of speaking rate on labial consonant-vowel articu- lation. J. Phonet., *2:* 47–63.

GEDDES, L. A. (1972) *Electrodes and the Measurement of Bioelectric Events,* John Wiley and Sons, Inc.

GELLHORN, E. (1947) Patterns of muscular activity in man. Arch. Phys. Med., *28:* 568–574.

GELLHORN, E. (1960) In *Science and Medicine of Exercise and Sports,* edited by W. R. Johnson, Chapter 7, pp. 108–122. Harper and Brothers, New York.

GERMANA, J. (1969) Patterns of autonomic and somatic activity during classical conditioning of a motor response. J. Compar. Physiol. Psychol., *69:* 173–178.

GETTRUP, E. (1966) Sensory regulation of wing twisting in locusts. Exper. Biol., *44:* 1–16.

GIBBS, C. H. (1975) Electromyographic activity during the motionless period of chewing. J. Prosth. Dent., *34:* 35–40.

GILLIATT, R. W. AND THOMAS, P. K. (1960) Changes in nerve conduction with ulnar lesions at the elbow. J. Neurol. Neurosurg. & Psychiat., *23:* 312–321.

GILLIES, J. D. (1972) Motor unit discharge patterns during isometric contraction in man. J. Physiol. (Lond.), *223:* 36–37P.

GILSON, A. S. AND MILLS, W. B. (1940) Single responses of motor units in consequence of volitional effort. Proc. Soc. Exper. Biol. & Med., *45:* 650–652.

GILSON, A. S. AND MILLS, W. B. (1941) Activities of single motor units in man during slight voluntary efforts. Am. J. Physiol., *133:* 658–669.

GIOVINE, G. P. (1959) Premesse al trattamento neurochirurgico della disfunzioni vescicali neurogene. I. Studi sulla funzione dello sfintere striato dell'uretra: l'elettrosfinterografia (Italian text). Chirurgia, *14:* 39–62.

GODAUX, E. AND DESMEDT, J. E. (1975a) Evidence for a monosynaptic mechanism in the tonic vibration reflex of the human masseter muscle. J. Neurol. Neurosurg. Psychiat., *38:* 161–168.

GODAUX, E. AND DESMEDT, J. E. (1975b) Exteroceptive suppression and motor control of the masseter and temporalis muscles in normal man. Brain Res., *85:* 447–458.

GODFREY, K. E., KINDIG, L. E. AND WINDELL, E. J. (1977) Electromyographic study of duration of muscle activity in sit-up variations. Arch. Phys. Med. Rehab., *58:* 132–135.

GOLDBERG, L. J. AND DERFLER, B. (1977) Relationship among recruitment order, spike amplitude, and twitch tension of single motor units in human masseter muscle. J. Neurophysiol., *40:* 879–890.

GOLSTEIN, I. BALSHAN. (1965) The relationship of muscle tension and autonomic activity to psychiatric disorders. Psychosom. Med., *27:* 39–52.

GOLDSTEIN, I. D. (see also Balshan, I. D.)

GOMEZ OLIVEROS, L. (1969) Estudios anatómicos y electromigráficos de la fonación. Acta Otorrinolar. Esp., *19(6):* 37–88.

GOSS, C. M. (1959) (editor) *Gray's Anatomy of the Human Body,* ed. 27. Lea & Febiger, Philadelphia.

GOTO, Y., KUMAMOTO, M. AND OKAMOTO, T. (1974) Electromyographic study of the function of the muscles participating in thigh elevation in the various planes. (Japanese text; English abstract). Res. J. Phys. Ed., *18:* 269–276.

GOTO, Y., MATSUSHITA, K., TSUJINO, A., AND OKAMOTO, T. (1976) Hadoru-Soh no Kineshiorogiteki Kohsatsu (A kinesiological study of the hurdle running). *The Science of Human Movement,* Kyorin Shorin, Tokyo, pp. 145–158.

GOUBEL, F. AND BOUISSET, S. (1967) Relation entre l'activité électromyographique intégrée et la travail mécanique effectué au cours d'un mouvement monoarticulaire simple. J. Physiol., *59:* 241.

GOUBEL, F., LESTIENNE, F. AND BOUISSET, S. (1968) Détermination dynamique de la compliance musculaire *in situ.* J. Physiol., *60:* 255.

GRANIT, R. (1958) Neuromuscular interaction in postural tone of the cat's isometric soleus muscle. J. Physiol., *143:* 387–402.

GRANIT, R. (1960) During discussion of his paper, Muscle tone and postural regulations, in "Muscle as a Tissue" International Conference, Lankenau Hospital, Philadelphia.

GRANIT, R. (1964) The gamma (γ) loop in the mediation of muscle tone. Clin. Pharmacol. & Therap., *5:* 837–847.

GRANIT, R., HENATSCH, H. D. AND STEG, G. (1956) Tonic and phasic ventral horn cells differentiated by post-tetanic potentiation in cat extensors. Acta Physiol. Scandinav., *37:* 114–126.

GRANIT, R., PHILLIPS, C. G., SKOGLUND, S. AND STEG, G. (1957) Differentiation of tonic from phasic alpha ventral horn cells by stretch, pinna and crossed extensor reflexes. J. Neurophysiol., *20:* 470–481.

GRANT, J. C. B. AND BASMAJIAN, J. V. (1965) *Grant's Method of Anatomy: By Regions Descriptive and Deductive*, ed. 7. The Williams & Wilkins Co., Baltimore.

GRANT, P. G. (1973) Lateral pterygoid: two muscles? Am. J. Anat., *138:* 1–10.

GRASSINO, A. E., WHITELAW, W. A. AND MILIC-EMILI, J. (1976). Influence of lung volume and electrode position on electromyography of the diaphragm. J. Appl. Physiol., *40:* 971–975.

GRAY, E. R. (1969) The role of leg muscles in variations of the arches in normal and flat feet. J. Am. Phys. Th. Assoc., *49:* 1084–1088.

GRAY, E. R. (1971a) Conscious control of motor units in a tonic muscle. Am. J. Phys. Med., *50:* 34–40.

GRAY, E. R. (1971b) Conscious control of motor units in a neuromuscular disorder. Electromyography, *11:* 515–517.

GRAY, E. R. AND BASMAJIAN, J. V. (1968) Electromyography and cinematography of leg and foot ("normal" and flat) during walking. Anat. Rec., *161:* 1–16.

GREEN, E. E., GREEN, A. M. AND WALTERS, E. D. (1970) Voluntary control of internal states: psychological and physiological. J. Transpersonal Psychol., *2:* 1–26.

GREEN, E. E., WALTERS, E. D., GREEN, A. M. AND MURPHY, G. (1969) Feedback technique for deep relaxation. Psychophysiology, *6:* 371–377.

GREEN, J. G. AND NEIL, E. (1955) The respiratory function of the laryngeal muscles. J. Physiol., *129:* 134–141.

GREENLAW, R. K. (1973) *Function of Muscles About the Hip During Normal Level Walking*. Ph.D. Thesis, Queen's University, Canada.

GREENWOOD, R. AND HOPKINS, A. (1976a) Muscle responses during sudden falls in man. J. Physiol., (Lond.) *254:* 507–518.

GREENWOOD, R. AND HOPKINS, A. (1976b) Landing from an unexpected fall and a voluntary step. Brain, *99:* 375–386.

GREENWOOD, R. AND HOPKINS, A. (1977) Monosynaptic reflexes in falling man. J. Neurol. Neurosurg. Psychiat., *40:* 448–454.

GREGG, R. A., MASTELLONE, A. F. AND GERSTEN, J. W. (1957) Cross exercise—a review of the literature and study utilizing electromyographic techniques. Am. J. Phys. Med., *36:* 269–280.

GRESCZYK, E. G. (1965) *Electromyographic Study of the Effect of Leg Muscles on the Arches of the Normal and Flat Foot.* Thesis for Master of Science Degree. University of Vermont, U. S. A.

GRIFFIN, C. J. AND MUNRO, R. R. (1969) Electromyography of the jaw-closing muscles in the open-close-clench cycle in man. Arch. Oral Biol., *14:2:* 141–150.

GRIM, P. (1971) Anxiety change produced by self-induced muscle tension and by relaxation with respiration feedback. Behavior Therapy, *2:* 11–17.

GRIMBY, L. (1963a) Normal plantar response: integration of flexor and extensor reflex components. J. Neurol. Neurosurg. & Psychiat., *26:* 39–50.

GRIMBY, L. (1963b) Pathological plantar response: disturbances of the normal integration of flexor and extensor reflex components. J. Neurol. Neurosurg. & Psychiat., *26:* 314–321.

GRIMBY, L. AND HANNERZ, J. (1968) Recruitment order of motor units on voluntary contraction: changes induced by proprioceptive afferent activity. J. Neurol. Neurosurg. Pyschiat., *31:* 565–573.

GRIMBY, L. AND HANNERZ, J. (1970) Differences in recruitment order of motor units in phasic and tonic flexion reflex in 'spinal man.' J. Neurol. Neurosurg. Psychiat., *33:* 562–570.

GRIMBY, L. AND HANNERZ, J. (1974a) Differences in recruitment order and discharge pattern of motor units in the early and late flexion reflex components in man. Acta Physiol. Scandinav. *90:* 555–564.

GRIMBY, L. AND HANNERZ, J. (1974b) Disturbances in the voluntary recruitment order of anterior tibial motor units in bradykinesia of Parkinsonism. J. Neurol. Neurosurg. Psychiat., *37:* 47–54.

GRIMBY, L. AND HANNERZ, J. (1977) Firing rate and recruitment order of toe extensor motor units in different modes of voluntary contraction. J. Physiol., *264:* 865–879.

GRIMBY, L., HANNERZ, J. AND RÅNLUND, T. (1974) Disturbances in the voluntary recruitment order of anterior tibial motor units in spastic paraparesis upon fatigue. J. Neurol. Neurosurg. Psychiat., *37:* 40–46.

GRØNBAEK, P. AND SKOUBY, A. P. (1960) The activity pattern of the diaphragm and some muscles of the neck and trunk in chronic asthmatics and normal controls: a comparative electromyographic study. Acta Med. Scandinav., *168:* 413–425.

GROSSMAN, W. I. AND WEINER, H. (1966) Some factors affecting the reliability of surface electromyography. Psychosom. Med., *28:* 78–83.

GRUNDY, M., TOSH, P. A., McLEISH, R. D. AND SMIDT, L. (1975) An investigation of the centres of pressure under the foot while walking. J. Bone & Joint Surg., *57-B;* 98–103.

GRYNBAUM, B. B., BRUDNY, J., KOREIN, J. AND BELANDRES, P. V. (1976) Sensory feedback therapy for stroke patients. Geriatrics, *31:* 43–47.

GUITAR, B. (1975) Reduction of stuttering frequency using analog electromyographic feedback. J. Speech & Hearing Res., *18:* 672–685.

GULD, C., ROSENFALCK, A. AND WILLISON, R. G. (1970) Report of the committee on emg instrumentation. Electroenceph. Clin. Neurophysiol., *28:* 399–413.

GURFINKEL', V. S., IVANOVA, A. N., KOTS, Y. M., PYATETSKII-SHAPIRO, I. M. AND SHIK, M. L. (1964) Quantitative characteristics of the work of motor units in the steady state. Biofizika, *9(5):* 636–638.

GURFINKEL', V. S. AND LEVIK, Y. S. (1976) Forming an unfused tetanus. Human Physiol. (Russian), *2:* 914–924.

GURKOW, H. J. AND BAST, T. H. (1958) Innervation of striated skeletal muscle. Am. J. Phys. Med., *37:* 269–277.

GUTTMANN, L. AND SILVER, J. R. (1965) Electromyographic studies on reflex activity of the intercostal and abdominal muscles in cervical cord lesions. Paraplegia, *3:* 1–22.

GYDIKOV, A. AND KOSAROV, D. (1972) Studies on the activity of alpha motoneurons in man by means of a new eletromyographic method, neurophysiology studied in man. Exerpta Medica, Amsterdam: 321–329.

GYDIKOV, A. AND KOSAROV, D. (1974a) Influence of various factors on the length of the summated depolarized area of the muscle fibres in voluntary activating of motor units and in electrical stimulation. Electromyogr. Clin.

Neurophysiol., *14:* 79–93.

GYDIKOV, A. AND KOSAROV, D. (1974b) Some features of different motor units in human biceps brachii. Pflugers Arch., *347:* 75–88.

HAGBARTH, K.-E., HELLSING, G. AND LÖFSTEDT, L. (1976) TVR and vibration-induced timing of motor impulses in the human jaw elevator muscles. J. Neurol. Neurosurg. Psychiat., *39:* 719–728.

HAINES, R. W. (1932) The laws of muscle and tendon growth. J. Anat., *66:* 578–585.

HAINES, R. W. (1934) On muscles of full and of short action. J. Anat., *69:* 20–24.

HÅKANSSON, C. H. (1956) Conduction velocity and amplitude of the action potential as related to circumference in the isolated fibre of frog muscle. Acta Physiol. Scandinav., *37:* 14–34.

HÅKANSSON, C. H. (1957a) Action potentials recorded intra- and extracellularly from the isolated frog muscle fibre in Ringer's solution and in air. Acta Physiol. Scandinav., *39:* 291–318.

HÅKANSSON, C. H. (1957b) Action potential and mechanical response of isolated cross striated frog muscle fibres at different degrees of stretch. Acta Physiol. Scandinav., *41:* 199–216.

HÅKANSSON, C. H. AND TOREMALM, N. G. (1967) Studies of the physiology of the trachea, part IV. Ann. Otolar. Rhinol. Laryngol., *76:* 873–885.

HALLÉN, L. G., AND LINDAHL, O. (1967) Muscle function in knee extension: an emg study. Acta Orthop. Scand. *38:* 434–444.

HALLETT, M., SHAHANI, B. T. AND YOUNG, R. R. (1975) EMG analysis of stereotyped voluntary movements in man. J. Neurol. Neurosurg. Psychiat., *38:* 1154–1162.

HAMILTON, W. J. AND APPLETON, A. B. (1956) in *Textbook of Human Anatomy*, by J. D. Boyd, W. E. Le Gros Clark, W. J. Hamilton, J. M. Yoffey, S. Zuckerman and A. B. Appleton, p. 206. Macmillan & Co. Ltd., London.

HANNAM, A. G. (1972) Effect of voluntary contraction of the masseter and other muscles upon the masseteric reflex in man. J. Neurol. Neurosurg. Psychiat., *35:* 66–71.

HANNAM, A. G. (1976) Computer analysis of the correlation between the activity of the masseter muscles during unilateral chewing in man. Electromyogr. Clin. Electrophysiol., *16:* 165–175.

HANNERZ, J. (1973) Discharge properties of motor units in man. Experientia, *29:* 45–46

HANNERZ, J. (1974) An electrode for recording single motor unit activity during strong muscle contractions. Electroenceph. Clin. Neurophysiol., *37:* 179–181.

HANSON, R. J., JR., SUSSMAN, H. M. AND MacNEILAGE, P. F. (1971) Single motor unit potentials in speech musculature. In *Proc. VII Intern. Congr. Phon. Sc.*, The Hague, Mouton, pp. 316–319.

Hardy, R. H. (1959) A method of studying muscular activity during walking. Med. & Biol. Illustration, *9:* 158–163.

HARDYCK, C. D., PETRINOVICH, L. F. AND ELLSWORTH, D. W. (1966) Feedback of speech muscle activity during silent reading; rapid extinction. Science, *154:* 67–1468.

HARRIS, K. S. (1971) Action of the extrinsic musculature in the control of tongue position: preliminary report. Report SR-25/26, pp 87–96, Haskins Laboratories, New Haven, Connecticut.

HARRIS, K. S., ROSOV, R., COOPER, F. S. AND LYSAUGHT, G. F. (1964) A multiple suction electrode system. Electroencephalog. & Clin. Neurophysiol., *17:* 698–700.

HARRIS, R. I., AND BEATH, T. (1948) Hypermobile flat-foot with short tendo achillis. J. Bone & Joint Surg., *30-A:* 116–140.

HARRISON, V. F. AND KOCH, W. B. (1972)

Voluntary control of single motor unit activity in the extensor digitorum muscle. Physical Ther., *52:* 267–272.

HARRISON, V. F. AND MORTENSEN, O. A. (1962) Identification and voluntary control of single motor unit activity in the tibialis anterior muscle. Anat. Rec., *144:* 109–116.

HART, B. L. AND KITCHELL, R. L. (1966) Penile erection and contraction of penile muscles in the spinal and intact dog. Am. J. Physiol., *210:* 257–262.

HARVEY, A. M. AND MASLAND, R. L. (1941) Method for study of neuromuscular transmission in human subjects. Bull. Johns Hopkins Hosp., *68:* 81–93.

HASKELL, B. AND ROVNER, H. (1970) Electromyography in the management of the incompetent anal sphincter. Dis. Colon & Rectum, *10:* 81–84.

HAYES, K. J. (1960) Wave analyses of tissue noise and muscle action potentials. J. Appl. Physiol., *15:* 749–752.

HEFFERLINE, R. F. AND PERERA, T. B. (1963) Proprioceptive discrimination of a covert operant without its observation by the subject. Science, *139:* 834–835.

HELLEBRANDT, F. A., HOUTZ, S. J., PARTRIDGE, M. J. AND WALTERS, C. E. (1956) Tonic neck reflexes in exercises of stress in man. Am. J. Phys. Med., *35:* 144–159.

HELLEBRANDT, F. A. AND WATERLAND, J. C. (1962a) Indirect learning: the influence of unimanual exercise on related muscle groups of the same and opposite side. Am. J. Phys. Med., *41:* 45–55.

HELLEBRANDT, F. A. AND WATERLAND, J. C. (1962b) Expansion of motor patterning under exercise stress. Am. J. Phys. Med., *41:* 56–66.

HENDERSON, R. L. (1952) Remote action potentials at the moment of response in a simple reaction-time situation. J. Exper. Psychol., *44:* 238–241.

HENNEMAN, E., SOMJEN, G. AND CARPENTER, D. O. (1965) Excitability and inhibitibility of motoneurons of different sizes. J. Neurophysiol., *28:* 599–620.

HENSON, O. W., JR. (1965) The activity and function of the middle-ear muscles in echo-locating bats. J. Physiol., *180:* 871–887.

HERBERTS, P. AND KADEFORS, R. (1976) A study of painful shoulder in welders. Acta Orthop. Scandinav., *47:* 381–387.

HERBERTS, P., KAISER, E., MAGNUSSON, R. AND PETERSÉN, I. (1969) Power spectra of myoelectric signals in muscles of arm amputees and healthy normal controls. Acta Orthop. Scandinav., *39:* 1–32.

HERMAN, R. AND BRAGIN, S. J. (1967) Function of gastrocnemius and soleus muscles. J. Am. Phys. Ther. Assoc., *47:* 105–113.

HERMANN, G. W. (1962) An electromyographic study of selected muscles involved in the shot put. Res. Quart., *33:* 1–9.

HERSHLER, C. AND MILNER, M. (1976) Kinematic analysis of gait by stroboscopic flash photography and computer. *Human Engineering Program Progress Report No. 1,* Chedoke Hospitals, Hamilton, Canada.

HICKS, J. H. (1951) The function of the plantar aponeurosis. J. Anat., *85:* 414–415.

HICKS, J. H. (1954) The mechanics of the foot. II. The plantar aponeurosis and the arch. J. Anat., *88:* 25–31.

HINSON, M. M. (1969) An electromyographic study of the push-up for women. Res. Quart., *40:* No. 2.

HIROSE, H. AND GAY, T. (1972) The activity of the intrinsic muscles in voicing control: an electromyographic study. Phonetica, *25:* 140–160.

HIROSE, H. AND GAY, T. (1973) Laryngeal control in vocal attack: an electromyographic study. Folio Phoniat., *25:* 203–213.

HIROSE, K., UONO, M. AND SOBUE, I. (1974) Quantitative electromyography comparison between manual values and computer ones on normal subjects. Electromyogr. Clin. Neurophysiol., *14:* 315–320.

HIROTO, I., HIRANO, M., TOYOZUMI, Y., AND SHIN, T. (1967) Electromyographic investigation of the intrinsic laryngeal muscles related to speech sounds. Otol. Rhinol. Laryngol., *76:* 861–873.

HIRSCHBERG, G. G. (1957) Electromyographic evidence of the role of intercostal muscles in breathing. News Letter, Am. A. of EMG and Electrodiag., *4:* 2–3.

HIRSCHBERG, G. G., ADAMSON, J. P., LEWIS, L. AND ROBERTSON, K. J. (1962) Patterns of breathing of patients (with) poliomyelitis and respiratory paralysis. Arch. Phys. Med., *43:* 529–533.

HIRSCHBERG, G. G. AND DACSO, M. M. (1953) The use of electromyography in the study of clinical kinesiology of the upper extremity. Am. J. Phys. Med., *32:* 13–21.

HIRSCHBERG, G. G. AND NATHANSON, M. (1952) Electromyographic recording of muscular activity in normal and spastic gaits. Arch. Phys. Med., *33:* 217–225.

HISHIKAWA, Y., SUMITSUJI, N., MATSUMOTO, K. AND KANEKO, Z. (1965) H-reflex and EMG of the mental and hyoid muscles during sleep, with special reference to narcolepsy. Electroencephalog. & Clin. Neurophysiol., *18:* 487–492.

HIXON, T. J., SIEBENS, A. A., AND MINIFIE, F. D. (1969) An EMG electrode for the diaphragm. J. Acoust. Soc. Am., *46:* 1588–1589.

HOBART, D. J., KELLEY, D. L. AND BRADLEY, L. S. (1975) Modification occurring during acquisition of a novel throwing task. Am. J. Phys. Med., *54:* 1–24.

HODES, R., GRIBETZ, L., MOSKOWITZ, J. A. AND WAGMAN, I. H. (1965) Low threshold associated with slow conduction velocity. Arch. Neurol., *12:* 510–526.

HODES, R., LARRABEE, M. G. AND GERMAN, W. (1948) The human electromyogram in response to nerve stimulation and the conduction velocity of motor axons. Arch. Neurol. & Psychiat., *60:* 340–365.

HOEFER, P. F. A. (1952) Physiological mechanisms in spasticity. Brit. J. Phys. Med., n.s. *15:* 88–90.

HOF, A. L. AND VAN DEN BERG J. W. (1977) Linearity between the weighted sum of the EMGs of the human triceps surae and the total torque. J. Biomech., *10:* 529–539.

HOGAN, N. J. (1976) Myoelectric prosthesis control: optimal estimation applied to EMG and the cybernetic considerations for its use in a man-machine interface. Ph.D. Dissertation, M.I.T., Cambridge, MA, U.S.A.

HUNTINGTON, D. A., HARRIS, K. S. AND SHOLES, G. N. (1968) An electromyographic study of consonant articulation in hearing-impaired and normal speakers. J. Speech Hearing Res., *11:* 147–158.

HUSSON, R. (1950) Etude des phénomenes physiologiques et acoustiques fondamentaux de la voix chantée. Thesis, Faculty of Sciences, Paris.

HUSTERT, R. (1975) Neuromuscular coordination and proprioceptive control of rhythmical abdominal ventilation in intact *locusta migratoria migratorioides*. J. Comp. Physiol., *97:* 159–179.

HOGUE, R. E. (1969) Upper-extremity muscular activity at different cadences and inclines in normal gait. J. Am. Phys. Therap. Assoc., *49:* 963–972.

HOLLIDAY, T. A., VAN METER J. R., JULIAN, L. M. AND ASMUNDSON, V. S. (1965) Electromyography of chickens with inherited muscular dystrophy. Am. J. Physiol., *209:* 871–876.

HOLLINSHEAD, W. H. (1958) *Anatomy for Surgeons,* vol. 3, p. 388. Hoeber-Harper, New York.

HOLT, K. S. (1966) Facts and fallacies about neuromuscular function in cerebral palsy as revealed by electromyography. Develop. Med. Child Neurol., *8:* 255–268.

HOOGMARTENS, M. J. AND BASMAJIAN, J. V. (1976) Postural tone in the deep spinal muscles of idiopathic scoliosis patients and their siblings: an etiologic study based on vibration-induced electromyography. Electromyogr. Clin. Neurophysiol., *16:* 93–114.

HOOVER, F. (1922) The functions and integration of the intercostal muscles. Arch. Int. Med., *30:* 1–33.

HOPF, H. C., SCHLEGEL, H. J. AND LOWITZSCH, K. (1974) Irridiation of voluntary activity to the contralateral side in movements of normal subjects and patients with central motor disturbances. Europ. Neurol., *12:* 142–147.

HORN, C. V. (1969) Electromyographic investigation of muscle imbalance in patients with paralytic scoliosis. Electromyography, *9:* 447–455.

HOSHIKAWA, T., MATSUI, H. AND MIYASHITA, M. (1973) In *Medicine and Sport, vol. 8: Biomechanics III,* edited by E. Jokl, pp. 342–348, Karger, Basel.

HOSHIKO, M. (1960) Sequence of action of breathing muscles during speech. J. Speech & Hearing Res., *3:* 291–297.

HOSHIKO, M. (1962) Electromyographic investigation of the intercostal muscles during speech. Arch. Phys. Med., *43:* 115–119.

HOUTZ, S. J. AND FISCHER, F. J. (1959) An analysis of muscle action and joint excursion during exercise on a stationary bicycle. J. Bone & Joint Surg., *41-A:* 123–131.

HOUTZ, S. J. AND FISCHER, F. J. (1961) Function of leg muscles acting on foot as modified by body movements. J. Appl. Physiol., *16:* 597–605.

HOUTZ, S. J. AND WALSH, F. P. (1959) Electromyographic analysis of the function of the muscles acting on the ankle during weight-bearing with special reference to the triceps surae. J. Bone & Joint Surg., *41-A:* 1469–1481.

HOYLE, G. (1964) Exploration of neuronal mechanisms underlying behaviour in insects. In *Neural Theory and Modeling,* ed. R. Reiss; p. 346 to 376. Stanford University Press.

HOYLE, G. AND WILLOWS, A. O. D. (1973) Neuronal basis of behavior in *tritonia.* J. Neurobiol., *4:* 239–254.

HRYCYSHYN, A. W. AND BASMAJIAN, J. V. (1972) Electromyography of the oral stage of swallowing in man. Am. J. Anat., *133:* 333–340.

HUBBARD, A. W. (1960) In *Science and Medicine of Exercise and Sports,* edited by W. R. Johnson, Chapter 2, pp. 7–39. Harper and Brothers, New York.

HUGHES, G. M. AND BALLINTIJN, C. M. (1968) Electromyography of the respiratory muscles and gill water flow in the dragonet. J. Exp. Biol., *49:* 583–602.

HUTCH, J. A. AND ELLIOTT, H. W. (1968) Electromyographic study of electrical activity in the paraurethral muscles prior to and during voiding. J. Urol., *99:* 759–765.

IHRE, T. (1974) Studies on anal function in continent and incontinent patients. Scand. J. Gastroent., Suppl. 25, 64 pp.

IIDA, M. AND BASMAJIAN, J. V. (1974) Electromyography of hallux valgus. Orthopaed. Rel. Res., *101:* 220–224.

IIDA, M. AND BASMAJIAN, J. V. (1975) Electromyography of plantaris muscle. Electromyogr. Clin. Neurophysiol., *15:* 311–316.

IIDA, M., VIEL, E., IWASAKI, T., ITO, H. AND YAZAKI, K. (1976) Activité E.M.G. des muscles superficiels et profonds du dos pendant les exercices de rééducation couramment utilisés. Electrodiagnostic-therapie, *13:* 55–67.

IKAI, M. (1956) Crossed reflexes of limbs observed in healthy man. Jap. J. Physiol., *6:* 29–39.

INGLIS, J., CAMPBELL, D. AND DONALD, M. W. (1976a) Electromyographic biofeedback and neuromuscular rehabilitation. Can. J. Behav. Sci., *8:* 299–323.

INGLIS, J., SPROULE, M., LEICHT, M., DONALD, M. W. AND CAMPBELL, D. (1976b) Electromyographic biofeedback treatment of residual neuromuscular disabilities after cerebrovascular accident. *28:* 260–264.

INMAN, V. T. (1947) Functional aspects of the abductor muscles of the hip. J. Bone & Joint Surg., *29:* 607–619.

INMAN, V. T., RALSTON, H. J., SAUNDERS, J. B. DEC. M., FEINSTEIN, B. AND WRIGHT, E. W., JR. (1951) Relation of human electromyogram to muscular tension, Advisory Committee on Artificial Limbs, N. R. C., Series 11, Issue 18.

INMAN, V. T., RALSTON, H. J., SAUNDERS, J. B. DEC. M., FEINSTEIN, B. AND WRIGHT, E. W., JR. (1952) Relation of human electromyogram to muscular tension. Electroencephalog. & Clin. Neurophysiol., *4:* 187–194.

INMAN, V. T., SAUNDERS, J. B. DEC. M. AND ABBOTT, L. C. (1944) Observations on the function of the shoulder joint. J. Bone & Joint Surg., *26:* 1–30.

INNOCENT, G. M. (1971) Electrical activity of single extraocular muscles during the oculocompensatory positions. Electromyography, *11:* 25–38.

INOUYE, T. AND SHIMIZU, A. (1970) The electromyographic study of verbal hallucination. J. Nerv. Ment. Dis., *151:* 415–422.

ISAKSSON, I., JOHANSON, B., PETERSÉN, I. AND SELLDEN, U. (1962) Electromyographic study of the Abbe- and fan flaps. Acta Chir. Scandinav., *123:* 343–350.

ISHIDA, H., KIMURA, T. AND OKADA, M. (1974) Symp. 5th Cong. Int. Primat. Soc. Japan Science Press, Tokyo.

ISLEY, C. L. AND BASMAJIAN, J. V. (1973) Electromyography of the human

cheeks and lips. Anat. Rec., *176:* 143–148.

ISMAIL, A. H., BARANY, J. W. AND MANNING, K. R. (1965) *Assessment and Evaluation of Hemiplegic Gait.* Technical Report for the National Instit. Health, Purdue University, Lafayette, Indiana.

ITO, H., IWASAKI, T., YAMADA, M., YASAKI, K., TANAKA, S. AND IIDA, M. (1976) Electromyography of latissimus dorsi muscle. J. Jap. Ph. Assoc., *3:* 23–36.

JACOB, P. P., HARIDAS, R. AND AMMAL, P. J. (1971) An electromyographic study of the behaviour of orbicularis oris and mentalis muscle. Indian J. Med. Res., *59:* 311–320.

JACOBS, A. AND FENTON, G. S. (1969) Visual feedback of myoelectric output to facilitate muscle relaxation in normal persons and patients with neck injuries. Arch. Phys. Med., *50:* 34–39.

JACOBS, L., FELDMAN, M. AND BENDER, M. B. (1971) Eye movements during sleep. I. The pattern in the normal human. Arch. Neurol., *25:* 151–159.

JACOBS, M. B. (1976) *Antagonist EMG Temporal Patterns During Rapid Voluntary Movement.* Ph.D. Dissertation, University of Toledo, Ohio

JACOBSON, A., KALES, A., LEHMANN, D. AND HOEDEMAKER, F. S. (1964) Muscle tonus in human subjects during sleep and dreaming. Exper. Neurol., *10:* 418–424.

JACOBSON, E. (1929) *Progressive Relaxation.* University of Chicago Press, Chicago, Ill.

JACOBSON, E. (1933) Electrical measurements concerning muscular contraction (tonus) and the cultivation of relaxation in man: studies on arm flexors. Am. J. Physiol., *107:* 230–248.

JAMPOLSKY, A. (1970) What can electromyography do for the ophthalmologist? Invest. Ophthalmol., *9:* 570–599.

JAMPOLSKY, A., TAMLER, E. AND MARG, E. (1959) Artifacts and normal variations in human ocular electromyography. A.M.A. Arch. Ophth., *61:* 402–413.

JANDA, V. AND KOZÁK, P. (1964) Zur Funktion der motorischen Einheit unter Ischämie. Deutsch Ztschr. Nervenh., *185:* 598–605.

JANDA, V. AND STARÁ, V. (1965) The role of thigh adductors in movement patterns of the hip and knee joint. Courrier (Centre Internat. de l'Enfance), *15:* 1–3.

JANDA, V. AND VÉLE, F. (1963) A polyelectromyographic study of muscle testing with special reference to fatigue. Proc. of IX World Rehab. Congress, Copenhagen, pp. 80–84.

JARCHO, L. W., EYZAGUIRRE, C., BERMAN, B. AND LILIENTHAL, J. L., JR. (1952) Spread of excitation in skeletal muscle: some factors contributing to the form of the electromyogram. Am. J. Physiol., *168:* 446–457.

JARCHO, L. W., VERA, C. L., McCARTHY, C. G., AND WILLIAMS, P. M. (1958) The form of motor-unit and fibrillation potentials. Electroencephalog. & Clin. Neurophysiol., *10:* 527–540.

JASPER, H. H. AND BALLEM, G. (1949) Unipolar electromyograms of normal and denervated human muscle. J. Neurophysiol., *12:* 231–244.

JASPER, H. H. AND FORDE, W. O. (1947) The R. C. A. M. C. electromyograph mark III. Canad. J. Res., *25:* 100–110.

JEFFERSON, N. C., OGAWA, T., SYLEOS, C., ZAMBETOGLOU, A. AND NECHELES, H. (1960) Restoration of respiration by nerve anastomosis. Am. J. Physiol., *198:* 931–933.

JEFFERSON, N. C., PHILLIPS, C. W. AND NECHELES, H. (1949) Observations on diaphragm and stomach of the dog following phrenicotomy. Proc. Soc. Exper. Biol. & Med., *72:* 482–485.

JEN, P. H.-S. AND SUGA, N. (1976) Coordinated activities of middle-ear and laryngeal muscles in echolocating bats. Science, *191:* 950–952.

JENERICK, H. (1964) An analysis of the striated muscle fibre action current. Biophysical J., *4:* 77–91.

JESEL, M., ISCH TREUSSARD, C. AND ISCH, F. (1970) EMG of the anal and urethral sphincters in the diagnosis of lesions of the cauda equina and lumbar section of the spinal cord. Rev. Neurol., *122:* 431–434.

JOHNSON, C. E., BASMAJIAN, J. V. AND DASHER, W. (1972) Electromyography of sartorius muscle. Anat. Rec., *173:* 127–130.

JOHNSON, C. P. (1976) Analysis of five tests commonly used in determining the ability to control single motor units. Am. J. Phys. Med., *55:* 113–121.

JOHNSON, D. R. (1970) *An Electromyographic Study of Extrinsic and Intrinsic Muscles of the Thumb.* M.Sc. Thesis. Queen's University, Canada.

JOHNSON, D. R. AND FORREST, W. J. (1970) An electromyographic study of the abductors and flexors of the thumb in man (abstract). Anat. Rec., *166:* 325.

JOHNSON, E. W. AND OLSEN, K. J. (1960) Clinical value of motor nerve conduction velocity determination. J. A. M. A., *172:* 2030–2035.

JOHNSON, H. E. AND GARTON, W. H. (1973) Muscle re-education in hemiplegia by use of electromyographic device. Arch. Phys. Med., *54:* 320–322.

JOHNSTON, R. AND LEE, K.-H. (1976) Myofeedback: a new method of teaching breathing exercises in emphysematous patients. Phys. Ther., *56:* 826–829.

JOHNSTON, T. B., DAVIES, D. V. AND DAVIES, F. (1958) *Gray's Anatomy: Descriptive and Applied,* ed. 32. Longmans, Green and Co., New York, Toronto, London.

JOHNSTON, T. B. AND WHILLIS, J. (1954) (editors) *Gray's Anatomy: Descriptive and Applied,* ed. 31. Longmans, Green and Co., London.

JONES, D. S., BEARGIE, R. J. AND PAULY, J. E. (1953) An electromyographic study of some muscles of costal respiration in man. Anat. Rec., *117:* 17–24.

JONES, D. S. AND PAULY, J. E. (1957) Further electromyographic studies on muscles of costal respiration in man. Anat. Rec., *128:* 733–746.

JONES, F. W. (1942) *The Principles of Anatomy as Seen in the Hand,* ed. 2, p. 258–259. Baillière, Tindall and Cox, London.

JONES, F. W. (1949) *Structure and Function as Seen in the Foot,* ed. 2, p. 246–265. Baillière, Tindall & Cox, London.

JONES, G. M. AND WATT, D. G. D. (1971) Observations on the control of stepping and hopping movements in man. J. Physiol., *219:* 709–727.

JONES, R. L. (1941) The human foot. An experimental study of its mechanics, and the role of its muscle and ligaments in the support of the arch. Am. J. Anat., *68:* 1–39.

JONES, R. L. (1945) The functional significance of the declination of the axis of the subtalar joint. Anat. Rec., *93:* 151–159.

JONSSON, B. (1970) The functions of individual muscles in the lumbar part of the erector spinae muscle. Electromyography, *10:* 5–21.

JONSSON, B. (1973) Electromyography of the erector spinae muscle. In *Medicine and Sport, vol. 8: Biomechanics III,* ed. by E. Jokl, pp. 294–300, Karger, Basel.

JONSSON, B. AND BAGGE, U. E. (1968) Displacement, deformation and fracture of wire electrodes for electromyography. Electromyography, *8:* 328–347.

JONSSON, B. AND HAGBERG, M. (1974) The effect of different working heights on the deltoid muscle: a preliminary methodological study. Scand. J. Rehab. Med. *3:* suppl. 26–32.

JONSSON, S. AND JONSSON, B. (1975) Function of the muscles of the upper limb in car driving, Part I. The deltoid muscle; Part II. The trapezius muscle; Part III. The brachialis, brachioradialis, biceps brachii and triceps brachii muscles. IV. The pectoralis major, serratus anterior and latissimus dorsi muscles. Ergonomics, *18:* 375–388 & 643–649.

JONSSON, S. AND JONSSON, B. (1976) Function of the muscles of the upper limb in car driving. Part V. The supraspinatus, infraspinatus, teres minor and teres major muscles. Ergonomics, *19:* 711–717.

JONSSON, B., OLOFSSON, B. M. AND STEFFNER, L. CH. (1972) Function of the teres major, latissimus dorsi and pectoralis major muscles: a preliminary study. Acta Morpol. Neerl.-Scand., *9:* 275–280.

JONSSON, B. AND REICHMANN, S. (1968) Reproducibility in kinesiologic EMG-investigations with intramuscular electrodes. Acta Morph. Neerl.-Scand., *7:* 73–90.

JONSSON, B. AND REICHMANN, S. (1970) Radiographic control in the insertion of emg electrodes in the lumbar part of the erector spinae muscle. Z. Anat. Entwickl.-Gesch., *130:* 192–206.

JONSSON, B. AND RUNDGREN, A. (1971) The peroneus longus and brevis muscles: a roentgenologic and electromyographic study. Electromyography, *11:* 93–103.

JONSSON, B. AND STEEN, B. (1966) Function of the gracilis muscle. An electromyographic study. Acta Morphol. Neer-Scandinav., *4:* 325–341.

JONSSON, B. AND SYNNERSTAD, B. (1967) Electromyographic studies of muscle function in standing: a methodological study. Acta Morphol. Neerl.-Scand., *6:* 361–370.

JOSEPH, J. (1960) *Man's Posture: Electromyographic Studies,* 88 pp. Charles C Thomas, Springfield, Ill.

JOSEPH, J. (1965) Electromyography of posture and gait in man (abstract). Bull. Am. Assoc. EMG and Electrodiag., *12:* 24.

JOSEPH, J. (1968) The pattern of activity of some muscles in women walking in high heels. Ann. Phys. Med. *9:* 295–299.

JOSEPH, J. AND NIGHTINGALE, A. (1952) Electromyography of muscles of posture: leg muscles in males. J. Physiol., *117:* 484–491.

JOSEPH, J. AND NIGHTINGALE, A. (1954) Electromyography of muscles of posture: thigh muscles in males. J. Physiol., *126:* 81–85.

JOSEPH, J. AND NIGHTINGALE, A. (1956) Electromyography of muscles of posture: leg and thigh muscles in women, including the effects of high heels. J. Physiol., *132:* 465–468.

JOSEPH, J., NIGHTINGALE, A. AND WILLIAMS, P. L. (1955) A detailed study of the electric potentials recorded over some postural muscles while relaxed and standing. J. Physiol., *127:* 617–625.

JOSEPH, J. AND WATSON, R. (1967) Telemetering electromyography of muscles used in walking up and down stairs. J. Bone & Joint Surg. *49B:* 774–780.

JOSEPH, J. AND WILLIAMS, P. L. (1957) Electromyography of certain hip muscles. J. Anat., *91:* 286–294.

JÜDE, H. D., DRECHSLER, F. AND NEUHAUSER, B. (1975) Elementare elektromyographische Analyse and Bewegungstmuster des M. myloglossus. Dtsch. Zahärztl. Z., *30:* 457–461.

KADEFORS, R., KAISER, E. AND PETERSÉN, I. (1968) Dynamic spectrum analysis of myo-potentials with special reference to muscle fatigue. Electromyography, *8:* 39–74.

KADEFORS, R. AND PETERSÉN, I. (1970) Spectral analysis of myo-electric signals from muscles of the pelvic floor during voluntary contraction and during reflex contractions connected with ejaculation. Electromyography, *10:* 45–68.

KADEFORS, R., PETERSÉN, I. AND TENGROTH, B. (1974) Quantitative analysis of EMG from m rectus lateralis oculi. Scand. J. Rehab. Med., *6: Suppl 3:* 115–120.

KAHN, S. D. (1971) Comparative advantages of bipolar abraded skin surface electrodes over bipolar intramuscular electrodes for single motor unit recording in psychophysiological research. Psychophysiology, *8:* 635–647.

KAISER, E. AND PETERSÉN, I. (1963) Frequency analysis of muscle action potentials during tetanic contraction. Electromyography. *3:* 5–17.

KAISER, E. AND PETERSÉN, I. (1965) Muscle action potentials studied by frequency analysis and duration measurement. Acta Neurol. Scandinav. *41:* 19–41.

KAISER, E. AND PETERSÉN, I. (1965) Muscle action potentials studied by frequency analysis and duration measurement. Acta Neurol. Scand. suppl. *13:* 213–236.

KALEN, F. C. AND GANS, C. (1970) How does the bat chew? Evidence from electromyography (abstract). Anat. Rec., *166:* 327.

KAMEI, S., MATSUI, H. AND MIYASHITA, M. (1971) An electromyographic analysis of Japanese archery (Japanese text: English abstract). Res. J. Phys. Ed., *15:* 39–46.

KAMMER, A. E. AND HEINRICH, B. (1972) Neural control of bumblebee fibrillar muscles during shivering. J. Comp. Physiol., 337–345.

KAMON, E. (1966) Electromyography of static and dynamic postures of the body supported on the arms. J. Appl. Physiol., *21:* 1611–1618.

KAMON, E. AND GORMLEY, J. (1968) Muscular activity pattern for skilled performance and during learning of a horizontal bar exercise. Ergonomics, *11:* 345–357.

KAMP, A. KOK, M. L. AND DE QUARTEL, F. W. (1965) A multiwire cable for recording from moving subjects. Electroencephalog. & Clin. Neurophysiol., *18:* 422–423.

KAPLAN, M. AND KAPLAN T. (1935) Flat foot. A consideration of the anatomy and physiology of the normal foot, the pathology and mechanism of flat foot, with the resulting Roentgen manifestations. Radiology, *25:* 485–491.

KARLINS, M. AND ANDREWS, L. M. (1973) *Biofeedback: Turning on the Power of Your Mind.* Warner Paperback Library, New York. Originally published by J. B. Lippincott Co., Inc., 1972.

KARLSSON, E. AND JONSSON, B. (1965) Function of the gluteus maximus muscle: an electromyographic study. Acta Morphol. Neer.-Scandinav., *6:* 161–169.

KASEDA, Y. AND NOMURA, S. (1973) Electromyographic studies on the swimming movement of carp: I. body movement. Jap. J. Vet. Sci., *35:* 335–342.

KASVAND, T., MILNER, M., QUANBURY, A. O. AND WINTER, D. A. (1976) Computers and the kinesiology of gait. Comput. Biol., Med., *6:* 111–120.

KASVAND, T. MILNER, M. AND RAPLEY, L. F. A computer-based system for the analysis of some aspects of human locomotion. *Transactions of Conference on Human Locomotor Engineering,* University of Sussex, 1971, Institute of Mechanical Engineers, Publishers, London, pp. 297–306.

KATO, M. AND TANJI, J. (1972a) Volitionally controlled single motor units in human finger muscles. Brain Res., *40:* 345–357.

KATO, M. AND TANJI, J. (1972b) Cortical motor potentials accompanying volitionally controlled single motor unit discharges in human finger muscles. Brain Res., *47:* 103–111.

KAWAMURA, Y. AND FUJIMOTO, J. (1957) Some physiologic considerations on measuring rest position of the mandible. Med. J. Osaka Univ., *8:* 247–255.

KAWASAKI, M., OGURA, J. H. AND TAKENOUCHI, S. (1964) Neurophysiologic observations of normal deglutition. Laryngoscope, *74:* 1747–1780.

KAZAI, N., KUMAMOTO, M., OKAMOTO, T., YAMASHITA, N., GOTO, Y., AND MARUYAMA, H. (1976) Yakyu no Toh-Dohsa (Oba-hando Suroh) ni okeru Johshi Johshitaikingun no Sayo-Kijyo (Electromyographic study of the overhand pitching in terms of the functional mechanism of the upper extremity and the shoulder girdle muscles). Res. J. Phys. Ed. *21:* 137–144.

KAZAI, N., OKAMOTO, T. AND KUMAMOTO, M. (1976) Electromyographic study of supported walking of infants in the initial period of learning to walk. In *Biomechanics V,* ed. by P. V. Komi *et al.* University Park Press, Baltimore.

KEAGY, R. D., BRUMLIK, J. AND BERGAN, J. J. (1966) Direct electromyography of the psoas major muscle in man. J. Bone & Joint Surg., *48-A:* 1377–1382.

KEAR, M. AND SMITH, R. N. (1975) A method of recording tendon strain in sheep during locomotion. Acta Orthop. Scandinav., *46:* 896–905.

KEITH, SIR ARTHUR. (1920) In a discussion of a paper by L. H. Buxton: The teeth and jaws of savage man. Tr. Brit. Soc. Orthodontists, 1916–20, 79–88.

KEITH, A. (1929) The history of the human foot and its bearing on orthopaedic practice. J. Bone & Joint Surg., *11:* 10–32.

KELLY, K. A., CODE, C. F. AND ELVEBACK, L. R. (1969) Patterns of canine gastric electrical activity. Am. J. Physiol., *217:* 461–470.

KELMAN, A. W. AND GATEHOUSE, S. (1975) A study of the electromyographic activity of the muscle orbicularis oris. Folia Phoniat., *27:* 177–189.

KELSEN, S. G., ALTOSE, M. D., STANLEY, N. N., LEVINSON, R. S., CHERNIACK, N. S. AND FISHMAN, A. P. (1976) Electromyographic response of respiratory muscles during elastic loading. Am. J. Physiol., *230:* 675–683.

KELTON, L. W. AND WRIGHT, R. D. (1949) The mechanism of easy standing by man. Australian J. Exper. Biol. & M. Sc., *27:* 505–515.

KENNEY, W. E. AND HEABERLIN, P. C., JR.

(1962) An electromyographic study of the locomotor pattern of spastic children. Clin. Orthop., *24:* 139–151.

KHAN, M. A. (1969) *Morphology and Electromyography of the Flexor Pollicis Brevis and Adductor Pollicis Muscles.* M.Sc. Thesis. Queen's University, Canada.

KIESSWETTER, H. (1970) EMG-patterns of pelvic floor muscles with surface electrodes. Urol. Internat., *31:* 60–69.

KIMM, J. AND SUTTON, D. (1973) Foreperiod effects on human single motor unit reaction times. Physiol. Behav., *10:* 539–542.

KINSMAN, R. A., O'BANION, K., ROBINSON, S. AND STAUDENMAYER, H. (1975) Continuous biofeedback and discrete posttrial verbal feedback in frontalis muscle relaxation training. *12:* 30–35.

KIRIKAE, I. (1960) *The Structure and Function of the Middle Ear.* The University of Tokyo Press, Tokyo.

KIRIKAE, I., HIROSE, H., KAWAMURA, S., SAWASHIMA, M. AND KOBAYASHI, T. (1962) An experimental study of central motor innervation of the laryngeal muscles in the cat. Ann. Otol., Rhin. & Laryng., *71:* 222–242.

KIVIAT, M. D., ZIMMERMANN, T. A. AND DONOVAN, W. H. (1975) Sphincter stretch: a new technique resulting in continence and complete voiding in paraplegics. J. Urol., *114:* 895–897.

KLAUSEN, K. (1965) The form and function of the loaded human spine. Acta Physiol. Scandinav., *65:* 176–190.

KLINEBERG, I. J., GREENFIELD, B. E. AND WYKE, B. D. (1970) Contributions to the reflex control of mastication from mechanoreceptors in the temporomandibular joint capsule. Dent. Pract., *21:* 73–83.

KLOPROGGE, M. J. G. M. (1975) Reflex control of the jaw muscles by stimuli from receptors in the periodontal membrane. J. Oral Rehab., *2:* 259–272.

KNUTSSON, B., LINDH, K. AND TELHAG, H. (1966) Sitting—an electromyographic and mechanical study. Acta Orthop. Scand., *37:* 415–428.

KNUTTSON, E., MÅRTENSSON, A. AND MARTENSSON, B. (1969) The normal electromyogram in human vocal muscles. Acta Oto-laryng., *68:* 526–536.

KOCZOCIK-PRZEDPELSKA, J., TOBOLA, S. AND GRUSZCZYŃSKI, W. (1966) Aktywność miolektryczna podczas czynności zautomatyzowanej oraz ruchn dowolnego (Polish text; English abstract). Acta Physiol. Polon., *17:* 593–599.

KOEPKE, G. H., SMITH, E. M., MURPHY, A. J. AND DICKINSON, D. G. (1958) Sequence of action of the diaphragm and intercostal muscles during respiration. I. Inspiration. Arch. Phys. Med., *39:* 426–430.

KOLLBERG, S., PETERSÉN, I. AND STENER, I. (1962) Preliminary results of an electromyographic study of ejaculation. Acta Chir. Scandinav., *123:* 478–483.

KOMI, P. V. (1973) Relationship between muscle tension, EMG and velocity of contraction under concentric and eccentric work. In *New Developments in Electromyography and Clinical Neurophysiology* ed. by J. E. Desmedt, Karger, Basel.

KOMI, P. V. AND BUSKIRK, E. R. (1970) Reproducibility of electromyographic measurements with inserted wire electrodes and surface electrodes. Electromyography, *10:* 357–367.

KOMI, P. V. AND RUSKO, H. (1974) Quantitative evaluation of mechanical changes during fatigue loading of eccentric and concentric work. Scand. J. Med., *3:* suppl. 121–126, 1974.

KOMI, P. AND VIITASALO, J. H. T. (1976) Signal characteristics of EMG at different levels of muscle tension. Acta Physiol. Scand., *96:* 267–276.

KOREIN, J., BRUDNY, J., GRYNBAUM, B., SACHS-FRANKEL, G., WEISINGER, M. AND LEVIDOW, L. (1976) Sensory feedback therapy of spasmodic torticollis and dystonia: results in treatment of

55 patients. Adv. Neurol., *14:* 375–402.

KOTBY, M. N. (1975) Percutaneous laryngeal electromyography: standardization of the technique. Folio Phoniat., *27:* 116–127.

KOTBY, M. N. AND HAUGEN, L. K. (1970a) The mechanics of laryngeal function. Acta Otolaryng., *70:* 203–211.

KOTBY, M. N. AND HAUGEN, L. K. (1967b) Critical evaluation of the action of the posterior crico-arytenoid muscle, utilizing direct emg-study. Acta Otolaryng., *70:* 260–268.

KOTBY, M. N. AND HAUGEN, L. K. (1970c) Attempts at evaluation of the function of various laryngeal muscles in the light of muscle and nerve stimulation experiments in man. Acta Otolaryng., *70:* 419–427.

KOTBY, M. N. AND HAUGEN, L. K. (1970d) Clinical application of electromyography in vocal fold mobility disorders. Acta Orolaryng., *70:* 428–437.

KOZMYAN, E. I. (1965) Time relations of excitation and inhibition of antagonist muscles (Russian text). Zh. Vssh. Nerv. Deiat. Pavlov, *17:* 125–133.

KRAMER, H., FRAUENDORF, H. AND KÜCHLER, G. (1972) Die Beeinflussung des mittels Oberflächenelektroden abgeleiteten Elektromyogramms durch ableittechnische Variablen. II. Zum Einfluss von Abstand, Flächengrösse und Andruck der Electroden auf die ableitbare elektrische Muskelaktivität. Acta Biol. Med. Germ., *28:* 489–496.

KRANZ, H. AND BAUMGARTNER, G. (1974) Human alpha motoneurone discharge, a statistical analysis. Brain Res., *67:* 324–329.

KRNJEVIĆ, K. AND MILEDI, R. (1958a) Motor units in the rat diaphragm. J. Physiol., *140:* 427–439.

KRNJEVIC, K. AND MILEDI, R. (1958b) Failure of neuromuscular propagation in rats. J. Physiol. (London), *140:* 440–461.

KUFFLER, S. W. AND VAUGHAN WILLIAMS, E. M. (1953) The distribution of small motor nerves to frog skeletal muscle, and the membrane characteristics of the fibres they innervate. J. Physiol., *121:* 289–317.

KUGELBERG, E. (1953) Clinical electromyography. Progr. Neurol. & Psychiat., *8:* 264–282.

KUKULKA, C. G., BROWN, D. M. AND BASMAJIAN, J. V. (1975) Biofeedback training for early finger joint mobilization. Am. J. Occup. Ther., *29:* 469–470.

KURODA, E., KLISSOURAS, V. AND MULSUM, J. H. (1970) Electrical and metabolic activities and fatigue in human isometric contraction. J. Appl. Physiol., *29:* 358–367.

KUROZUMI, TASHIRO, T. AND HARADY, Y. (1971) Laryngeal responses to electrical stimulation of the medullary respiratory centers in the dog. Laryngoscope, *81:* 1960–1967.

LABAN, M. M., RAPTOU, A. D. AND JOHNSON, E. W. (1965) Electromyographic study of function of iliopsoas muscle. Arch. Phys. Med., *46:* 676–679.

LADD, H., JONSSON, B. AND LINDEGREN, U. (1972) The learning process for fine neuromuscular controls in skeletal muscles of man. Electromyog. Clin. Neurophysiol., *12:* 213–223.

LADD, H. W. AND SIMARD, T. G. (1972) Bilaterally controlled neuromuscular activity in congenitally malformed children—an electromyographic study. Interclinic Information Bull., *11:* 9–16.

LAFRATTA, C. W. AND SMITH, O. H. (1964) A study of the relationship of motor nerve conduction velocity in the adult to age, sex and handedness. Arch. Phys. Med., *45:* 407–412.

LAGO, P. AND JONES, N. B. (1977) Effect of motor-unit firing time statistics on e.m.g. spectra. Med. Biol. Eng. Comput., *15:* 648–655.

LA JOIE, W. J., COSGROVE, M. D. AND JONES, W. G. (1976) Electromyographic evaluation of human detrusor muscle activity in relation to abdominal muscle activity. Arch. Phys. Med. Rehab., *57:* 382–386.

LAKE, N. (1937) The arches of the foot. Lancet, *2:* 872–873.

LAKE, L. F. (1954) An electromyographic study of the action of the second and third dorsal interosseous muscles and their interaction with extensor digitorum communis and flexor digitorum sublimis of the normal hand. M. A. Thesis, Department of Anatomy, Washington University, St. Louis, Missouri.

LAKE, L. F. (1957) An electromyographic study of finger movement. Anat. Rec., *127:* 322–323.

LAMARRE, Y. AND LUND, J. P. (1975) Load compensation in human masseter muscles. J. Physiol. (Lond.), *253:* 21–35.

LANDA, J. (1974) Shoulder muscle activity during selected skills on the uneven parallel bars. Res. Quart., *45:* 120–127.

LANDAU, W. M. (1951) Comparison of different needle leads in EMG recording from a single site. Electroencephalog. & Clin. Neurophysiol., *3:* 163–168.

LANDAU, W. M. AND CLARE, M. H. (1959) The plantar reflex in man: with special reference to some conditions where the extensor response is unexpectedly absent. Brain, *82:* 321–355.

LANE, R. H. (1975) Clinical application of anorectal physiology. Proc. Roy. Soc. Med., *68:* 28–30.

LAPIDES, J., AJEMIAN, E. P., STEWART, B. H., BREAKEY, B. A. AND LICHTWARDT, J. R. (1960) Further observations on the kinetics of the urethrovesical sphincter. J. Urol., *84:* 86–94.

LAPIDES, J., SWEET, R. B. AND LEWIS, L. W. (1957) Role of striated muscle in urination. J. Urol., *77:* 247–250.

LARSON, J. D. AND FOULKES, D. (1969) Electromyographic suppression during sleep, dream recall, and orientation time. Psychophysiology, *5:* 548–555.

LARSSON, L. E., LINDERHOLM, H. AND

RINGQVIST, T. (1965) The effect of sustained and rhythmic contractions on the electromyogram (EMG). Acta Physiol. Scandinav., *65:* 310–318.

LAST, R. J. (1954) *Anatomy Regional and Applied,* 665 pp. J. & A. Churchill, Ltd., London.

LATIF, A. (1957) An electromyographic study of the temporalis muscle in normal persons during selected positions and movements of the mandible. Am. J. Orthodontics, *43:* 577–591.

LAUERMA, K. S. L., HARVEY, J. E. AND OGURA, J. H. (1972) Cricopharyngeal myotomy in subtotal supraglottic laryngectomy: an experimental study. Laryngoscope, *82:* 447–453.

LEANDERSON, R. (1972) *On the Functional Organization of Facial Muscles in Speech,* Published in book form by Acta Laryngol., Stockholm.

LEANDERSON, R., MEYERSON, B. A. AND PERSSON, A. (1971) Effect of L-dopa on speech in Parkinsonism: an EMG study of labial articulatory function. J. Neurol. Neurosurg. Psychiat., *34:* 679–681.

LEANDERSON, R. PERSSON, A. AND ÖHMAN, S. (1970) Electromyographic studies of the function of the facial muscles in dysarthria. Acta Otolaryng., *263:* 89–94.

LEAVITT, L. A. AND BEASLEY, W. C. (1964) Clinical application of quantitative methods in the study of spasticity. Clin. Pharmacol. & Therap., *5:* 918–941.

LE FEVER R. S. AND DE LUCA C. J. (1976) The contribution of individual motor units to the EMG power spectrum. Proc *29th* ACEMB: 56.

LEHR, R. P., BLANTON, P. L. AND BIGGS, N. L. (1971) An electromyographic study of the mylohyoid muscle. Anat. Rec., *169:* 651–660.

LEIBRECHT, B. E., LLOYD, A. J. AND POUNDER, S. (1973) Auditory feedback and conditioning of the single motor unit. Psychophysiology, *10:* 1–7.

LEIFER, L. J. (1969) Characterization of single muscle fiber discharge during voluntary isometric contraction of the biceps brachii muscle in man. Ph. D. Thesis, Stanford University, California, U.S.A.

LENMAN, J. A. R. AND POTTER, J. L. (1966) Electromyographic measurement of fatigue in rheumatoid arthritis and neuromuscular disease. Ann. Rheumat. Dis., *25:* 76–84.

LESAGE, Y. AND LE BAR, R. (1970) Étude électromyographique simultanée des différent chefs du quadriceps. Ann. Méd. Phys. *13:* 292–297.

LESOINE, W. AND PAULSEN, H-J. (1968) Elektromyographische Unterschungen am Musculus cricothyreodeus und Musculus digastricus venter anterior. Z. Laryng. Rhinol. Otol., *47:* 543–55.

LESTIENNE, F. AND BOUISSET, S. (1968) Pattern temporel de la mise en jeu d'un agoniste et d'un antagoniste en fonction de la tension de l'agoniste, Rev. Neurol., Paris, *118:* 550–554.

LESTIENNE, F. AND GOUBEL, F. (1969) Contribution relatif de deux agonistes à un travail avec et sans raccourcissement. J. Physiol., *61:* 342–343.

LEVENS, A. S., INMAN, V. T. AND BLOSSER, J. A. (1948) Transverse rotation of the segments of the lower extremity in locomotion. J. Bone & Joint Surg., *30-A:* 859–872.

LEVINE, I. M., JOSSMANN, P. B., TURSKY, B., MEISTER, M. AND DE ANGELIS, V. (1964) Telephone telemetry of bioelectric information. J. A. M. A., *188:* 794–798.

LEVITT, M. N., DEDO, H. H. AND OGURA, J. H. (1965) The cricopharyngeus muscle, an electromyographic study in the dog. Laryngoscope, *75:* 122–136.

LEVY, R. (1963) The relative importance of the gastrocnemius and soleus muscles in the ankle jerk of man. J. Neurol. Neurosurg. Psychiat., *26:* 148–150.

LEWIS, R. S. AND BASMAJIAN, J. V. (1959) Electromyographic studies in embryos and fetuses. Proc. Canad. Fed. Biol. Soc., *2:* 41–42.

LI, C.-L. AND LUNDERVOLD, A. (1958) Electromyographic study of cleft palate. Plast. & Reconstruct. Surg., *21:* 427–432.

LIBERSON, W. T. (1936) Une nouvelle application de Quartz pièzoélectrique: Pièzoélectrographie de la marche et des mouvements volontaires. Le Travail Humain (Paris), *4:* 1–7.

LIBERSON, W. T. (1960) New development in the study of conduction time in peripheral neuropathies. Electroenceph. Clin. Neurophysiol., *12:* 264.

LIBERSON., W. T. (1965a) Normal and pathological gaits (abstract). Bull. Am. Assoc. EMG and Electrodiag., *12:* 25.

LIBERSON, W. T. (1965b) Biomechanics of gait: a method of study. Arch. Phys. Med., *46:* 37–48.

LIBERSON, W. T., DONDEY, M. AND ASA, M. M. (1962) Brief repeated isometric maximal exercises. Am. J. Phys. Med., *41:* 3–14.

LIBERSON, W. T., GRATZER, M., ZALIS. A. AND GRABINSKI, B. (1966) Comparison of conduction velocities of motor and sensory fibers determined by different methods. Arch. Phys. Med., *47:* 17–23.

LIBERSON, W. T. AND KIM, K. C. (1963) Mapping evoked potentials elicited by stimulation of the median and peroneal nerves. Electroenceph. Clin. Neurophysiol. *15:* 721.

LIBERSON, W. T., HOLMQUEST, H. J. AND HALLS, A. (1962) Accelerographic study of gait. Arch. Phys. Med., *43:* 547–551.

LIBKIND, M. S. (1968) II. Modelling of interference bioelectrical activity. Biofizika, *13:* 685–693.

LIEB, F. J., AND PERRY, J. (1968) Quadriceps function: an anatomical and mechanical study using amputated limbs. J. Bone & Joint Surg., *50-A:* 1535–1538.

Lieb, F. J. and Perry, J. (1971) Quadriceps function: an electromyographic study under isometric conditions. J. Bone & Joint Surg., *53-A:* 749–758.

Lindqvist, C. (1959) The motor unit potential in severely paretic muscles after acute anterior poliomyelitis: an electromyographic study using fatigue experiments. Acta Psychiat. Neurol. Scandinav., *34:* suppl. 131, 1–72.

Lindsley, D. B. (1935) Electrical activity of human motor units during voluntary contraction. Am. J. Physiol., *114:* 90–99.

Lindström, L. R. (1970) *On the frequency spectrum of EMG signals,* Technical Report-Research Laboratory of Medical Electronics, Chalmers University of Technology, Göteborg, Sweden.

Lindström, L. R., Magnusson, R. Petersén, I. (1970) Muscular fatigue and action potential conduction velocity changes studied with frequency analysis of EMG signals. Electromyography, *4:* 341–353.

Lippold, O. C. J. (1952) The relation between integrated action potentials in a human muscle and its isometric tension. J. Physiol., *117:* 492–499.

Lippold, O. C. J. (1971) Physiological tremor. Sci. Amer., *224:* 65–73.

Lippold, O. C. J., Redfearn, J. W. T. and Vučo, J. (1957) The rhythmical activity of groups of motor units in the voluntary contraction of muscle. J. Physiol., *137:* 473–487.

Lippold, O. C. J., Redfearn, J. W. T. and Vučo, J. (1970) The electromyography of fatigue. Ergonomics, *3:* 121–131.

Littler, J. W. (1960) The physiology and dynamic function of the hand. Surg. Clin. N. A., *40:* 259–266.

Livingston, R. B., Paillard, J., Tournay, A. and Fessard, A. (1951) Plasticité d'une synergie musculaire dans l'exécution d'un mouvement volontaire chez l'homme. J. Physiol., Paris, *43:* 605–619.

Lloyd, A. J. (1968) Muscle activity and kinesthetic position responses. J. Appl. Physiol., *25:* 659–663.

Lloyd, A. J. and Caldwell, L. S. (1965) Accuracy of active and passive positioning of the leg on the basis of kinesthetic cues. J. Comp. Physiol. Psychol., *60:* 102–106.

Lloyd, A. J. and Leibrecht, B. C. (1971) Conditioning of a single motor unit. J. Exper. Psychol., *88:* 391–395.

Lloyd, A. J. and Shurley, J. T. (1976) The effect of sensory perceptual isolation on single motor unit conditioning. Psychophysiol., *13:* 340–344.

Lloyd, A. J. and Voor, J. H. (1973) The effect of training on performance efficiency during a competitive isometric exercise. J. Motor Behav., *5:* 17–24.

Lockhart, R. D. (1951) in *Cunningham's Textbook of Anatomy,* ed. 9, edited by J. C. Brash. Oxford University Press, London, New York, Toronto.

Lockhart, R. D., Hamilton, G. F. and Fyfe, F. (1959) *Anatomy of the Human Body,* pp. 92, 216. Faber and Faber Ltd., London.

Long, C. and Brown, M. E. (1962) Electromyographic kinesiology of the hand: Part III. Lumbricalis and flexor digitorum profundus to the long finger. Arch. Phys. Med., *43:* 450–460.

Long, C. and Brown, M. E. (1964) Electromyographic kinesiology of the hand: muscles moving the long finger. J. Bone & Joint Surg., *46-A:* 1683–1706.

Long, C., Brown, M. E. and Weiss, G. (1960) Electromyographic kinesiology of the hand. Part II. Third dorsal interosseus and extensor digitorum of the long finger. Arch. Phys. Med., *42:* 559–565.

Long, C., Conrad, P. W., Hall, E. W. and Furler, S. L. (1970) Intrinsic-extrinsic muscle control of the hand in power grip and precision handling. J. Bone & Joint Surg., *52-A:* 853–867.

Lopata, M., Evanich, M. J. and Lourenço, R. V. (1976) The electromyogram of the diaphragm in the investigation of human regulation of ventilation. Chest, *70: suppl.* 162S-165S.

LOUIS-SYLVESTRE, J. AND MACLEOD, P. (1968) Le muscle vocal humain est-il asynchrone? J. Physiol., Paris, *60:* 373-389.

LOOFBOURROW, G. N. (1948) Electrographic evaluation of mechanical response in mammalian skeletal muscle in different conditions. J. Neurophysiol., *11:* 153-168.

LOURENÇO, R. V., CHERNIAK, N. S., MALM, J. R. AND FISHMAN, A. P. (1966) Nervous output from the respiratory center during obstructed breathing. J. Appl. Physiol., *21:* 527-533.

LOW, M. D., BASMAJIAN, J. V. AND LYONS, G. M. (1962) Conduction velocity and residual latency in the human ulnar nerve and the effects on them of ethyl alcohol. Am. J. Med. Sc., *244:* 720-730.

LUBKER, J. F. (1968) An electromyographic-cinefluorographic investigation of velar function during normal speech. Cleft Palate J., *5:* 1-18.

LUBKER, J. F. (1975) Normal velopharyngeal function in speech. Clin. Plastic Surg., *2:* 249-259.

LUBKER, J., FRITZELL, B. AND LINDQVIST, J. (1970) Speech production. Velopharyngeal function: an electromyographic study, privately printed.

LUBKER, J. AND MAY, K. (1973) Palatoglossus function in normal speech production. In *Papers from the Institute of Linguistics, 1973.* pp. 17-26. University of Stockholm, Sweden.

LUBKER, J. F. AND PARRIS, P. J. (1970) Simultaneous measurements of intraoral pressure, force of labial contact, and labial electromyographic activity during production of the stop consonant cognates /p/ and /b/. J. Acoust. Soc. Am., *47:* 625-633.

LUGARESI, E., COCCAGNA, G., FARNETI, P., MONTOVANI, M. AND CIRIGNOTTA, F. (1975) Snoring. Electroenceph. Clin. Neurophysiol., *39:* 59-64.

LUKAS, J. S., PEELER, D. J. AND KRYTER, K. D. (1970) Effects of sonic booms and subsonic jet flyover noise on skeletal muscle tension and a paced tracing task. Report to N.A.S.A. (CR-1522), 35 pp.

LUND, J. P. (1976) Oral-facial sensation in the control of mastication and voluntary movements of the jaw. In *Mastication and Swallowing,* ed. by B. J. Sessle and A. G. Hannam, University of Toronto Press, Toronto.

LUND, J. P., MCLACHLAN, R. S. AND DELLOW, P. G. (1971) A lateral jaw movement reflex. Exper. Neurol., *31:* 189-199.

LUNDERVOLD, A. J. S. (1951) *Electromyographic Investigations of Position and Manner of Working in Typewriting,* 171 pp. A. W. Brøggers Boktrykkeri A/S, Oslo.

LUNDERVOLD, A., BRULAND, H. AND STENSRUD, P. (1965) Conduction velocity in peripheral nerves: a general introduction. Acta Neurol. Scandinav., *41:* suppl. 13, 259-262.

LUNDERVOLD, A. AND LI, C.-L. (1953) Motor units and fibrillation potentials as recorded with different kinds of needle electrodes. Acta Psychiat. Neurol., *28:* 201-211.

LUSCHEI, E., SASLOW, C. AND GLICKSTEIN, M. (1967) Muscle potentials in reaction time. Exper. Neurol., *18:* 429-442.

MACCONAILL, M. A. (1946) Some anatomical factors affecting the stabilizing functions of muscles. Irish J. M. Sc., *6:* 160-164.

MACCONAILL, M. A. (1949) The movements of bones and joints. 2. Function of the musculature. J. Bone & Joint Surg., *31-B:* 100-104.

MACCONAILL, M. A. AND BASMAJIAN, J. V. (1969) *Muscles and Movements: A Basis for Human Kinesiology.* Williams & Wilkins Co., Baltimore.

MACDOUGALL, J. D. B. AND ANDREW, B. L. (1953) An electromyographic study of the temporalis and masseter muscles. J. Anat., *87:* 37-45.

MACHADO, see de Sousa, O. Machado.

MACNEILAGE, P. F. (1973) Preliminaries to the study of single motor unit activity in speech musculature. J. Phonet., *1:* 55–71.

MACNEILAGE, P. F., ROOTES, T. O. AND CHASE, R. A. (1967) Speech production and perception in a patient with severe impairment of somesthetic perception and motor control. J. Speech Hearing Res., *10:* 449–467.

MACNEILAGE, P. F. AND SZABO, R. K. (1972) Frequency control of single motor units in upper articulatory musculature. Paper presented at 83rd Meeting, Acoustical Soc. of Am., Buffalo, New York.

MACNEILAGE, P. F., SUSSMAN, H. M. AND SUSSMAN, R. J. (1972) Parametric study of single motor unit waveforms in upper articulatory musculature. Paper presented at 83rd Meeting, Acoustical Soc. of Am., Buffalo, New York.

MAGLADERY, J. W. AND MCDOUGAL, D. B. JR. (1950a) Electrophysiological studies on nerve and reflex activity in normal man (Part I). Bull. Johns Hopkins Hosp., *86:* 265–290.

MAGLADERY, J. W., MCDOUGAL, D. B. JR. AND STOLL, J. (1950b,c) Electrophysiological studies of nerve and reflex activity in normal man: Parts II and III. Bull. Johns Hopkins Hosp., *86:* 291–312 and 313–339.

MAGORA, A., ROZIN, R., ROBIN, G. C., GONEN, B. AND SIMKIN, A. (1970) Investigation of gait. I. A technique of combined recording. Electromyography, *10:* 385–396.

MAGOUN, H. W. AND RHINES, R. (1947) *Spasticity: the Stretch-Reflex and Extrapyramidal Systems,* Charles C Thomas, Springfield, Ill.

MALMO, R. B., SHAGASS, C. AND DAVIS, J. F. (1951) Electromyographic studies of muscular tension in psychiatric patients under stress. J. Clin. Exper. Psychopath., *12:* 45–66.

MALMO, R. B. AND SMITH, A. A. (1955) Forehead tension and motor irregularities in psychoneurotic patients under stress. J. Personality, *23:* 391–406.

MANN, R. AND INMAN, V. T. (1964) Phasic activity of intrinsic muscles of the foot. J. Bone & Joint Surg., *66-A:* 469–481.

MARCINIAK, W. (1975) Czynność bioelektryczna mięśni goleni w chodzie poziomym u dziece zdrowyck (Bioelectric activity of leg muscles of healthy children during level walking). Chir. Narz. Ruchu Ortop. Pol., *15:* 643–650.

MAREY, E. J. (1895) *Movement.* Tr. by E. Pritchard. D. Appleton and Co., New York.

MARG, E. (1951) Development of electrooculography. A.M.A. Arch. Ophth., *45:* 169–185.

MARG, E., TAMLER, E. AND JAMPOLSKY, A. (1962) Activity of a human oculorotary muscle unit. Electroencephalog. & Clin. Neurophysiol., *14:* 754–757.

MARINACCI, A. A. (1959) Dynamics of neuromuscular diseases. Arch. Neurol., *1:* 243–257.

MARINACCI, A. A. (1968) *Applied Electromyography.* Lea & Febiger, Philadelphia.

MARKEE, J. E., LOGUE, J. T., WILLIAMS, M., STANTON, W. B., WRENN, R. N. AND WALKER, L. B. (1955) Two-joint muscles of the thigh. J. Bone & Joint Surg., *37-A:* 125–142.

MARSDEN, C., MEADOWS, J. AND HODGSON, H. (1969) Observations on the reflex response to muscle vibration in man and its voluntary control. Brain, *92:* 829–846.

MARSDEN, C. D., MEADOWS, J. C., LANGE, G. W. AND WATSON, R. S. (1969) (a) The role of the ballistocardiac impulse in the genesis of physiological tremor. EEG and Clin. Neurophysiol., *27:* 169–178. (b) Variations in human physiological finger tremor, with particular reference to changes with age. Brain, *92:* 647–662.

MARSDEN, C. D., MEADOWS, J. C. AND

MERTON, P. A. (1976) Fatigue in human muscle in relation to the number and frequency of motor impulses. J. Physiol., (Lond.), *258:* 94P.

MARSDEN, C. D., MERTON, P. A. AND MORTON, H. B. (1976) Servo action in the human thumb. J. Physiol. (Lond.), 1–44.

MÅRTENSSON, A. AND SKOGLUND, C. R. (1964) Contraction properties of internal laryngeal muscles. Acta Physiol. Scandinav., *60:* 318–336.

MARTIN, H. N. AND HARTWELL, E. M. (1879) On the respiratory function of the internal intercostal muscles. J. Physiol., *2:* 24–27.

MARTONE, A. L. (1962) Anatomy of facial expression and its prosthodontic significance, J. Pros. Den., *12:* 1020–1042.

MASLAND, W. S. SHELDON, D. AND HERSHEY, C. D. (1969) The stochastic properties of individual motor unit interspike intervals. Am. J. Physiol., *217 (5):* 1384–1388.

MATHEWS, A. M. AND GELDER, M. G. (1960) Psycho-physiological investigations of brief relaxation training. J. Psychosomat. Res., *13:* 1–12.

MATON, B. AND BOUISSET, S. (1975) Motor unit activity and preprogramming of movement in man. Electroenceph. Clin. Neurophysiol., *38:* 658–660.

MATSUSHITA, K., GOTO, Y., OKAMOTO, T., TSUJINO, A. AND KUMAMOTO, M. (1974) Soh no Kindenzuteki Kenkyu (An electromyographic study of spring running). Res. J. Phys. Ed., *19:* 147–156.

MAWDSLEY, C. AND MAYER, R. F. (1965) Nerve conduction in alcohol polyneuropathy. Brain, *88:* part II, 335–356.

MAYER, R. F. (1963) Nerve conduction studies in man. Neurology, *13:* 1021–1030.

McALLISTER, R., LUBKER, J. AND CARLSON, J. (1974) An emg study of some characteristics of the Swedish rounded vowels. J. Phonet., *2:* 267–278.

McCOLLUM, B. B. (1943) Oral diagnosis.

J. Am. Dent. A., *30:* 1218–1233.

McFARLAND, G. B., KRUSEN, U. L. AND WEATHERSBY, H. T. (1962) Kinesiology of selected muscles acting on the wrist: electromyographic study. Arch. Phys. Med., *43:* 165–171.

McGLONE, R. E. AND SHIPP, T. (1971) Some physiologic correlates of vocal-fry phonation. J. Speech Hearing Res., *14:* 769–775.

McGREGOR, A. L. (1950) *Synopsis of Surgical Anatomy,* ed. 7. The Williams & Wilkins Co., Baltimore.

McGUIGAN, F. J. (1966) Covert oral behavior and auditory hallucinations. Psychophysiology, *3:* 73–80.

McGUIGAN, F. J. (1970) Covert oral behavior during the silent performance of language tasks. Psychol. Bull., *74:* 309–326.

McGUIGAN, F. J. AND RODIER, W. I., III. (1968) Effects of auditory stimulation on covert oral behavior during silent reading. J. Exper. Psychol., *76:* 649–655.

McLEOD, W. D., NUNNALLY, H. N. AND CANTRELL, P. E. (1976) Dependence of emg power spectra on electrode type. IEEE Trans. Biomed. Eng., (March) 172–175.

McLEOD, W. D. AND THYSELL, R. (1973) Cortically evoked potentials co-related with single motor units, to be published.

McNAMARA, J. A. (1973) The independent functions of the two heads of the lateral pterygoid muscle. Am. J. Anat., *138:* 197–206.

McPHEDRAN, A. M., WUERKER, R. B. AND HENNEMAN, E. (1965) Properties of motor units in a homogeneous red muscle (soleus) of the cat. J. Neurophysiol., *28:* 71–84.

MEIJERS, L. M. M., TEULINGS, J. L. H. M. AND EIJKMAN, E. G. J. (1976) Model of the electromyographic activity during brief isometric contractions. Biol. Cybernetics, *25:* 7–16.

MERTON, P. A. (1954) Voluntary strength and fatigue. J. Physiol., *123:* 553–564.

METZ, H. S. AND SCOTT, A. B. (1970) Innervational plasticity of the oculomotor system. Arch. Ophthal., *84:* 86–91.

MIDDAUGH, S. J. (1976) Single motor unit control with emg feedback, non-feedback performance, and discrimination of motor unit level events. (abstract); EMG feedback as a muscle reeducation technique: a controlled study (abstract). Two papers presented to the Society of Psychophysiological Research, 1976. Psychophysiology, *14:* 102–103.

MILES, M., MORTENSEN, O. A. AND SULLIVAN, W. E. (1947) Electromyography during normal voluntary movements. Anat. Rec., *98:* 209–218.

MILLER, A. J. (1972a) Characteristics of the swallowing reflex induced by peripheral nerve and brain stem stimulation. Exper. Neurol., *34:* 210–222.

MILLER, A. J. (1972b) Significance of sensory inflow to the swallowing reflex. Brain Res., *43:* 147–159.

MILLER, J. E. (1958) Electromyographic pattern of saccadic eye movements. Am. J. Ophth., *46:* (No. 5, special Part II (not I) of the November issue) 183–186.

MILLER, J. E. (1959) The electromyography of vergence movement. A. M. A. Arch. Ophth., *62:* 790–794.

MILNER, M., BASMAJIAN, J. V. AND QUANBURY, A. O. (1971) Multifactorial analyses of walking by electromyography and computer. Am. J. Phys. Med., *50:* 235–258.

MILNER-BROWN, H. S. AND BROWN, W. F. (1976) New methods of estimating the number of motor units in a muscle. J. Neurol. Neurosurg. Psychiat., *39:* 258–265.

MILNER-BROWN, H. S. AND STEIN, R. B. (1975) The relation between the surface electromyogram and muscular force. J. Physiol. (Lond.), *246:* 549–569.

MILNER-BROWN, H. S., STEIN, R. B. AND LEE, R. G. (1975) Synchronization of human motor units: possible roles of exercise and supraspinal reflexes. Electroencephal. Clin. Neurophysiol., *38:* 245–254.

MILNER-BROWN, H. S., STEIN, R. B. AND YEMM, R. (1972) Mechanisms for increasing force during voluntary contractions. J. Physiol., Lond., *226:* 18–19P.

MILNER-BROWN, H. S., STEIN, R. B. AND YEMM, R. (1973a) Changes in firing rate of human motor units during linearly changing voluntary contractions. J. Physiol., *230:* 371–390.

MILNER-BROWN, H. S., STEIN, R. B. AND YEMM, R. (1973b) The contractile properties of human motor units during voluntary isometric contractions. J. Physiol., *228:* 285–306.

MILOJEVIC, B. (1965) Electronystagmography. Laryngoscope, *75:* 243–258.

MILOJEVIC, B. AND HAST, M. (1964) Cortical motor centers of the laryngeal muscles in the cat and dog. Ann. Otol., Rhin. & Laryng., *73:* 979–989.

MIRANDA, J. M. AND LOURENÇO, R. V. (1968) Influence of diaphragm activity on the movement of total chest compliance. J. Appl. Physiol., *24:* 741–746.

MISSIURO, W. (1963) Studies on developmental stages of children's reflex reactivity. Child Developm., *34:* 33–41.

MISSIURO, W. AND KOZLOWSKI, S. (1961) Investigations on adaptative changes in reciprocal innervation of muscle. Arch. Phys. Med., *44:* 37–41.

MISSIURO, W., KIRSCHNER, H. AND KOZLOWSKI, S. (1962a) Electromyographic manifestations of fatigue during work of different intensity. Acta Physiol. Polon., *13:* 11–23.

MISSIURO, W., KIRSCHNER, H. AND KOZLOWSKI, S. (1962b) Contribution à l' étude de la fatigue au cours de travaux musculaires d'intensités différentes. J. Physiol., Paris, *54:* 717–727.

MITOLO, M. (1956) Studio elettromiografico nel corso dell'allenamento all'esercizio fisico (Italian text). Boll. Soc. Ital. Biol. Sper., *32:* 1413–1415.

MITOLO, M. (1957) Studio elettromiografico nel corso dell'allenamento all'esercizio fisico (Italian text). Lav. Umano, *9:* 2–23.

MITOLO, M. (1964) L'allenamento del muscolo all'esercizio fisico in vecchiaia (ricerche dinamometriche, ergografiche ed elettromiografiche). Lavoro Umano, *16:* 371–390.

MIWA, N. AND MATOBA, M. (1959) Relation between the muscle strength and electromyogram (Japanese text). Orthop and Traumatol., *8:* 121–124.

MIWA, N., TANAKA, T. AND MATOBA, M. (1963) Electromyography in kinesiologic evaluations: subjects on the two joint muscle and the relation between the muscular tension and electromyogram. J. Jap. Orthop. Assoc., *36:* 1025–1035.

MIYASHITA, M., MIURA, M., MATSUI, H. AND MINAMITATE, K. (1972) Measurement of the reaction time of muscular relaxation. Ergonomics, *15:* 555–562.

MIYOSHI, K. (1966) The mechanism of arch support in the foot (Japanese text; English summary). J. Jap. Orthop. Assoc., *40:* 977–984.

MØLLER, A. R. (1961) Bilateral contraction of the tympanic muscles in man: examined by measuring acoustic impedence-change. Ann. Otol. Rhin. & Laryng., *70:* 735–752.

MØLLER, E. (1975) Quantitative features of masticatory muscle activity. In *Occlusion: Research in Form and Function,* ed. by N.H. Rowe. University of Michigan Press, Ann Arbor.

MØLLER, E. (1976) Evidence that the rest position is subject to servo-control. In *Mastication,* ed. by D.J. Anderson and B. Mathews, Wright and Sons, Bristol.

MOORE, A. D. (1966) Synthetic EMG waves. In *Muscles Alive,* 3rd ed. by J.V. Basmajian, Baltimore, pp. 53–70.

MOORE, J. C. (1966) Fabrication of suction-cup electrodes for electromyography. Electroencephalog. & Clin. Neurophysiol., *20:* 405–406.

MOORE, J. C. (1975) Excitation overflow: an electromyographic investigation. Arch. Phys. Med. Rehab., *56:* 115–120.

MORI, S. (1973) Discharge patterns of soleus motor units with associated changes in force exerted by foot during quiet stance in man. J. Neurophysiol. *36:* 458–471.

MORI, S. AND ISHIDA, A. (1976) Synchronization of motor units and its simulation in parallel feedback system. Biol. Cybernetics, *21:* 107–11.

MORIN, C., PIERROT-DESEILLIGNY, E. AND BUSSEL, B. (1976) Role of muscular afferents in the inhibition of the antagonist motor nucleus during a voluntary contraction in man. Brain Res., *103:* 373–376.

MOROSOVA, I. A. AND SHIK, L. L. (1957) Action potentials of respiratory muscles in patients with respiratory deficiencies (Russian text). Byull. Eksper. Biol. Med., *63:* 61–65.

MORRIS, J. M., BENNER, G. AND LUCAS, D. B. (1962) An electromyographic study of the intrinsic muscles of the back in man. J. Anat., *96:* 509–520.

MORRIS, J. M., LUCAS, D. B. AND BRESLER, B. (1961) The role of the trunk in stability of the spine. Biomechanics Laboratory, University of California. Publication no. 42.

MORTIMER, J. T. MAGNUSSON, R. AND PETERSÉN, I. (1970) Isometric contraction, muscle blood flow, and the frequency spectrum of the electromyogram. In *Proceedings of the First Nordic Meeting on Medical and Biological Engineering.* E. Spring, T. Jauhiainen and T. Honkavaara, Eds.; Helsinki, Finland, Soc. Med. & Biol. Eng., Publishers.

MORTON, D. J. (1935) *The Human Foot,* p. 119. Columbia University Press, New York.

MORTON, D. J. (1952) *Human Locomotion and Body Form. A Study of Gravity and*

Man. The Williams & Wilkins Co., Baltimore.

MOSHER, C. B., GERLACH, R. L. AND STUART, D. G. (1972) Soleus and anterior tibial motor units of the cat. Brain Res., *44:* 1–11.

MOYERS, R. E. (1949) Temporomandibular muscle contraction patterns in angle class II, division 1 malocclusions: an electromyographic analysis. Am. J. Orthodontics, *35:* 837–857.

MOYERS, R. E. (1950) An electromyographic analysis of certain muscles involved in temporomandibular movement. Am. J. Orthodontics, *36:* 481–515.

MUELLNER, S. R. (1958) The voluntary control of micturition in man. J. Urol., *80:* 473–478.

MUNRO, R. R. (1974) Activity of the digastric muscle in swallowing and chewing. J. Dent. Res., *53:* 530–537.

MUNRO, R. R. AND ADAMS, C. (1971) Electromyography of the intercostal muscles in connected speech. Electromyography, *11:* 365–378.

MUNRO, R. R. AND ADAMS, C. (1973) The patterns of internal intercostal muscle activity used by some non-native speakers of English. Electromyogr. Clin. Neurophysiol., *13:* 271–288.

MUNRO, R. R. AND BASMAJIAN, J. V. (1971) The jaw opening reflex in man. Electromyography, *11:* 191–206.

MURAKAMI, Y., FUKUDA, H. AND KIRSCHNER, J. A. (1972) The cricopharyngeus muscle: an electrophysiological and neuropharmacological study. Acta Otolaryngol., Suppl. 311.

MURAKAMI, Y. AND KIRSCHNER, J. A. (1971) Electrophysiological properties of laryngeal reflex closure. Acta Otolaryng., *71:* 416–425.

MURAKAMI, Y. AND KIRSCHNER, J. A. (1972) Mechanical and physiological properties of reflex laryngeal closure. Ann. Otol. Rhinol. Laryngol., *81:* 59–72.

MURPHEY, D. L., BLANTON, P. L. AND

BIGGS, N. L. (1971) Electromyographic investigation of flexion and hyperextension of the knee in normal adults. Am. J. Phys. Med., *50:* 80–90.

MURPHY, A. J., KOEPKE, G. H., SMITH, E. M. AND DICKINSON, D. G. (1959) Sequence of action of the diaphragm and intercostal muscles during respiration. II. Expiration. Arch. Phys. Med., *40:* 337–342.

MURRAY, M. P., DROUGHT, A. B. AND KORY, R. C. (1964) Walking patterns of normal men. J. Bone & Joint Surg., *46-A:* 335–360.

MURRAY, M. P., SEIRGE, A. A. AND SEPIC, S. B. (1975) Normal postural stability and steadiness: quantitative assessment. J. Bone & Joint Surg., *57-A:* 510–516.

MUSTARD, W. T. (1952) Iliopsoas transfer for weakness of the hip abductors. J. Bone & Joint Surg., *34-A:* 647–650.

MUSTARD, W. T. (1958) Personal communications.

MUYBRIDGE, E. (1887) *The Human Figure in Motion.* Reprinted in 1955 by Dover Publications Inc., New York.

MYERS, S. J. AND SULLIVAN, W. P. (1968) Effect of circulatory occlusion on time of muscular fatigue. J. Appl. Physiol., *24:* 54–59.

NACHEMSON, A. (1966) Electromyographic studies of the vertebral portion of the psoas muscle. Acta Orthop. Scandinav., *37:* 177–190.

NACHMANSOHN, D. (1959) *Chemical and Molecular Basis of Nerve Activity.* Academic Press, New York.

NAFPLIOTIS, H. (1976) Electromyographic feedback to improve ankle dorsiflexion, wrist extension, and hand grasp. Phys. Ther., *56:* 821–825.

NAGASAWA, T., SASAKI, H. AND TSURU, H. (1976) Masseteric silent period after tooth contact in full denture wearers. J. Dent. Res., *55:* 314.

NAKAARAI, K. (1968) Electromyographic study on the bulbocavernosus muscle of male subjects with normal and abnormal urination. Electromyography, *8:* 367–381.

NAPIER, J. R. (1957) The foot and the shoe. Physiotherapy, *43:* 65–74.

NAPONIELLO, L. V. (1957) An electromyographic study of certain muscles in the easy standing position (abstract). Anat. Rec., *127:* 339.

NEGUS, V. E. (1949) *The Comparative Anatomy and Physiology of the Larynx.* William Heinemann, Ltd., London.

NESBIT, R. M. AND LAPIDES, J. (1959) The physiology of micturition. J. Michigan M. Soc., *58:* 384–388.

NETSELL, R. AND CLEELAND, C. S. (1973) Modification of lip hypertonia in dysarthria using EMG feedback. J. Speech Hearing Dis., *38:* 131–140.

NIELSON, V. K. (1973) Sensory and motor nerve conduction in the median nerve in normal subjects. Acta Med. Scand., *194:* 435–443.

NIEPORENT, H. J. (1956) An electromyographic study of the function of the respiratory muscles in normal subjects (Dissertation for M.D. degree) University of Zurich, 30 pp.

NORDH, E., JONSSON, B. AND LADD, H. (1974) The learning process for fine neuromuscular controls in skeletal muscles of man: VII. interelectrode distance. Electromyogr. Clin. Neurophysiol., *14:* 475–483.

NORRIS, E. H., JR. AND GASTEIGER, E. L. (1955) Action potentials of single motor units in normal muscle. Electroencephalog. & Clin. Neurophysiol., *7:* 115–126.

NORRIS, F. H., JR., GASTEIGER, E. L. AND CHATFIELD, P. O. (1957) An electromyographic study of induced and spontaneous muscle cramps. Electroencephalog. & Clin. Neurophysiol., *9:* 139–147.

NORRIS, F. H., JR. AND IRWIN, R. L. (1961) Motor unit area in a rat muscle. Am. J. Physiol., *200:* 944–946.

NORRIS, A. H., SHOCK, N. W. AND WAGMAN, I. H. (1953) Age changes in the maximum conduction velocity of motor fibers of human ulnar nerves. J. Appl. Physiol., *5:* 589–593.

NOVOZAMSKY, V. AND BUCHBERGER, J. (1970) Die Fusswölbung nach Belastung durch einin 100 km-Marsch. Z. Anat. Entwickl.-Gesch., *131:* 243–248.

OBRIST, P. A. (1968) Heart rate and somatic coupling during classical aversive conditioning in humans. J. Exper. Psychol., *77:* 180–193.

O'CONNELL, A. L. (1958) Electromyographic study of certain leg muscles during movements of the free foot and during standing. Am. J. Phys. Med., *37:* 289–301.

O'CONNELL, A. L. (1971) Effect of sensory deprivation on postural reflexes. Electromyography, *11:* 519–527.

O'CONNELL, A. L. AND GARDNER, E. B. (1963) The use of electromyography in kinesiological research. Res. Quart., *34:* 166–184.

O'CONNELL, A. L. AND MORTENSEN, O. A. (1957) An electromyographic study of the leg musculature during movements of the free foot and during standing. Anat. Rec., *127:* 342.

O'DELL, N. L., TODD, G. L., III, AND BERNARD, G. R. (1970) Musculoskeletal arangements for lateral mandibular movements in the rabbit and rat: electromyographic and other analyses. J. Dent. Res., *49:* 1111–1117.

OGANISYAN, A. A. AND IVANOVA, S. N. (1964) A new method of implantation of electrodes into the muscles of a dog's extremities for EMG recording during free movement (Russian text). Byull. Eksper. Biol. Med., *57:* 136–138.

OGAWA, T., JEFFERSON, N. C., TOMAN, J. E., CHILES, T. ZAMBETOGLOU, A. AND NECHELES, H. (1960) Action potentials of accessory respiratory muscles in dogs. Am. J. Physiol., *199:* 569–572.

OHYAMA, M., UEDA, N., HARVEY, J. E., MOGI, G. AND OGURA, J. H. (1972) Electrophysiologic study of reinnervated laryngeal motor units. Laryngoscope, *82:* 237–251.

OKA, H., OKAMOTO, T. AND KUMAMOTO, M. (1976) Electromyographic and cinematographic study of the volleyball spike. *International Series on Biomechanics, Vol. 1A, Biomechanics, Vol. V-A,* University Park Press, Baltimore, pp. 326–331.

OKADA, M. (1972) An electromyographic estimation of the relative muscular load in different human postures. J. Human Ergol., *1:* 75–93.

OKADA, M. (1975) Quantitative studies on the bearing of the anti-gravity muscles in human postures with special references to electromyographic estimation of the postural muscle load. J. Fac. Sc., Univ. Tokyo, Sec. V, *4:* 471–530.

OKAMOTO, T. (1968a) Electromyographic study of the function of m. rectus femoris. Res. J. Phys. Ed. (Japan) (English text), *12:* 175–182.

OKAMOTO, T. (1968b) A study of the variation of discharge pattern during flexion of the upper extremity. J. Lib. Arts Dept., Kansai Med. School, *2:* 111–122.

OKAMOTO, T. (1973) Electromyographic study of the learning process of walking in 1- and 2-year-old infants. Medicine and Sports: Biomechanics III, *8:* 328–333.

OKAMOTO, T. (1976) Yoshoji no Suiei-Shidokatei no Bunseki (A kinesiological study of swimming). J. Health Phys. Ed. Rec., *26:* 409–414.

OKAMOTO, T., AND KUMAMOTO, M. (1971) Electromyographic study of the process of acquisition of proficiency in gymnastic kip. Res. J. Phys. Ed., *17:* 385–394.

OKAMOTO, T. AND KUMAMOTO, M. (1972) Electromyographic study of the learning process of walking in infants. Electromyography, *12:* 149–158.

OKAMOTO, T., KUMAMOTO, M. AND TAKAGI, K. (1966) Electromyographic study of the function of m. adductor longus and m. adductor magnus. Jap. J. Phys. Fitness, *15:* 43–48.

OKAMOTO, T., TAKAGI, K. AND KUMAMOTO, M. (1967) Electromyographic study of elevation of the arm. Res. J. Phys. Ed. (Jap. Soc. Phys. Ed.), *11:* 127–136.

OKAMOTO, T., TOKUYAMA, H., YOSHIZAWA, M., DODAIRA, A., TSUJINO, A. AND KUMAMOTO, M. (1976) Yohshoji no Suiei no Kindenzuteki Kenkyu (An electromyographic study of swimming in children). *The Science of Human Movement,* Kyorin Shoin, Tokyo, pp. 115–126.

OKHNYANSKAYA, L. G., YUSEVICH, YU. S. AND NIKIFOROVA, N. A. (1974) Some questions relating to electromyographic investigations of synergies in man. Electromyogr. Clin. Neurophysiol., *14:* 391–414.

OLSEN, C. B., CARPENTER, D. O. AND HENNEMAN, E. (1968) Orderly recruitment of muscle action potentials: motor unit threshold and EMG amplitude. Arch. Neurol., *19:* 591–597.

ONGERBOER DE VISSER, B. W. AND GOOR, C. (1976) Cutaneous silent period in masseter muscles: a clinical and diagnostic evaluation. J. Neurol. Neurosurg. Psychiat., *39:* 674–679.

ONO, K. (1958) Electromyographic studies of the abdominal wall muscles in visceroptosis. I. Analysis of patterns of activity of the abdominal wall muscles in normal adults. Tohoku J. Exper. Med., *68:* 347–354.

OOTA, Y. (1956a) Electromyography in the study of clinical kinesiology. Kyushu J. M. Sc., *7:* 75–91.

OOTA, Y. (1956b) Electromyography in the practice of orthopedic surgery. Kyushu J. M. Sc., *7:* 49–62.

ÖWALL, B. AND ELMQVIST, D. (1975) Motor pauses in emg activity during chewing and biting. Odont. Rev., *26:* 17–38.

PANIN, N., LINDENAUER, H. J., WEISS, A.

A. AND EBEL, A. (1961) Electromyographic evaluation of the "cross exercise" effect. Arch. Phys. Med., *42:* 47–52.

PARTRIDGE, M. J. AND WALTERS, C. E. (1959) Participation of the abdominal muscles in various movements of the trunk in man; an electromyographic study. Phys. Therapy Rev., *39:* 791–800.

PATON, W. D. M. AND WAND, D. R. (1967) The margin of safety of neuromuscular transmission. J. Physiol. (London), *191:* 59–90.

PATTON, N. J. AND MORTENSEN, O. A. (1970) A study of some mechanical factors affecting reciprocal activity in one-joint muscles. Anat. Rec., *166:* 360.

PATTON, N. J. AND MORTENSEN, A. O. (1971) An electromyographic study of reciprocal activity of muscles. Anat. Rec., *170:* 255.

PAUL, G. L. (1969) Physiological effects of relaxation training and hypnotic suggestion. J. Abnorm. Psychol., *74:* 425–437.

PAULY, J. E. (1957) Electromyographic studies of human respiration. Chicago M. School Quart., *18:* 80–86.

PAULY, J. E. (1966) An electromyographic analysis of certain movements and exercises. Part I: some deep muscles of the back. Anat. Rec., *155:* 223–234.

PAULY, J. E. AND STEELE, R. W. (1966) Electromyographic analysis of back exercises for paraplegic patients. Arch. Phys. Med., *47:* 730–736.

PAYTON, O. D. (1974) Electrical correlates of motor skill development in an isolated movement: the triceps and anconeus as agonists. Proc. World Conf. Phys. Ther., Montreal, pp. 132–139.

PAYTON, O. D. AND KELLEY, D. L. (1972) Electromyographic evidence of the acquisition of a motor skill. Phys. Therapy, *52:* 261–266.

PAYTON, O. D., SU, S. AND MEYDRICH, E.

F. (1976) Abductor digiti shuffleboard: a study in motor learning. Arch. Phys. Med. Rehab., *57:* 169–174.

PEARSON, K. (1976) The control of walking. Sci. Amer., *235:* 72–87.

PEAT, M. AND GRAHAME, R. E. (1977) Shoulder function in hemiplegia: a method of electromyographic and electrogoniometric analysis. Physioth. Can., *29:* 1–7.

PEDOTTI, A. (1977) A study of motor coordination and neuromuscular activities in human locomotion. Biol. Cyber., *26:* 53–62.

PERRITT, R. Q. AND MILNER, M. (1976) A simple, inexpensive 8-channel multiplexer for electromyography in human locomotion. Med. & Biol. Eng., (Jan.) 104–106.

PERSON, R. S. (1958) Electromyographical study of coordination of the activity of human antagonist muscles in the process of developing motor habits (Russian text). Jurn. Vys'cei Nervn. Dejat., *8:* 17–27.

PERSON, R. S. (1965) *Antagonistic Muscles in Human Movements.* (Russian test; English Abstract). Moscow, Acad. Science, U.S.S.R. (publishers).

PERSON, R. S. (1969) *Electromyography in Human Study* (Russian text). Moscow, Acad. Science, U.S.S.R. (publishers).

PERSON, R. S. AND KUDINA, L. P. (1971) Pattern of human motoneuron activity during voluntary muscular contraction. Neurophysiology, *3(6):* 455–462.

PERSON, R. S. AND KUDINA, L. P. (1972) Discharge frequency and discharge pattern in human motor units during voluntary contractions of muscle. Electroenceph. Clin. Neurophysiol., *32:* 471–483.

PERSON, R. S. AND MISHIN, L. N. (1964) Auto- and cross-correlation analysis of the electrical activity of muscles. Med. Electron. Biol. Engng., *2:* 155–159.

PERSON, R. S. AND ROSHTCHINA, N. A. (1958) Electromyographic investigation of coordinated activity of antagonistic muscles in movements of fingers of the human hand (Russian text). J. Physiol U.S.S.R., *94:* 455–462.

PERSSON, A., LEANDERSON, R. AND ÖHMAN, S. (1969) Electromyographic studies of facial muscle activity in speech. Proc. 6th IFSECN Congress, in Electroenceph. Clin. Neurophysiol., *27:* 641–735.

PERTUZON, E. AND LESTIENNE, F. (1968) Caractère électromyographique d'un mouvement monoarticulaire executé à vitesse maximale. J. Physiol., *60:* 513.

PETAJAN, J. H. AND PHILIP, B. A. (1969) Frequency control of motor unit action potentials. Electroceph. Clin. Neurophysiol., *27:* 66–72.

PETAJAN, J. H. AND WILLIAMS, D. D. (1972) Behavior of single motor units during pre-shivering tone and shivering tremor. Am. J. Phys. Med., *51:* 16–22.

PETERS, J. F. (1971) Eye movement recording: a brief review. Psychophysiology, *8:* 414–416.

PETERSÉN, I. AND FRANKSSON, C. (1955) Electromyographic study of the striated muscles of the male urethra. Brit. J. Urol., *27:* 148–153.

PETERSÉN, I., FRANKSSON, C. AND DANIELSSON, C.-O. (1955) Electromyographic study of the muscles of the pelvic floor and urethra in normal females. Acta Obst. Gynec. Scandinav., *34:* 273–285.

PETERSÉN, I. AND KUGELBERG, E. (1949) Duration and form of action potential in the normal human muscle. J. Neurol. Neurosurg. & Psychiat., *12:* 124–128.

PETERSÉN, I. AND STENER, B. (1959) Experimental evaluation of the hypothesis of ligamento-muscular protective reflexes. III. A study in man using the medial collateral ligament of the knee joint. Acta Physiol. Scandinav., *48:* suppl. 166, 51–61.

PETERSEN, I. AND STENER, I. (1970) An electromyographical study of the striated urethral sphincter, the striated anal sphincter, and the levator ani muscle during ejaculation. Electromyography, *10:* 23–44.

PETERSÉN, I. STENER, I., SELLDÉN, U. AND KOLLBERG, S. (1962) Investigation of urethral sphincter in women with simultaneous electromyography and micturition urethro-cystography. Acta Neurol. Scandinav., *38:* suppl. 3, 145–151.

PETIT, J. M., DELHEZ, L., DEROANNE, R., PIRNAY, F. AND BOCCAR, M. (1968) Régulation de la ventilation pendant l'exercice musculaire. Trav. Soc. Méd. Belge d'Educ. Phys. Sports, *20:* 82–87.

PETIT, J. M., DELHEZ, L. AND TROQUET, J. (1965) Aspects actuels de la mécanique ventilatoire. J. Physiol., Paris, *57:* 7–113.

PETIT, J. M., MILIC-EMILI, G. AND DELHEZ, L. (1960) Role of the diaphragm in breathing in conscious normal man: an electromyographic study. J. Appl. Physiol., *15:* 1101–1106.

PHILLIPS, C. G. (1975) In Laying the ghost of 'muscles versus ligaments' a lecture to the Xth Can. Congr. of Neurol. Science, 1975.

PHUON-MONICH. (1963) La capacité de travail statique intermittent. (M.D. thesis) Paris, Editions A.G.E.M.P. (Publishers).

PIERCE, D. S. AND WAGMAN, I. H. (1964) A method of recording from single muscle fibers or motor units in human skeletal muscle. J. Appl. Physiol., *19:* 366–368.

PIERCE, J. M., JR., ROBERGE, J. T. AND NEWMANN, M. M. (1960) Electromyographic demonstration of bulbocavernosus reflex. J. Urol., *83:* 319.

PIETTE, P. (1974) L'exploration électromyographique du muscle transversaire épinaux. Electromyogr. Clin. Neurophysiol., *14:* 69–77.

PIHKANEN, T., HARENKO, A. AND HUHMAR, E. (1965) Observations on the conduction velocity in peripheral nerves in states of drug intoxication: studies of the ulnar nerve in acute drug intoxication. Acat Neurol. Scandinav., *41:* suppl. 13, 267–271.

PIPER, H. (1912) *Electrophysiologie Menschlicher Muskeln* (German text). Springer-Verlag, Berlin.

PISHKIN, V. AND SHURLEY, J. T. (1968) Electrodermal and electromyographic parameters in concept identification. Psychophysiology, *5:* 112–118.

PISHKIN, V., SHURLEY, J. T. AND WOLFGANG, A. (1968) Assessment of drugs in decision-making under stress. Dis. Nerv. Syst., *29:* 841–843.

PIVIK, T. AND DEMENT, W. C. (1970) Phasic changes in muscular and reflex activity during non-REM sleep. Exper. Neurol., *27:* 115–124.

POCOCK, G. S. (1963) Electromyographic study of the quadriceps during resistive exercises. J. Amer. Phys. Ther. Assoc., *43:* 427–434.

PODIVINSKÝ, F. (1964) Factors affecting the course and the intensity of crossed motor irradiation during voluntary movement in healthy human subjects. Physiologica Bohemoslov., *13:* 172–178.

POLLOCK, L. J. AND DAVIS, L. (1930) Reflex activities of the decerebrate animal. J. Comp. Neurol., *50:* 377–411.

POMMERANKE, W. T. (1928) A study of the sensory areas eliciting the swallowing reflex. Am. J. Physiol., *84:* 36–41.

PORTER, N. H. (1960) (unpublished) Reported in Todd, I. P. (1962) Some aspects of the physiology of continence and defaecation. Arch. Dis. Childhood, *37:* 181–183.

PORTMANN, G. (1956) Myographies des cordes vocales chez l'homme (French text). Rev. Laryng., *77:* 1–10.

PORTMANN, G. (1957) The physiology of phonation (the Semon lecture for 1956)

J. Laryng. & Otol., *71:* 1–15.

PORTMANN, G., ROBIN, J.-L., LAGET, P. AND HUSSON, R. (1956) La myographie des cordes vocales (French text). Acta Otolaryng., *46:* 250–263.

PORTNOY, H. AND MORIN, F. (1956) Electromyographic study of postural muscles in various positions and movements. Am. J. Physiol., *186:* 122–126.

POUDRIER, C. AND KNOWLTON, G. C. (1964) Command-force relations during voluntary muscle contraction. Am. J. Phys. Med., *43:* 109–116.

POWERS, W. R. (1969) *Conscious Control of Single Motor Units in the Preferred and Non-Preferred Hand.* Ph.D. Thesis. Queen's University Kingston, Ontario, Canada.

PÓZNIAK-PATEWICZ, E. (1975) Electromyographic investigations of nape and temporal muscles in headaches of various origin (Polish text; English abstract). Neur. Neurochir. Pol., *9:* 461–468.

PRABHU, V. G. AND OEXTER, Y. T. (1971) Electromyographic studies of skeletal muscle of rat given cortisone. Arch. Neurol., *24:* 253–258.

PRECHTL, H. F. R. AND LENARD, H. G. (1967) A study of eye movements in sleeping newborn infants. Brain Res., *5:* 477–493.

PRUZANSKY, S. (1952) The application of electromyography to dental research. J. Am. Dent. A., *44:* 49–68.

RADCLIFFE, C. W. (1962) The biomechanics of below-knee prostheses in normal, level, bipedal walking. Artif. Limbs. *6:* 16–24.

RAINAUT, J.-J. (1971) Étude de la marche par electromyographie télémétrique. Rev. Chir. Orthop. Repat. App. Moteur (Paris), *57:* 427–437.

RALSTON, H. J. (1961) Uses and limitations of electromyography in the quantitative study of skeletal muscle function. Am. J. Orthodont. *47:* 521–530.

RALSTON, H. J. (1964) Effects of immobilization of various body segments on the energy cost of human locomotion. Ergonomics, *7:* 53–60.

RALSTON, H. J. AND LIBET, B. (1953) The question of tonus in skeletal muscle. Amer. J. Phys. Med., *32:* 85–92.

RALSTON, H. J., TODD, F. N. AND INMAN, V. T. (1976) Comparison of electrical activity and duration of tension in the human rectus femoris muscle. Electromyogr. Clin. Neurophysiol., *16:* 277–286.

RAMSEY, R. W. (1960) Some aspects of the biophysics of muscle, in *The Structure and Function of Muscle*, Vol. II., p. 323–327, G. H. Bourne, Ed. Academic Press, New York.

RANNEY, D. A. AND BASMAJIAN, J. V. (1960) Electromyography of the rabbit fetus. J. Exper. Zool., *144:* 179–186.

RAO, V. R. (1963) Interesting electromyographic changes induced by smoking. J. Postgrad. Med., *9:* 138–139.

RAO, V. R. (1965) Reciprocal inhibition: inapplicability to tendon jerks. J. Postgrad. Med., *11:* 123–125.

RAO, V. R. AND RINDANI, T. H. (1962) The influence of smoking on electromyograms. J. Postgrad. Med., *8:* 170–172.

RAPER, A. J., THOMPSON, W. T., JR., SHAPIRO, W. AND PATTERSON, J. L., JR. (1966) Scalene and sternomastoid muscle function. J. Appl. Physiol., *21:* 497–502.

RAPHAEL, L. J. (1975) The physiological control of durational differences between vowels preceding voiced and voiceless consonants in English. J. Phonet., *3:* 25–33.

RAPHAEL, L. J. AND BELL-BERTI, F. (1975) Tongue musculature and the feature of tension in English vowels. Phonetica, *32:* 61–73.

RASCH, P. J. AND BURKE, R. K. (1974) *Kinesiology and Applied Anatomy: The Science of Human Movement*, ed. 5. Lea & Febiger, Philadelphia.

RATTNER, W. H., GERLAUGH, B. L., MURPHY, J. J. AND ERDMAN, W. J., II. (1958) The bulbocavernosus reflex. I. Electromyographic study of normal patients. J. Urol., *80:* 140–141.

RAVAGLIA, M. (1957) Sulla particolare attività dei capi muscolari del quadricipite femorale nell'uomo (indagine elettromiografica) (Italian text). Circ. Org. Movimento, *44:* 498–504.

RAVAGLIA, M. (1958) Indagine elettromiografica della funzione dei muscoli pettorali nella meccanica respiratoria (Italian text). Ginnastica Med., *5:* 1–3.

RAY, R. D., JOHNSON, R. J. AND JAMESON, R. M. (1951) Rotation of the forearm: an experimental study of pronation and supination. J. Bone & Joint Surg., *33-A:* 993–996.

REDFORD, J. B., BUTTERWORTH, T. R. AND CLEMENTS, E. L. (1969) Use of electromyography as a prognostic aid in the management of idiopathic scoliosis. Arch. Phys. Med., *50:* 433–438.

REINKING, R. M., STEPHENS, J. A. AND STUART, D. G. (1975) The motor units of cat medial gastrocnemius: problem of their categorisation on the basis of mechanical properties. Exp. Brain Res., *23:* 301–313.

REIS, F. P. AND CARVALHO, C. A. F. (1973) Electromyographic study of the popliteus muscle. Electromyogr. Clin. Neurophysiol., *13:* 445–455.

RICHARDS, C. AND KNUTSSON, E. (1974) Evaluation of abnormal gait patterns by intermittent-light photography and electromyography. Scand. J. Rehab. Med., *3:* suppl. 61–68.

RICHTER, J. (1966) Vorläufige Mitteilung über elektromyographische Untersuchungen am unbelasteten und belasteten Fuss. Arch. Orthop. Unfall-Chir., *59:* 168–176.

RIDDLE, H. F. V. AND ROAF, R. (1955) Muscle imbalance in the causation of scoliosis. Lancet, *1:* June 18, 1245–1247.

ROBINSON, D. A. (1965) The mechanics of human smooth pursuit eye movement. J. Physiol., *180:* 569–591.

ROMAN, C. AND CAR, A. (1970) Déglutitions et contractions oesophagiennes réflexes obtenues par la stimulation des nerfs vague et laryngé supérieur. Exp. Brain Res., *11:* 48–74.

ROSEMEYER, B. (1971) Elektromyographische Untersuchungen der Rücken- und Schultermuskulatur im Stehen und Sitzen unter Berücksichtigung der Haltung des Autofahrers. Arch. Orthop. Unfall-Chir., *69:* 59–70.

ROSENFALCK, A. (1960) Evaluation of the electromyogram by mean voltage recording. In *Medical Electronics: Proceedings of the Second International Conference on Medical Electronics, Paris, 24–27 June, 1959,* ed. by C. N. Smyth, pp. 9–12. Iliffe & Sons, Ltd., London.

ROSENFALCK, P. (1969) *Intra- and Extracellular Potential Fields of Active Nerve and Muscle Fibers,* Academisk Forlag, Kobenhavn.

RUBIN, H. J. (1960) Further observations on the neurochronaxic theory of voice production. A.M.A. Arch. Otolaryng., *72:* 207–211.

RUCH, T. C., PATTON, H. D., WOODBURY, J. W. AND TOWE, A. L. (1963) *Neurophysiology.* W. B. Saunders Company, Philadelphia.

RUËDI, L. (1959) Some observations on the histology and function of the larynx (Semon lecture for 1958). J. Laryng. & Otol., *73:* 1–20.

RUMMEL, R. M. (1974) An electromyographic analysis of patterns used to reproduce muscular tension. Res. Quart. Am. Assoc. Health Phys. Ed., *45:* 64–71.

RUSKIN, A. P. (1970) Anal sphincter electromyography. Electromyography, *10:* 425–428.

SACCO, G., BUCHTHAL, F. AND ROSENFALCK, P. (1962) Motor unit potentials at different ages. Arch. Neurol., *6:* 366–373.

SALA, E. (1959) Studio elettromiografico dell'innervazione dei muscoli flessore breve ed opponente del pollice (Italian text). Riv. Pat. Nerv., *80:* 131–147.

SALOMON, G. AND STARR, A. (1963) Electromyography of middle ear muscles in man during motor activities. Acta Neurol. Scandinav., *39:* 161–168.

SAMILSON, R. L. AND MORRIS, J. M. (1964) Surgical improvement of the cerebral-palsied upper limb: electromyographic studies and results in 128 operations. J. Bone & Joint Surg., *46-A:* 1203–1216.

SAMSON, J. (1971) *The Acquisition, Retention and Transfer of Single Motor Unit Control.* Ph.D. Dissertation, Univ. of Illinois at Urbana-Champaign.

SANO, E., ANDO, K., KATORI, I., YAMADA, H., SAMPERI, H. AND SUGAHARA, R. (1977) Electromyographic studies on the forearm muscle activities during finger movements. J. Jap. Orthop. Assoc., *51:* 331–337.

SANT'AMBROGIO, G., FRAZIER, M. F., WILSON, M. F. AND AGOSTONI, E. (1963) Motor innervation and pattern of activity of cat diaphragm. J. Appl. Physiol., *18:* 43–46.

SANT'AMBROGIO, G. AND WIDDICOMBE, J. G. (1965) Respiratory reflexes acting on the diaphragm and inspiratory intercostal muscles of the rabbit. J. Physiol., *180:* 766–779.

SATO, M. (1963) Electromyographical study of skilled movement. J. Fac. Sci. Univ. Tokyo, *Sec. V, Vol. II, Part 4:* 323–369.

SATO, M. (1966) Muscle fatigue in the half rising posture (English text). Zinruigaku Zassi; J. of Anthropol. Soc. (Nippon), *74:* 13–19.

SATO, M., HAYAMI, A. AND SATO, H. (1965) Differential fatiguability between the one- and two-joint muscles. Zinruigaku Zassi (J. Anthropol. Soc., Nippon), *73:* 82–90.

SAUERLAND, E. K. AND HARPER, R. M.

(1976) The human tongue during sleep; electromyographic activity of the genioglossus muscle. Exper. Neurol., *51:* 160–170.

SAUERLAND, E. K. AND MITCHELL, S. P. (1975) Electromyographic activity of intrinsic and extrinsic muscles of the human tongue. Texas Reports Biol. & Med., *33:* 445–455.

SAUNDERS, J. B. DeC. M., INMAN, V. T. AND EBERHART, H. D. (1953) The major determinations in normal and pathological gait. J. Bone & Joint Surg., *35-A:* 543–558.

SCHÄRER, P. AND PFYFFER, G. L. (1970) Comparison of habitual and cerebrally stimulated jaw movements in the rabbit. Helv. Odont. Acta, *14:* 6–10.

SCHERR, M. S., CRAWFORD, P. L., SERGENT, B., SCHERR, C. A. (1975) Effect of bio-feedback techniques on chronic asthma in a summer camp environment. Ann. Allergy, *35:* 289–295.

SCHERRER, J. AND BOURGUIGNON, A. (1959) Changes in the electromyogram produced by fatigue in man. Am. J. Phys. Med., *38:* 170–180.

SCHERRER, J., BOURGUIGNON, A. AND MARTY, R. (1957) Evaluation électromyographique du travail statique (French text). J. Physiol., Paris, *49:* 376–378.

SCHERRER, J., BOURGUIGNON, A., SAMSON, M. AND MARTY, R. (1956) Sur un caractère particulier de la contraction isométrique maximum au cours de la fatigue chez l'homme (French text). J. Physiol., Paris, *48:* 704–707.

SCHERRER, J., LEFEBVRE, J. AND BOURGUIGNON, A. (1957) Activité électrique du muscle strié squelettique et fatigue (French text). In *Fourth International Congress of Electroencephalography and Clinical Neurophysiology*, Brussels, pp. 99–123.

SCHERRER, J. AND MONOD, H. (1960) Le travail musculaire local et la fatigue chez l'homme (French text). J. Physiol., Paris, *52:* 419–501.

SCHEVING, L. E. AND PAULY, J. E. (1959) An electromyographic study of some muscles acting on the upper extremity of man. Anat. Rec., *135:* 239–246.

SCHLAPP, M. (1973) Observations on a voluntary tremor—violinist's vibrato. Quart. J. Exper. Physiol., *58:* 357–368.

SCHLOON, H., O'BRIEN, M. J., SCHOLTEN, C. A. AND PRECHTL, H. F. R. (1976) Muscle activity and postural behaviour in newborn infants. Neuropädiatrie, *7:* 384–415.

SCHLOSSBERG, L. AND HARRIS, S. C. (1956) An electromyographic investigation of the functioning perioral and suprahyoid musculature in normal occlusion and Class I division I dysplasia cases. Am. J. Orthodont., *42:* 153.

SCHMIDT, R. S. (1972) Action of intrinsic laryngeal muscles during release calling in leopard frog. J. Exper. Zool., *181:* 233–244.

SCHOOLMAN, A. AND FINK, B. R. (1963) Permanently implanted electrode for electromyography of the diaphragm in the waking cat. Electroencephalog. & Clin. Neurophysiol., *15:* 127–128.

SCHUBERT, H. A. (1963) A study of motor nerve conduction: determination of velocity. South. M. J., *56:* 666–668.

SCHUBERT, H. A. (1964) Conduction velocities along course of ulnar nerve. J. Appl. Physiol., *19:* 423–426.

SCHULTE, F. J. AND SCHWENZEL, W. (1965) Motor control and muscle tone in the newborn period. Electromyographic studies. Biol. Neonat., *8:* 198–215.

SCHUSTER, M. M. (1975) The riddle of the sphincters. Gastroenterology, *69:* 249–262.

SCHWARTZ, G. E., FAIR, P. L., SALT, P., MANDEL, M. R. AND KLERMAN, G. L. (1976a) Facial expression and imagery in depression: an electromyographic study. Psychosom. Med., *38:* 337–347.

SCHWARTZ, G. E., FAIR, P. H., SALT, P., MANDEL, M. R. AND KLERMAN, G. L.

(1976b) Facial muscle patterning to affective imagery in depressed and non-depressed subjects. Science, *192:* 489–491.

SCHWARTZ, R. P., TRAUTMAN, O. AND HEATH. A. L., (1936) Gait and muscle function recorded by electrobasograph. J. Bone & Joint Surg., *18:* 445–454.

SCOTT, R. N. (1965) A method of inserting wire electrodes for electromyography. IEEE Tr. Bio-Med. Engng., *BME-12:* 46–47.

SCOTT, R. N. AND THOMPSON, G. B. (1969) An improved biopolar wire electrode for electromyography. Med. & Biol. Engng., *7:* 677–678.

SCRIPTURE, E. W., SMITH, T. L. AND BROWN, E. M. (1894) On the education of muscular control and power. Studies from the Yale Psychological Lab., *2:* 114–119.

SCULLY, H. E. AND BASMAJIAN, J. V. (1969) Motor-unit training and influence of manual skill. Psychophysiology, *5:* 625–632.

SEARS, M. L., TEASDALL, R. D. AND STONE, H. H. (1959) Stretch effects in human extraocular muscle: an electromyographic study. Bull. Johns Hopkins Hosp., *104:* 174–178.

SERRA, C. (1964) Neuromuscular studies on various oto-laryngological problems. Electromyography, *4:* 254–294.

SERRA, C. (1968) Motor driving aptitude evaluation: an electromyographic contribution. Europa Medicophys., *4:* 1–5.

SERRA, C. (1970) Electromyographic contributions to human visual perception mechanisms. Internat. J. Neurol., *8:* 62–69.

SERRA, C. AND COVELLO, L. (1959) Elettromiografia clinica (Italian text). Acta Neurol. (Naples), *20:* 1–507.

SERRA, C. AND LAMBIASE, M. (1957) Fumo e sistema nervoso; II. Modificazioni dei potenziali d'azione muscolari da fumo (Italian text). Acta Neurol. (Naples), *8:* 494–506.

SERRA, C., PASANISI, E. J. AND DE NATALE, G. (1963) L'attività elettrica muscolare del coniglio in condizione di ipotermia controllata. Il Cardarelli (Rev. Ospedali Riun. di Napoli), *5:* 1–8.

SETTINERI, L. I. C. AND RODRIGUEZ, R. B. (1974). Estudo eletromiográfico da mobilizacão ativa e passiva do cotovelo. Med. Esporte, Porto Alegre (Brazil), *1:* 161–166.

SEYFFARTH, H. (1940) *The Behaviour of Motor-Units in Voluntary Contraction,* I. Kommisjon Hos Jacob Dybwad, A. W. Brøggers Boktrykkeri A/S, Oslo.

SHAGASS, C. (1961) Evoked cortical potentials and sensations in man. J. Neuropsychiat., *2:* 262–270.

SHAFIK, A. (1975) A new concept of the anatomy of the anal sphincter mechanism and the physiology of defecation. Investig. Urol., *12:* 412–419.

SHAHANI, B. T. AND YOUNG, R. R. (1976) Physiological and pharmacological aid in the differential diagnosis of tremor. J. Neurol. Neurosurg. Psychiat., *39:* 772–783.

SHAMBES, G. M. (1976) Static postural control in children. Am. J. Phys. Med., *55:* 221–252.

SHANKWEILER, D., HARRIS, K. S. AND TAYLOR, M. L. (1968) Electromyographic studies of articulation in aphasia. Arch. Phys. Med., *49:* 1–8.

SHARP, J. T., DRUZ, W., DANON, J. AND KIM, M. J. (1976) Respiratory muscle function and the use of respiratory muscle electromyography in the evaluation of respiratory regulation. Chest, *70:* suppl. 150S–154S.

SHARRARD, W. J. W. (1964) Posterior iliopsoas transplantation in treatment of paralytic dislocation of the hip. J. Bone & Joint Surg., *46-B:* 426–44.

SHEFFIELD, F. J. (1962) Electromyographic study of the abdominal muscles in walking and other movements. Am. J. Phys. Med., *41:* 142–147.

SHEFFIELD, F. J., GERSTEN, J. W. AND

MASTELLONE, A. F. (1956) Electromyographic study of the muscles of the foot in normal walking. Am. J. Phys. Med., *35:* 223–236.

SHERRINGTON, C. S. (1929) Ferrier Lecture. Some functional problems attaching to convergence. Proc. Roy. Soc., *105B:* 332–362.

SHEVLIN, M. G., LEHMANN, J. F. AND LUCCI, J. A. (1969) Electromyographic study of the function of some muscles crossing the glenohumeral joint. Arch. Phys. Med., *50:* 264–270.

SHIAVI, R. (1974) A wire multielectrode for intramuscular recording. Med. & Biol. Eng., (Sept.), 721–723.

SHIAVI, R. AND NEGIN, M. (1975) Stochastic properties of motoneuron activity and the effect of muscular length. Biol. Cybernetics, *19:* 231–237.

SHIK, M. L. AND ORLOVSKY, G. N. (1976) Neurophysiology of locomotor automatism. Physiol. Rev., *56:* 465–501.

SHIMAZU, H., HONGO, T., KUBOTA, K. AND NARABAYASHI, H. (1962) Rigidity and spasticity in man: electromyographic analysis with reference to the role of the globus pallidus. Arch. Neurol., *6:* 10–17.

SHIPP, T. (1968) A technique for examination of laryngeal muscles during phonation. Proceedings of I.S.E.K., Montreal, 1968. Electromyography, *8:* suppl. 21–26.

SHIPP, T. (1970) EMG of pharyngoesophageal musculature during alaryngeal voice production. J. Speech Hearing Res. *13:* 184–192.

SHIPP, T., DEATSCH, W. W. AND ROBERTSON, K. (1968) A technique for electromyographic assessment of deep neck muscle activity. Laryngoscope, *78:* 418–432.

SHIPP, T., FISHMAN, B. V. AND MORRISSEY, P. (1970) Method and control of laryngeal emg electrode placement in man. J. Acoust. Soc. Am., *48:* 429–430.

SHIPP, T. AND MCGLONE, R. E. (1971) Laryngeal dynamics associated with voice frequency change. J. Speech Hearing Res., *14:* 761–768.

SILLS, F. D. AND OLSEN, A. L. (1958) Action potentials in unexercised arm when opposite arm exercised. Res. Quart A.A.H.P.E., *29:* 213–221.

SIMARD, T. (1969) Fine sensorimotor control in healthy children: an electromyographic study. Pediatrics, *43:* 1035–1041.

SIMARD, T. (1971) Unusual electromyographic motor unit action potentials in human healthy subjects. Electromyography, *11:* 83–91.

SIMARD, T. G. AND BASMAJIAN, J. V. (1965) Factors influencing motor unit training in man. Proc. Can. Fed. Biol. Soc., *8:* 63.

SIMARD, T. G. AND BASMAJIAN, J. V. (1967) Methods in training the conscious control of motor units. Arch. Phys. Med., *48:* 12–19.

SIMARD, T. G., BASMAJIAN, J. V. AND JANDA, V. (1968) Effect of ischemia on trained motor units. Am. J. Phys. Med., *47:* 64–71.

SIMARD, T. G. AND LADD, H. W. (1969) Conscious control of motor units with thalidomide children: an electromyographic study. Develop. Med. Child Neurol., *11:* 743–748.

SIMPSON, H. M. AND CLIMAN, M. H. (1971) Pupillary and electromyographic changes during an imagery task. Psychophysiology, *8:* 483–490.

SIMPSON, J. A. (1964) Fact and fallacy in measurement of conduction velocity in motor nerves. J. Neurol. Neurosurg. & Psychiat., *27:* 381–385.

SLOAN, A. W. (1965) Electromyography during fatigue of healthy rectus femoris. South African M. J., *39:* 395–398.

SMIDT, G. L. (1974) Methods of studying gait. Phys. Ther., *54:* 14–17.

SMIDT, G. L., ARORA, J. S. AND JOHNSTON, R. C. (1971) Accelerographic analysis of several types of walking. Am. J. Phys. Med., *50:* 285–300.

SMITH, H. M., JR., BASMAJIAN, J. V. AND VANDERSTOEP, S. F. (1974) Inhibition of neighboring motoneurons in conscious control of single spinal motoneurons. Science, *183:* 975–976.

SMITH, J. W. (1954) Muscular control of the arches of the foot in standing: an electromyographic assessment. J. Anat., *88:* 152–163.

SMITH O. C. (1934) Action potentials from single motor units in voluntary contraction. Am. J. Phsyiol., *108:* 629–638.

SMITH, R. P., JR. (1973) Frontalis muscle tension and personality. Psychophysiology, *10:* 311–312.

SMORTO, M. P. AND BASMAJIAN, J. V. (1972) *Clinical Electroneurography: an Introduction to Nerve Conduction Tests.* Williams & Wilkins Co., Baltimore.

SOECHTING, J. F. AND ROBERTS, W. J. (1975) Transfer characteristics between emg activity and muscle tension under isometric conditions in man. J. Physiol., Paris, *70:* 779–793.

SOTO, R. A., SANZ, O. P., SICA, R. E. P. AND CHORNY, D. (1974) Facilitation of muscle activity by contralateral homonymous muscle action in man. Medicina, *34:* 481–484.

SPIEGEL, M. H. AND JOHNSON, E. W. (1962) Conduction velocity in the proximal and distal segments of the motor fibers of the ulnar nerve of human beings. Arch. Phys. Med., *43:* 57–61.

SPOOR, A. AND VAN DISHOECK, H. A. E. (1960) Electromyography of the human vocal cords and the theory of Husson. Pract. Oto-rhinolaryng., *20:* 353–360.

SPRUIT, R. (1965) *Een Analyse van Vorm en Ligging van de MM. Glutaei en de Adductoren* (An Analysis of Form and Location of Gluteal and Adductor Muscles) (Dutch text with English summary).

Drukkerij Albani-Den. Haag, Leiden.

STÅLBERG, E. (1967) Propagation velocity in human muscle fibers in situ. Acta Physiol. Scandinav., *70:* suppl. 287.

STÅLBERG, E. AND EKSTEDT, J. (1973) Single fiber EMG and microphysiology of the motor unit in normal and diseased muscle. In *New Developments in Electromyography and Clinical Neurophysiology,* ed. by J. E. Desmedt, Karger, *1:* 113–129, Karger, Basel.

STÅLBERG, E., SCHWARTZ, M. S., THIELE, B. AND SCHILLER, H. H. (1976) the normal motor unit in man. J. Neurol. Sci., *27:* 291–301.

STÅLBERG, E., SCHWARTZ, M. S. AND TRONTELJ, J. V. (1975) Single fiber electromyography in various processes affecting the anterior horn cell. J. Neurol. Sci., *24:* 403–415.

STEENDIJK, R. (1948) On the rotating function of the ilio-psoas muscle. Acta Neerl. Morphol., *6:* 175–183.

STEIN, R. B. AND BAWA, P. (1976) Reflex responses of human soleus muscle to small preterbations. J. Neurophysiol., *39:* 1105–1116.

STEINER, J. E., MICHMAN, J. AND LITMAN, A. (1974) Time sequences of the activity of the temporal and masseter muscles in healthy young human adults during habitual chewing of different test foods. Arch. Oral Biol., *19:* 29–34.

STEINER, T. J., THEXTON, A. J. AND WEBER, W. V. (1972) An EMG electrode system for selected-site recording from a muscle. J. Appl. Physiol., *32:* 531–532.

STEINDLER, A. (1955) *Kinesiology of the Human Body under Normal and Pathological Conditions,* Charles C Thomas, Springfield, Ill.

STENER, B. (1959) Experimental evaluation of the hypothesis of ligamento-muscular protective reflexes. I. A method of adequate stimulation of tension receptors in the medial collateral ligament of the knee joint of the cat, and studies of the innervation of the ligament. Acta Physiol. Scandinav., *48:* suppl. 166, 5–26.

STEPHENS, J. A. AND TAYLOR, A. (1972) Fatigue of maintained voluntary muscle contraction in man. J. Physiol., *220:* 1–18.

STERN, J. T. (1971) Investigations concerning the theory of 'spurt' and 'shunt' muscles. J. Biomechanics, *4:* 437–453.

STERN, M. M. (1971) A model that relates low frequency EMG power to fluctuations in the number of active motor units: application to local muscle fatigue. Ph. D. Dissertation, University of Michigan.

STETSON, R. H. (1933) Speech movements in action. Tr. Am. Laryng. A., *55:* 29–41.

STILES, R. N. (1976) Frequency and displacement amplitude relations for normal hand tremor. J. Appl. Physiol., *40:* 44–54.

STOLOV, W. C. (1966) The concept of normal muscle tone, hypotonia and hypertonia. Arch. Phys. Med., *47:* 156–168.

STOREY, A. T. (1968a) Laryngeal initiation of swallowing. Exper. Neurol., *20:* 359–365.

STOREY, A. T. (1968b) A functional analysis of sensory units innervating epiglottis and larynx. Exper. Neurol., *20:* 366–383.

STOTZ, S. AND HEIMSTÄDT, P. (1970) Uber die Rollwirking des M. Iliopsoas in Abhängigkeit von Schenkelhalsund Antetorsionswinkel bie der infantilen Zerebralparese. Zeit. Orthop. Grenz., *108:* 229–243.

STRAUS, W. L. AND WEDDELL, G. (1940) Nature of the first visible contractions of the forelimb musculature in rat fetuses. J. Neurophysiol., *3:* 358–369.

STRIBLEY, R. F., ALBERTS, J. W., TOURTELLOTTE, W. W. AND COCKRELL, J. L. (1974) A quantitative study of stance in normal subjects. Arch. Phys. Med. Rehab., *55:* 74–80.

STRUPPLER, A. (1965) Silent period (s.p.) Electromyog. Clin. Neurophysiol., *5:* 163–168.

STUART, D. G., ELDRED, E., HEMINGWAY, A. AND KAWAMURA, Y. (1963) Neural regulation of the rhythm of shivering. In *Temperature—Its Measurement and Control in Science and Industry*, p. 545–557 Rheinhold Publishing Co., New York.

STYCK, J. AND HOOGMARTENS, M. (1975) A suitable type of wire for wire-electrodes. Electromyogr. Clin. Neurophysiol., *15:* 291.

SULLIVAN, W. E., MORTENSEN, O. A., MILES, M. AND GREENE, L. S. (1950) Electromyographic studies of m. biceps brachii during normal voluntary movement at the elbow. Anat. Rec., *107:* 243–252.

SUMITSUJI, N., MATSUMOTO, K., TANAKA, M., KASHIWAGI, T. AND KANEKO, Z. (1967) Electromyographic investigation of the facial muscles. Electromyography, *7:* 77–96.

SUNDERLAND, S. (1945) The actions of the extensor digitorum communis, interosseous and lumbrical muscles. Am. J. Anat., *77:* 189–209.

ŠURINA, I. AND JÁGR, J. Action potentials of levator and tensor muscles in patients with cleft palate. Acta Chir. Plast., *11:* 21–30.

SUSSET, J. G., RABINOVITCH, H. AND MACKINNON, K. J. (1965) Parameters of micturition: clinical study. J. Urol., *94:* 113–121.

SUSSMAN, H. M., HANSON, R. J. AND MACNEILAGE, P. F. (1972) Studies of single motor units in the speech musculature: methodology and preliminary findings. J. Acoust. Soc. Am., *51:* 1372–1374.

SUTHERLAND, D. H. (1966) An electromyographic study of the plantar flexors of the ankle in normal walking on the level. J. Bone & Joint Surg., *48-A:* 66–71.

SUTHERLAND, D. H., SCHOTTSTAEDT, E. R., LARSEN, L. J., ASHLEY, R. K., CAL-

LANDER, J. N. AND JAMES, P. M. (1969) Clinical and electromyographic study of seven spastic children with internal rotation gait. J. Bone & Joint Surg., *51-A:* 1070–1082.

SUTTON, D. L. (1962) Surface and needle electrodes in electromyography. Dental Progress, *2:* 127–131.

SUTTON, D. AND KIMM, J. (1969) Reaction time of motor units in biceps and triceps. Exper. Neurol., *23:* 503–515.

SUTTON, D. AND KIMM, J. (1970) Alcohol effects on human motor unit reaction time. Physiol. Behav., *5:* 889–892.

SUTTON, D., LARSON, C. R. AND FARRELL, D. M. (1972) Cricothyroid motor units. Acta Otolaryng., *74:* 145–151.

SUZUKI, R. (1956) Function of the leg and foot muscles from the viewpoint of the electromyogram (English text). J. Jap. Orthop. Surg. Soc., *30:* 775–786.

SWEZEY, R. L. AND FIEGENBERG, D. S. (1971) Inappropriate intrinsic muscle action in the rheumatoid hand. Ann. Rheum. Dis., *30:* 619–625.

TAKEBE, K. AND BASMAJIAN, J. V. (1976) Gait analysis in stroke patients to assess treatment of drop-foot. Arch. Phys. Med. & Rehab., *57:* 305–310.

TAKEBE, K., VITTI, M. AND BASMAJIAN, J. V. (1974) The functions of semispinalis capitis and splenius capitis muscles: an electromyographic study. Anat. Rec., *179:* 477–480.

TAMLER, E., JAMPOLSKY, A. AND MARG, E. (1958) An electromyographic study of asymmetric convergence. Am. J. Ophth., *46:* (No. 5, special Part II (not I) of the November issue) 174–181.

TAMLER, E., MARG, E. AND JAMPOLSKY, A. (1959) An electromyographic study of coactivity of human extraocular muscles in following movements. A. M. A. Arch. Ophth., *61:* 270–273.

TAMLER, E., MARG, E., JAMPOLSKY, A. AND NAWRATZKI, I. (1959) Electromyography of human saccadic eye movements. A. M. A. Arch. Ophth., *62:* 657–661.

TANJI, J. AND KATO, M. (1971) Volitionally controlled single motor unit discharges and cortical motor potentials in human subjects. Brain Res., *29:* 243–246.

TANJI, J. AND KATO, M. (1972) Discharges of single motor units at voluntary contraction of abductor digiti minimi muscle in man. Brain Res., *45:* 590–593.

TANJI, J. AND KATO, M. (1973a) Recruitment of motor units in voluntary contraction of a finger muscle in man. Exper. Neurol., *40:* 759–770.

TANJI, J. AND KATO, M. (1973b) Firing rate of individual motor units in voluntary contraction of abductor digiti minimi muscle in man. Exper. Neurol., *40:* 771–783.

TATHAM, M. A. A. AND MORTON, K. (1968a) Some electromyography data towards a model of speech production. "Occasional Papers", Language Centre, Univ. of Essex, Colchester, Engl., *1:* 1–24.

TATHAM, M. A. A. AND MORTON K. (1968b) Further electromyography data towards a model of speech production. "Occasional Papers," Language Centre, Univ. of Essex, Colchester, Engl., *1:* 25–59.

TATHAM, M. A. A. (1971) Speech synthesis and models of speech production. Interim Report to Science Research Council. Essex Univ., Colchester, Engl., privately printed.

TAUBER, E. S., COLEMAN, R. M. AND WEITZMAN, E. D. (1977) Absence of tonic electromyographic activity during sleep in normal and spastic non-mimetic skeletal muscles in man. Ann. Neurol., *2:* 66–68.

TAYLOR, A. (1960) The contribution of the intercostal muscles to the effort of respiration in man. J. Physiol., *151:* 390–402.

TEIG, E. (1973) Differential effect of graded contraction of middle ear muscles on the sound transmission of the ear. Acta Physiol. Scandinav., *88:*

382–391.

TENG, E. L., McNEAL, D. R., KRALJ, A. AND WATERS, R. L. (1976) Electrical stimulation and feedback training: effects on the voluntary control of paretic muscles. Arch. Phys. Med. & Rehab., *57:* 228–233.

TERGAST, P. (1873) Ueber das Verhaltnis von Nerve und Muskel (German text). Arch. Mikr. Anat., *9:* 36–46.

THOM, H. (1965) Elektromyographische Untersuchungen zur Funktion des M. rrapezius. Elektromedizin, *10:* 65–72.

THOMAS, G. (1969) The function of the extensor muscles of the back. German Med. Monthly, *14:* 564–566.

THOMAS, J. S., SCHMIDT, E. M. AND HAMBRECHT, F. T. (1976) Limitations of volitional control of single motor unit recruitment sequence (abstract). Society of Neuroscience, 1976, Annual Meeting, Toronto, Canada, p. 536.

THOMAS, L. J., TIBER, N. AND SCHIRESON, S. (1973) The effects of anxiety and frustration of muscular tension related to the temporomandibular joint syndrome. Oral Surg., Oral Med., Oral Pathol., *36:* 763–768.

THOMAS, P. K. (1961) Recent advances in the clinical electrophysiology of muscle and nerve. Postgrad. M. J., *37:* 377–384.

THOMAS, P. K., SEARS, T. A., AND GILLIATT, R. W. (1959) The range of conduction velocity in normal motor nerve fibers to the small muscles of the hand and foot. J. Neurol. Neurosurg. & Psychiat., *22:* 175–181.

THOMSON, S. A. (1960) Hallux varus and metatarsus varus. In *Clinical Orthopaedics,* No. 16, pp. 109–118. Lippincott, Philadelphia and Montreal.

THYSELL, R. V. (1969) Reaction time of single motor units. Psychophysiology, *6:* 174–185.

TICHAUER, E. R. (1971) A pilot study of the biomechanics of lifting in simulated industrial work situations. J. Safety Res., *3:* 98–115.

TOKITA, T., TASHIRO, K. AND KATO, K. (1970) Electromyography of the esophagus and its clinical applications. Acta Otolaryng., *70:* 269–278.

TOKIZANE, H. (1954) Electromyographic study of facial muscles. Ochyanomizu Med. J., *2:* 1.

TOKIZANE, T., KAWAMATA, K. AND TOKIZANE, H. (1952) Electromyographic studies on the human respiratory muscles. Jap. J. Physiol., *2:* 232–247.

TOKIZANE, T. AND SHIMAZU, H. (1964a) *Functional Differentiation of Human Skeletal Muscle.* Charles C Thomas, Springfield, Ill.

TOKIZANE, T. AND SHIMAZU, H. (1964b) *Functional Differentiation of Human Skeletal Muscle,* University of Tokyo Press.

TOKUYAMA, H., OKAMOTO, T. AND KUMAMOTO, M. (1976) Electromyographic study of swimming in infants and children. *International Series on Biomechanics, Vol. 1B, Biomechanics V-B.* University Park Press, Baltimore, pp. 215–220.

TOKUYAMA, H., OKAMOTO, T., YOSHIZAWA, M. AND KUMAMOTO, M. (1977) Suieiundoh no Kisoteki Kenkyu (A basic study of swimming—An electromyographic study of the flutter kick in childhood). *Reports of the Japanese Research Committee of Sports Sciences, No. IV:* 25–27.

TOMASZEWSKA, J. (1964) Badania patokinetyki mięśni po amputacji uda (Studies on the pathokinesiology of muscles after above-knee amputation) (Polish text with French and English summaries). Roczniki Nauk. WSWF W Poznan., *8:* 3–55.

TÖNNIS, D. (1966a) Elektromyographische Untersuchungen über die Rollwirkung des M. iliopsoas. Zeit. Orthop. Grenzg., *102:* 61–68.

TÖNNIS, D. (1966b) Ekektromyographische Untersuchungen über die Beterlegung des M. iliopsoas an der Ad- und Abduktion des Oberschenkels. Zeit. Orthop. Grenzg., *102:* 68–74.

TÖRÖK, Z. AND HAMMOND, P. H. (1971)

On the performance of single motor units as sources of control signals. Proceedings of the Conf. on Human Locomotor Engineering, U. of Sussex, pp. 35.

TOURNAY, A. AND FESSARD, A. (1948) Etude électromyographique de la synergie entre l'abducteur du pouce et le muscle cubital postérieur (French text). Rev. Neurol., *80:* 631.

TOURNAY, A. AND PAILLARD, J. (1953) Electromyographie des muscles radiaux à l'état normal (French text). Rev. Neurol., *89:* 277–279.

TOWNSEND, R. E., HOUSE, J. F. AND ADDARIO, D. (1975) A comparison of biofeedback-mediated relaxation and group therapy in the treatment of chronic anxiety. Am. J. Psychiat., *132:* 589–601.

TOYOSHIMA, S., MATSUI, H. AND MIYASHITA, M. (1971) An electromyographic study of the upper arm muscles involved in throwing (Japanese text, English abstract). Res. J. Phys. Ed., *15:* 103–110.

TRAVILL, A. A. (1962) Electromyographic study of the extensor apparatus of the forearm. Anat. Rec., *144:* 373–376.

TRAVILL, A. AND BASMAJIAN, J. V. (1961) Electromyography of the supinators of the forearm. Anat. Rec., *139:* 557–560.

TROJABORG, W. (1964) Motor nerve conduction velocities in normal subjects with particular reference to the conduction in proximal and distal segments of median and ulnar nerve. Electroencephalog. & Clin. Neurophysiol., *17:* 314–321.

TROJABORG, W. (1976) Motor and sensory conduction in the musculocutaneous nerve. J. Neurol. Neurosurg. Psychiat., *39:* 890–899.

TSURUMI, N. (1969) An electromyographic study of the gait of children. (Japanese text; English abstract). J. Jap. Orthop. Assoc., *43:* 611–628.

TURSKY, B. (1964) Integrators as measuring devices of bioelectric output. Clin.

Pharmacol. and Therap., *5:* 887–892.

TUSIEWICZ, K., MOLDOFSKY, H., BRYAN, A. C. AND BRYAN, M. H. (1977) Mechanics of the rib cage and diaphragm during sleep. J. Appl. Physiol., *43:* 600–602.

TUTTLE, R. AND BASMAJIAN, J. V. (1972) Electromyography of knuckle-walking: results of four experiments on the forearm of *Pan gorilla.* Am. J. Phys. Anthrop., *37:* 255–266.

TUTTLE, R. AND BASMAJIAN J. V. (1974) Electromyographic studies of brachial muscles in *Pan gorilla* and hominoid evolution. Am. J. Phys. Anthropol., *41:* 71–90.

TUTTLE, R. AND BASMAJIAN, J. V. (1976) Electromyography of the pongid shoulder muscles and hominid evolution. A. J. Phys. Anthrop., *44:* 212.

TUTTLE, R., BASMAJIAN, J. V. AND ISHIDA, H. (1975) Electromyography of the gluteus maximus muscle in gorilla and the evolution of bipedalism. In *Antecedents of Man*, Vol. 1 ed. by R. H. Tuttle, Mouton, the Hague.

UEDA, N., OHYAMA, M., HARVEY, J. E., MOGI, G. AND OGURA, J. H. (1971) Subglottal pressure and induced live voices of dogs with normal, reinnervated and paralyzed larynges. I. On voice function of the dog with a normal larynx. Laryngoscope, *81:* 1948–1959.

UEDA, N., OHYAMA, M., HARVEY, J. E. AND OGURA, J. H. (1972) Influence of certain extrinsic laryngeal muscles on artificial voice production. Laryngoscope, *82:* 468–482.

USHIJIMA, T. AND HIROSE, H. (1974) Electromyographic study of the velum during speech. J. Phonet., *2:* 315–326.

VAN ALLEN, M. W. AND BLODI, F. C. (1962) Electromyographic study of reciprocal innervation in blinking. Neurology, *12:* 371–377.

VAN DER STRAATEN, J. H. M., LOHMAN, A. H. M. AND VAN LINGE, B. (1975) A combined electromyographic & photographic study of the muscular control of the knee during walking. J. Human Mov. Studies, *1:* 25–32.

VANDERSTOEP, S. F. (1971) *A Comparison of the Ability to Control Single Motor Units in Selected Human Skeletal muscles.* Ph.D. Dissertation, Univ. of Southern California.

VAN GIJN, J. (1975) Babinski response: stimulus and effector. J. Neurol. Neurosurg. Psychiat., *38:* 180–186.

VAN GIJN, J. (1976) Equivocal plantar responses: a clinical and electromyographic study. J. Neurol. Neurosurg. Psychiatr., *39:* 275–282.

VAN GOOL, J. D., DE RIDDER, A. L. A., KUIJTEN, R. H., DONCKERWOLCKE, R. A. M. G. AND TIDDENS, H. A. (1976) Measurement of intravesical and rectal pressures simultaneously with electromyography of anal sphincter in children with myelomeningocele. Develop. Med. Child Neurol., *18:* 287–301.

VAN HARREVELD, A. (1946) The structure of the motor units in the rabbit's m. sartorius. Arch. Néerl. Physiol., *28:* 408–412.

VAN HARREVELD, A. (1947) On the force and size of motor units in the rabbit's sartorius muscle. Am. J. Physiol., *151:* 96–106.

VAN LINGE, B. AND MULDER, J. D. (1963) Function of the supraspinatus muscle and its relation to the supraspinatus syndrome. J. Bone & Joint Surg., *45-B:* 750–754.

VASILESCU, C. AND DIECKMANN, G. (1965) Electromyographic investigations in torticollis. Appl. Neurophysiol., *38:* 153–160.

VEREECKEN, R. L., DERLUYN, J. AND VERDUYN, H. (1975) Electromyography of the perineal striated muscles during cystometry. Urol. Internat., *30:* 92–98.

VIITASALO, J. H. T. AND KOMI, P. V. (1977) Signal characteristics of EMG during fatigue. Europ. J. Appl. Physiol., *37:* 111–121.

VILJANEN, A. A. (1967) The relation between the electrical and mechanical activity of human intercostal muscles during voluntary inspiration. Acta Physiol. Scandinav., Suppl. 296.

VILJANEN, A. A., POPPIUS, H., BERGSTRÖM, R. M. AND HAKUMÄKI, M. (1964) Electrical and mechanical activity in human respiratory muscles. Acta Neurol. Scandinav., *41:* suppl. 13, 237–239.

VISSER, S. L. AND DE RIJKE, W. (1974) Influence of sex and age on emg contraction pattern. Europ. Neurol., *12:* 229–235.

VITTI, M. (1965) Estudo electromiográfico dos músculos mastigadores no cão. Folia Clin. Biol., *34:* 101–114.

VITTI, M. (1969a) Análise electromiográfica do m. temporal no posição de repousa da mandíbula. O. Hospital (São Paulo) *76:* 339–349.

VITTI, M. (1969b) Comportamento electromiográfica do m. temporal no repuxamento da comissura bucal. O. Hospital (São Paulo), *75:* 349–354.

VITTI, M. (1970) Comportamento eletromiográfica do músculo temporal nas mordidas incisiva e molares homo e heterolaterais. O. Hospital (São Paulo), *78:* 207–214.

VITTI, M. (1971) Electromyographic analysis of the musculus temporalis in basic movements of the jaw. Electromyography, *11:* 389–403.

VITTI, M. AND BASMAJIAN, J. V. (1975) Muscles of mastication in small children: an electromyographic analysis. Am. J. Orthodont., *68:* 412–419.

VITTI, M. AND BASMAJIAN, J..V. (1976) Electromyographic investigation of procerus and frontalis muscles, Electromyogr. Clin. Neurophysiol., *16:* 227–236.

VITTI, M. AND BASMAJIAN, J. V. (1977) Integrated actions of masticatory muscles: simultaneous EMG from eight intramuscular electrodes. Anat. Rec., *187:* 173–189.

VITTI, M., BASMAJIAN, J. V., OUELETTE,

P. L., MITCHELL, D. L., EASTMAN, W. P. AND SEABORN, R. D. (1975) Electromyographic investigations of the tongue and circumoral muscular sling with fine-wire electrodes. J. Dent. Res., *54:* 844–849.

VITTI, M., CORRÊA, A. C. F., FORTINGUERRA, C. R. H., BÉRZIN, F. AND KÖNIG, B., JR. (1972) Electromyographic study of the "muscle depressor anguli oris." Electromyography Clin. Neurophysiol., *12:* 119–125.

VITTI, M., FORTINGUERRA, C. R. H., CORRÊA, A. C. F., KÖNIG, B., JR., AND BÉRZIN, F. (1974) Electromyographic behavior of the levator labii superioris alaeque nasi. Electromyography Clin. Neurophysiol., *14:* 37–43.

VITTI, M., FUJIWARA, M., IIDA, M. AND BASMAJIAN, J. V. (1973) The integrated roles of longus colli and sternocleidomastoid muscles: an electromyographic study. Anat. Rec., *177:* 471–484.

VOGT, A. T. (1975) Electromyograph responses and performance success estimates as a function of internal-external control. Percept. Motor Skills, *41:* 977–978.

VON MEYER, H. (1868) *Die Wechslnde Lage des Schwerpunktes im Menschlichen Körper,* Engelmann, Leipzig.

VRBOVÁ, G. (1963) The effect of motoneurone activity on the speed of contraction of striated muscle. J. Physiol, *169:* 513–526.

VREDENBREGT, J. AND RAU, G. (1973) Surface electromyography in relation to force, muscle length and endurance. In *New Developments in EMG and Clinical Neurophysiology ed. by J. E. Desmedt. 1:* 607–622, S. Karger, Basel.

VREELAND, R. W., SUTHERLAND, D. H., DORSA, J. J., WILLIAMS, L. A., COLLINS, C. C. AND SCHOTTSTEADT, E. R. (1961) A three-channel electromyography with synchronized slow-motion photography. IRE Transactions Bio-med.

Electronics, *BME-8:* 4–6.

WACHHOLDER, K., AND MCKINLEY, C. (1929) Über die innervation und tätigheit der atemmuskeln. Arch. Ges. Physiol., *222:* 575–588.

WAGMAN, I. H., PIERCE, D. S., AND BURGER, R. E. (1965) Proprioceptive influence in volitional control of individual motor units. Nature, *207:* 957–958.

WAGNER, A. L. AND BUCHTHAL, F. (1972) Motor and sensory conduction in infancy and childhood: reappraisal. Develop. Med. Child Neurol., *14:* 189–216.

WAIT, J. V., ATWATER, A. E. AND WETZEL, M. C. (1974) Computer-aided approach to gait analysis. Am. J. Phys. Med., *53:* 229–233.

WALMSLEY, R. P. (1977) Electromyographic study of the phasic activity of peroneus longus and brevis. Arch. Phys. Med. & Rehab., *58:* 65–69.

WALSHE, F. M. R. (1923) On certain tonic or postural reflexes in hemiplegia with special reference to the so-called "associated movements." Brain, *46:* 1–37.

WALTERS, C. E. AND PARTRIDGE, M. J. (1957) Electromyographic study of the differential action of the abdominal muscles during exercise. Am. J. Phys. Med., *36:* 259–268.

WATERLAND, J. C. AND HELLEBRANDT, F. A. (1964) Involuntary patterning associated with willed movement performed against progressively increasing resistance. Am. J. Phys. Med., *43:* 13–29.

WATERLAND, J. C. AND MUNSON, N. (1964a) Reflex association of head and shoulder girdle in nonstressful movements of man. Am. J. Phys. Med., *43:* 98–108.

WATERLAND, J. C. AND MUNSON, N. (1964b) Involuntary patterning evoked by exercise stress: radioulnar pronation and supination. J. Am. Phys. Therap.

A., *44:* 91–97.

WATERLAND, J. C. AND SHAMBES, G. M. (1970) Head and shoulder linkage: stepping in place. Am. J. Phys. Med., *49:* 279–289.

WATERS, R. L. AND MORRIS, J. M. (1970) Effect of spinal support on the electrical activity of muscles of the trunk. J. Bone & Joint Surg., *52-A:* 51–60.

WATERS, R. L. AND MORRIS, J. M. (1972) Electrical activity of muscles of the trunk during walking. J. Anat., *111:* 191–199.

WATT, D. AND JONES, G. M. (1966) On the functional role of the myotatic reflex in man. Proc. Canad. Fed. Biol. Soc., *9:* 13.

WAYLONIS, G. W. AND ASEFF, J. N. (1973) Anal sphincter electromyography in the first two years of life. Arch. Phys. Med. and Rehab., *54:* 525–527.

WEATHERSBY, H. T. (1957) Electromyography of the thenar muscles. Anat. Rec., *127:* 386.

WEATHERSBY, H. T. (1966) The forearm stabilisers of the thumb: an electromyographic study. Anat. Rec., *154:* 439.

WEATHERSBY, H. T., SUTTON, L. R. AND KRUSEN, U. L. (1963) The kinesiology of muscles of the thumb: an electromyographic study. Arch. Phys. Med., *44:* 321–326.

WEDDELL, G., FEINSTEIN, B. AND PATTLE, R. E. (1944) The electrical activity of voluntary muscle in man under normal and pathological conditions. Brain, *67:* 178–257.

WEISS, M. A. (1959) Electromyography in the selection of muscle reeducation methods. Polish Medical History and Science, July 1959 Bulletin.

WEISS, M. A. (1966) The neurophysiological aspects of the immediate post myoplastic amputation fitting from the neurophysiological point of view. (Unpublished manuscript of talk given in Washington.)

WELLS, J. G. AND MOREHOUSE, L. E. (1950) Electromyographic study of the effect of various headward accelerative forces upon the pilot's ability to perform standardized pulls on the aircraft control stick. J. Aviation Med., *21:* 48–54.

WERSÄLL, R. (1958) The tympanic muscles and their reflexes: physiology and pharmacology with special regard to noise generation by the muscles. Acta Otolaryng., *139:* suppl. 1–112.

WERTHEIMER, L. G. AND FERRAZ, E. C. D. F. (1958) Observações electromiográficas sòbre as funções dos músculos supra-espinhal e deltóide nos movimentos do ombro, (Portuguese text). Folia Clin. Biol., *28:* 276–289.

WHATMORE, G. B. AND KOHLI, D. R. (1968) Dysponesis: a neurophysiologic factor in functional disorders. Behav. Sci., *13:* 102–124.

WHEATLEY, M. D. AND JAHNKE, W. D. (1951) Electromyographic study of the superficial thigh and hip muscles in normal individuals. Arch. Phys. Med., *32:* 508–515.

WHITE, E. R. AND BASMAJIAN, J. V. (1973) Electromyography of lip muscles and their role in trumpet playing. J. Appl. Physiol., *35:* 892–897.

WIDMALM, S.-E. (1976) *Reflex Activity of the Masseter Muscle in Man: and EMG Study* (Suppl. 72 vol. 34 of Acta Odontol. Scandinav.) Doctoral Thesis, Faculty of Odontology, Göteborg University.

WIDMALM, S.-E. AND HEDEGÅRD, B. (1976) Reflex activity in the masseter muscle of young individuals. Part I. Experimental procedures—results: & Part II. (no subtitle). J. Oral Rehab., *3:* 41–55 and 167–180.

WIEDENBAUER, M. M. AND MORTENSEN, O. A. (1952) An electromyographic study of the trapezius muscle. Am. J. Phys. Med., *31:* 363–372.

WIESENDANGER, M., SCHNEIDER, P. AND

VILLOZ, J.-P. (1967) Elektromyographische Analyse der raschen Wilkürbewegung. Schweiz. Arch. Neurol. Neurochir. Psychiat., *100:* 88–99.

WIESENDANGER, M., SCHNEIDER, P. AND VILLOZ, J.-P. (1969) Electromyographic analysis of a rapid volitional movement. Am. J. Phys. Med., *48:* 17–24.

WILKINS, B. R. (1964) A theory of position memory. J. Theor. Biol., *7:* 374–387.

WILLIAMS, J. G. L. (1963) A resonance theory of "microvibrations." Psychol. Rev., *70:* 547–558.

WILSON, A. AND WILSON, A. S. (1970) Psychophysiological and learning correlates of anxiety and induced muscle relaxation. Psychophysiology, *6:* 740–748.

WILSON, D. M. (1965) Proprioceptive leg reflexes in cockroaches. J. Exper. Biol., *43:* 397–409.

WINDLE, W. F. (1940) *Physiology of the Fetus.* W. B. Saunders Co. Philadelphia and London.

WINDLE, W. F., MINEAR, W. L., AUSTIN, M. F. AND ORR, D. M. (1935) The origin and early development of somatic behaviour in the albino rat. Physiol. Zool., *8:* 156–185.

WINTER, D. A. AND QUANBURY, A. O. (1975) Multichannel biotelemetry system for use in emg studies, particularly in locomotion. Am. J. Phys. Med., *54:* 142–147.

WINTER, D. A., QUANBURY, A. O., HOBSON, D. A., SIDWALL, H. G., REIMER, G., TRENHOLM, B. G., STEINKE, T. AND SHLOSSER, H. (1974) Kinematics of normal locomotion—statistical study based on T.V. data. J. Biomech., *7:* 479–486.

WOELFEL, J. B., HICKEY, J. C., STACY, R. W. AND RINEAR, L. (1960) Electromyographic analysis of jaw movements. J. Prosthet. Dent., *10:* 688–697.

WÓJTOWICZ, S. (1966) Reciprocal innervation of directly and indirectly synergic and antagonistic external eye muscles. Polish Med. J., *5:* 656–661.

WÓJTOWICZ, S. AND GWÓŹDŹ, E. (1968) Protective function of the orbicularis oculi muscle in electromyographic investigations. Polish Med. J., *7:* 465–470.

WOLF, S. L. AND BASMAJIAN, J. V. (1973) A rapid cooling device for controlled cutaneous stimulation. Am. Phys. Therap. Assoc. J., *53:* 25–27.

WOLF, S. L. AND LETBETTER, W. D. (1975) Effect of skin cooling on spontaneous emg activity in triceps surae of the decerebrate cat. Brain Res., *91:* 151–155.

WOLF, S. L., LETBETTER, W. D. AND BASMAJIAN, J. V. (1976) Effects of a specific cutaneous cold stimulus on single motor unit activity of medial gastrocnemius muscle in man. Am. J. Phys. Med., *55:* 177–183.

WOLPERT, R. AND WOOLDRIDGE, C. P. (1975) The use of electromyography as biofeedback therapy in the management of cerebral palsy: a review and case study. Physioth. Can., *27:* 5–10.

WOODBURNE, R. T. (1957) *Essentials of Human Anatomy.* Oxford University Press, New York.

WRIGHT, G., WASSERMAN, E., POTTALA, E. AND DUKES-DOBOS, F. (1976) The effect of general fatigue on isometric strength-endurance measurements and the electromyogram of the biceps brachii. Am. Indust. Hyg. Assoc. J., *37:* 274–279.

WYKE, B. (1968) Effects of anesthesia upon intrinsic laryngeal reflexes. J. Laryng. Otol., *82:* 603–612.

WYKE, B. (1969) Deus ex machina vocis: an analysis of the laryngeal reflex mechanisms of speech. Br. J. Disorders Communic., *4:* 3–25.

WYMAN, R. (1970) Patterns of frequency variation in dipteran flight motor units. Comp. Biochem. Physiol., *35:* 1–16.

WYRICK, W. AND DUNCAN, A. (1974) Electromyographical study of reflex, premotor, and simple reaction time of relaxed muscle to joint displacement. J. Motor Behav., *6:* 1–10.

YABE, K. AND TAMAKI, Y. (1976) Inhibitory effect of unilateral contraction on the contralateral arm. Percep. Motor Skills, *43:* 979–982.

YAMAJI, K. AND MISU, A. (1968) Kinesiologic study with electromyography of low back pain. Electromyography, *8:* 189.

YAMASHITA, N. (1975) The mechanism of generation and transmission of forces in leg extension. J. Human Ergol., *4:* 43–52.

YAMASHITA, N., KUMAMOTO, M. AND OKAMOTO, T. (1972) Electromyographic study of the giant swing on the horizontal bar. *Sixth International Congress of Physical Medicine, 2:* 361–363.

YAMASHITA, N., TAKAGI, K., OKAMOTO, T. (1971) Tetsubo-Undoh ni okeru Junte-Sharin no Kindenzuteki Kenkyu (Electromyographic study on the giant swing [forward] on the horizontal bar). Res. Phys. Ed. *15:* 93–102.

YAMAZAKI, A. (1968) Electrophysiological study on "flick" eye movements during fixation (Japanese text; English abstract), *71:* 2446–2459.

YAMSHON, L. J. AND BIERMAN, W. (1948) Kinesiologic electromyography. II. The trapezius. Arch. Phys. Med., *29:* 647–651.

YAMSHON, L. J. AND BIERMAN, W. (1949) Kinesiologic electromyography. III. The deltoid. Arch. Phys. Med., *30:* 286–289.

YEMM, R. (1969a) Variations in the electrical activity of the human masseter muscle occurring in association with emotional stress. Arch. Oral Biol., *14:* 873–878.

YEMM, R. (1969b) Masseter muscle activity in stress: adaptation of response to a repeated stimulus in man. Arch. Oral Biol., *14:* 1437–1439.

YEMM, R. (1969c) Temporomandibular dysfunction and masseter muscle response to experimental stress. Br. Dent. J., *127:* 508–510.

YOSHI, H., TAKAGI, K., KUMAMOTO, M., ITO, M., ITO, K., YAMASHITA, N., OKAMOTO, T. AND NAKAGAWA, H. (1974) Suiso ni yoru Kayak Soho no Kindenzugakuteki Kenkyu (Electromyographic study of kayak paddling in the paddling tank). Res. J. Phys. Ed., *18:* 191–198.

YOSHIZAWA, M., TOKUYAMA, H., OKAMOTO, T. AND KUMAMOTO, M. (1976) Electromyographic study of the breaststroke. *International Series on Biomechanics, Vol. B, Biomechanics V-B.* University Park Press, Baltimore, pp. 222–229.

YOUMANS, W. B., MURPHY, Q. R., TURNER, J. K., DAVIS, L. D., BRIGGS, D. I. AND HOYE, A. S. (1963) *The Abdominal Compression Reaction: Activity of Abdominal Muscles Elicited form the Circulatory System.* Williams & Wilkins Co., Baltimore.

YOUMANS, W. B., TJIOE, D. T. AND TONG, E. Y. (1974) Control of involuntary activity of abdominal muscles. Am. J. Phys. Med., *53:* 57–74.

YOUNG, R. R. AND SHAHANI, B. (1969) An EMG study of cutaneous blink reflexes (abstract) Electroenceph. Clin. Neurophysiol., *26:* 630–636.

ZAPPALÁ, A. (1970) Influence of training and sex on the isolation and control of single motor units. Am. J. Phys. Med., *49:* 348–361.

ZERNICKE, R. F. AND WATERLAND, J. C. (1972) Single motor unit control in m. biceps brachii. Electromyog. Clin. Neurophysiol., *12:* 225–241.

ZHUKOV, E. K. AND ZAKHARYANTS, J. Z. (1960) Electrophysiological data concerning certain mechanisms of overcoming fatigue (Russian text) J. Physiol., U. S. S. R., *46:* 819–827.

Żuk, T. (1960) Badania elektromiograficzne w skoliozach (Polish text; Russian and English summaries) Chir. Narz. Ruchu i Ortop. Polska, *25:* 589–595.

Żuk, T. (1962a) Etiopatogeneze skoliózy na podkladě elektromyografických záznamů (Czech test; Russian and English summaries). Acta Chir. Orthop. et Traumat. Čechoslov., *24:* 69–74.

Żuk, T. (1962b) The role of spinal and abdominal muscles in the pathogenesis of scoliosis. J. Bone & Joint Surg., *44-B:* 102–105.

Index

A

Abdomen, 185, 319–338
  compression reaction, 355
  wall, 319
Acceleration, 152, 153
  centrifugal, 153
Accelerometer, 297, 301
Acetylcholine, 148
Action potentials, 12–21, 45, 46, 47, 48, 53–78,
    86, 109–111
  intracellular, 16–17
Activity, antagonistic, see Antagonist
  resting, 79–83
  reflex, intermittent, 179
  spinal reflex, 333
Afferent discharge, 167
Agonists, 93–102, 138, 155, 402
Amplifiers, 37–39, 419
  miniature, 38, 39, 419
Amputees, 296, 301
Analysis, computer, 45, 46
Anesthesia, surgical, 404
  eye position during, 404
Ankle, 176, 177, 179, 256, 260–266, 271–275,
    305–307
  dorsiflexion of, 264, 305
  free movements of, 264, 273
  plantar flexion, 265, 308
Antigravity torque, 97
Aponeurosis, plantar, 166
Antagonist, 93–102, 138, 155, 203, 205, 214,
    257, 285, 402–407
Anterior horn, spinal cord, 6
Anti-gravity mechanisms, 175–188
Anxiety, effect on EMG, 80, 132–133, 134
Apnea, 354
Apparatus, 23–51, 417–419
Arches, longitudinal, of foot, 181, 266–271
Arm, 186, 189–207
  swinging, 314
Artifact, movement, 38
Asphyxia, 356
Asymmetric convergence, 405
Asynchrony of motor unit contractions, 10
Atonia, 79–83
Atrophy, disuse, 104
Autocorrelation, 89
Axon, 6, 7, 12
  impulses, 7, 12

B

Babies, hypertonic, 79, 111
  premature, 110

Babinski sign, 98, 279
Back, 183–185, 281–293
Backache, 284
Bats, 412
Bicycle, human, 312
Bicycle-pedalling, 243, 244, 248, 274
Biofeedback, 115–140
Bite, 171, 380–385
Bladder, spinal reflex activity, 331
  urinary, 331–337
Blinking, 410
Breath-holding, 352
Breathing, 326–329, 339–357
  positive pressure, 329

C

Cable, multiwire, 39
Cage, shielded, 49
Capsule, 84, 166, 186
Camera, 45
Carbon monoxide poisoning, 105
Cathode-ray oscilloscope, 419
Cats, 18, 82, 84, 104, 369, 381
Cavity, orbital, 401
Central nervous system, 84, 102
Chest, pressure gradient, 354
Chewing, 379–390
Childbirth, 334–336
Circulation, arterial, 179
  venous, 179
Cinematography, 296
Chlorpromazine, effects of, 80
Clavicle, 190–191
Cleft palatal repairs, 362
Cockroaches, limb muscles of, 113
Cocontraction, 93–100, 155, 407
Computers, 46, 297
Conduction velocity, 141–149
Conductor, superflex, 40
Contraction, continuous, 88
  isometric, 14, 15, 90, 102, 106, 169, 203, 205,
    258
  isotonic, 169
  localized, involuntary, sustained, 104
  maximum, voluntary, 90, 169
  reflex, muscular, 167
  static, 258
  voluntary, 14, 169
Convergence, angular, 406
  asymmetric, 405–407
Coordination, 93–103
Co-reflex phenomenon, 100
Cortex, cerebral, 109

Cortical stimulus, 109
Counter, electronic, 122, 170
Coughing, 285, 326, 328, 330, 352, 374
Cramps, 103
Cross-correlation, 89
Cross exercise, 102
  education, 102
Cues, 117
Current, ionic, 16
Cyclograms, 296

D

Deceleration, 151
Defecation, 330
Delay, residual, 147–148
Denervation, 109
Diaphragm, electromyography of, 345–356
  esophageal hiatus, 352
  intercostal interrelationship, 353
  motor units of, 356
  movement of, 345–352
  rat, single motor units of, 356
Disc, intervertebral, 290
Disease, Parkinson's, 314
Dislocation, of humerus, 196–200
  of joint, 186
Divergence, innervational, 404
Dogs, 339, 354, 355, 358, 369, 371
Dorsiflexion, 264, 305
Ductus (vas) deferens, 326
Dyspnea, 356

E

ECG vector analysis, 48
  equipment, 37
EEG, 37
  recording paper, 43
Ejaculation, 337
Elasticity of muscle, natural, 79
Elbow, 93, 152, 153, 154, 155, 186, 200–205,
  210
Electrobasograph, 296
Electrodes, 23–37
  Beckman, 25
  bipolar concentric-needle, 27
  chronic intramuscular, 31
  construction, 32–35
  clip, 31, 346
  connector, 36
  fine-wire, 12, 29, 32–35, 117, 194, 211, 216,
    220, 236, 242, 250, 289, 331, 359,
    361, 362, 366, 374, 376, 386, 392,
    393, 394

jelly, 24
multielectrode, 10
NASA-type, 395
needle, 12, 27
paint-connector, 36
plastic suction cup, 25
pliable, indwelling, multiple wire, 32–39
rubber suction cup, 25
silver/silver chloride, 25
skin, see "surface" (below)
spray-on, 395
spring-connector, 36
stainless steel wire, 29
surface, 23–26, 132
unipolar needle, 27
wire, self-retaining, very fine, 32–35
Electrogoniogram, 301
Electromyogram, effect of age, 18–21
  of sex, 18–21
  of smoking, 105
  of sustained contraction, 17
  of temperature, 104
  feedback, 115–140
  integrated, 48
Electromyograph, 37–42, 417–419
Electromyography, apes, 242, 317
  birds, 111
  diaphragmatic, 345–352
  feedback, 115–140
  fetal, 109–111, 346
  fish, 111
  future, 413
  gymnastics, 189
  insects, 113
  integrated, 48
  laryngeal, 369–377
  mandibular, 379
  model, 66
  newborn, 109–111
  ocular, 401–410
  pharyngeal, 365–369
  postural, 175–188
  reptiles, 111
  sleep, 138
  surface, 139
  synthetic, 66
  vector, 48
Electroneurography, 141–149
Electronic counting device, 122, 170
Electronystagmography, 401
Electro-oculography, 401
Elephant, limbs of, 178
Embryology, 109–111
Endplate, 12, 90

Energy, internal, 176
  kinetic, 170
Equilibrium, 176, 284
Esodeviation, 404
Esophageal hiatus, 352, 369
Esophagus, 352, 369
Esotropia, 404
Eversion, 256, 273
Evolution, 317
Exercises, bilateral leg-raising, 320
  contralateral, 98, 102
  fitness, 288
  non-resistive, 102
  reeducation, spinal, 292
  stress, 399
  trunk, 288, 292, 323
Exertion, 87
Exodeviation, 404
Exotropia, intermittent, 404
Expiration, 323, 328, 343–351
  forced, 322, 329, 355
Eyeball, 401–409
  convergence, 405
  position of, 402
  infraduction, 407
  position, 402
  stationary, electromyogram of, 406
  supraduction, 407
Eye-head coordination, 409
Eyelids, 410

F

Fabrica, of Vesalius, 1
Facial palsies, 390
Faradization, 5
Fasciae, posterior, 282
Fasciculation, 82
Fatigue, 83–91, 102, 166, 178, 413
  extreme, subjective, tremors, 84
  general, 83, 84
  local, 88, 166
  of muscle fibres, 84, 90
  peripheral neuromuscular, 83, 84, 86, 90
  spectral analysis, 89
  threshold of, 7
Feedback EMG, 115–140
Feedback loops, 91, 115–140, 167–169, 396
Feet, 179–183, 266–280, 301–314
Femur, dislocation of, 257
  rotation of, 244, 249
FET follower, 38
Fetus, EMG of, 109–111
  muscles, 109–111
Fibre potentials, 12–15, 53

Fibrillation, 82, 104, 109
Fingers, 213–230
Fist, opening and closing, 167, 213–219
Fitness, 288
Fixation, ocular, 406
Flat feet, 181, 266–271, 275
Flexor response, 279
Foot, see Feet
Force, kinetic, 301
Force-plate, 296
Force-transducer, 296
Forearm, 153, 167, 205–212
  pronation, 205
  supination, 205
Frequency, propagation, 7, 14, 117
  range, 37, 43

G

Gait, 41, 235–250, 260–266, 295–301
Galvanometers, 45, 418
Gamma loop, 82, 92
Ganglion, pterothoracic, 114
Glans penis, 337
Glenoid cavity or fossa, 186, 198–200
Glottis, 352, 369–377
Goat, fetus of, 109
Goniometer, neck, 398
Gravity, 97, 175, 185, 295
  centre of, 97, 265
  line of, 175–177, 265, 292
  torque, 97
Great toe, 276–280
Grip, 217

H

Hallux valgus and varus, 278
Hand, 213–233
Harvard step test, 88
Head positioning, 398
Head-stands, 96
Head-turning, 398
Heel-contact, 299–304
Heel-off, 299–304
Hemiplegia, 103, 137, 200
Hernia, 322, 325
Hinge joint, whip-like motion, 94, 155
Hip, 181, 182, 244–250, 295–304, 297, 301, 308
Hook-lying, 238
Humerus, 186, 192–200
Hypopharynx, 369
Hyperpnea, 329, 355
Hypotonia, 79–82
Hysterics, 80

I

Immobilization, 104
Impedance, 32
Impulses, nervous, 7
    propagated, 16
Incontinence, urinary, 336
Infants, 108, 109
Inguinal canal, 325
Inhibition, autogenic or reciprocal, 94, 101
Inspiration, 326, 339–358
Integration, 48, 169, 171, 173
Interference pattern, emg, 13, 113
Interphalangeal joint, 214–217
Infraduction of eye, 407
Inversion, 272, 273
Irradiation, crossed motor, 102
Isometric contractions, 14, 15, 90, 102, 169,
    172, 203, 205, 258
Isotonic contractions, 169

J

Jaw, movements, 379–390
    muscles, 379–390
    protraction, 383, 386, 388
    rest position, 379
    retraction, 383, 384, 388
Joint, ankle, (see Ankle)
    elbow, (see Elbow)
    hip, (see Hip)
    interphalangeal, 214–217
    intervertebral, 178
    knee, (see Knee)
    metacarpophalangeal, 213–219
    movement-centres, 178
    protection, 94
    shoulder, (see Shoulder)
    stability, maintaining, 196
    subtalar, 275
    temporomandibular, 379–390
    transverse tarsal, 275
Junction, neuromuscular, 90, 104

K

Kinesiology, 5, 26, 32
Knee, 94, 167, 182, 245–260, 272
    of cat, 167

L

Larynx, 10, 120, 369–377
Latency, residual, 141, 147
Laughing, 352
Learning, motor, 106
    precision, 107
    skill, 107

Leg, 179, 260–266, 273
Ligament, 84, 155, 164–169, 183, 269
    articular, 166
    of back, 282, 285
    of foot, 166, 182
    inguinal, 319, 326
    injured, 167
    interspinous, 183–184, 282, 285
    of knee, 167
        posterior cruciate, 182, 257
    torn, 167
Ligamentum flavum, 282
Light beam galvanometer, 45, 418
Limb, hanging, 164–167, 186
    lower, 179, 235–280, 295
    upper, 186, 189–233, 314
Lips, 416
Locomotion, 295–317
Loop, gamma, 82, 92
Loudspeaker, 118, 419
Lungs, 339–358

M

Mandible, 101, 379–390
    rest position, 379
Mastication, 379–390
Membrane, ionic current, 12
    voltage, 12
Microelectrodes, 10
Microvibrations, 49
Micturition, 329–337
Modulation, 92
Momentum, 153, 155
Monkeys, 132, 363
Motion pictures, 45, 346
Motor cortex, 107
    end-plates, 10
    neurons, 115–130
    unit, 5–21, 115, 356
        control, 115–130
        fast-fibre (fast-twitch), 12, 14
        fatigability, 14
        fibres, lateral spread of, 356
        frequency, 14, 61, 120
        intermingling, 10, 356
        isolation, 115–130
        jitter, 17
        potential, 12–16, 53, 83–93, 109–111, 116
        recruitment, 15–16, 61–64
        rhythm, 120
        size, 7–11
        slow-fibre (slow-twitch), 12, 14, 18
        temperature effect, 127
        training, 15,106, 115–130

Mouth, 359-365
  angle, 392
  opening of, 384
Movement, ballistic, 155
  -centres, 178
  sign of life, 1
Multielectrode, 12-lead, 10
Muscle, abdominal, 185, 314, 319
    postural rôle of, 185
    respiration, 326, 356
  abductor, digiti minimi (quinti), 14, 224, 261
    hallucis, 267, 276-279, 308
    paralysis, 236
    pollicis, 118, 220-230, 232
    of vocal cords, 372-377
  action, antagonistic, 93
    ballistic, 155
    concentric, 257
  adductor, hallucis, 278
    of hip, 244, 309
    pollicis, 101, 220-231
  of amputees, 169
  anal sphincter, 329, 333
  anconeus, 210, 219
  antagonist, (see Antagonist)
  anterior tibial, 99, 123, 179, 260-266, 305, 311, 312
  anti-gravity, 175-188
  arytenoideus, 372
  of back, 134, 183-185, 281-293, 314
  ballistic, 155
  biceps, 10, 12, 95, 96, 102, 120, 128, 153-155, 165, 186, 187, 193, 197, 200-205, 209-211
    femoris, 94, 245-247, 309
  brachialis, 152, 153, 155, 165, 186, 200-205
  brachioradialis, 152, 153, 154, 155, 165, 187, 200-205, 209
  buccinator, 120, 136, 362, 392
  bulbocavernosus, 332, 337
  calf, 179, 180, 260-269, 271-275, 305-308
  cell, 6
  cloacal, 333
  coarse-acting, 7
  cocontraction, 155, 407
  concentric, 257
  contraction, isometric and isotonic, 169-174
  constrictors of pharynx, 8, 365-369
  coordination, 93
  corrugator supercibii, 135
  cramps, 103
  cricoarytenoid, 372-377
  cricopharyngeus, 372-377
  cricothyroid, 135, 372-377
  cross-education, 103, 257

  crural anterior, 260-275
  deltoid, 136, 164, 186, 192-194, 196-199
    of dogs, 358
  denervated, 82, 110
  detrusor, 337
  diaphragm, see Diaphragm
    urogenital, 329-336
  digastric, 101, 171, 361, 386
    of dogs, 358
  dorsiflexors, 273, 305
  ear, 411-412
  erector spinae, 183-185, 281-293, 314
  -evaluation, 22
  of expiration, 339-358
  of expression, 390-395
  extensor carpi radialis brevis, 213
    longus, 213-214
    ulnaris, 213-214
    digitorum, 213-219
      brevis, 279
      longus, 274, 279
    of great toe, 98, 276-280
    hallucis, 98, 276-280
    of wrist, 98
  external anal sphincter, 329, 333
    oblique, 319-326
      in dogs, 358
  extraocular, 8, 9, 401-410
  eye, 401-410
  facial, 13, 362, 390-397
  fatigue, 83-91, 166, 174
  fetal, 109-111
  fibre, 6-20
    crab, 17
    density, 10
    fast, 17
    slow, 17
    soleus, 18
    of vertebrates, 18
    of fingers, 213-222
    of fish, 111-113
  flexor carpi radialis, 213
    digitorum brevis, 269, 275
      profundus, 214-218
      superficialis, 216-218
    of elbow, 153-155, 186, 187, 200-211
    of finger, 97, 214-218
    hallucis brevis, 275
      longus, 99, 267, 307
    hip joint, 235-240
    pollicis brevis, 220-233
  floor of mouth, 359
  of foot, 181, 266-280, 305-314
  of forearm, 86, 97, 187, 205-216
  frontalis, 80, 140, 394
  gastrocnemius, 9, 81, 93, 99, 151, 156, 181,

Muscle—*continued*
  gastrocnemius—*continued*
    263, 265, 266, 308
    of rabbit, 79
  genioglossus, 360
  geniohyoid, 360
  gluteal, 181, 240–244
  gluteus maximus, 158, 181, 240, 241–242,
    303, 309
      paralysis of, 241
    medius, 181, 242, 308
    minimus, 181, 242, 308
  gracilis, 11, 156, 244, 248
  hamstrings, 158–164, 245–248, 309, 310
  of hand, 11, 13, 213–233
  of head, 359–412
  of hip, 182, 235–245, 308
  of hyoid bone in monkeys, cats, dogs, 369
  hypothenar, 220–230
  hyperirritability, 168, 169
  iliopsoas, 158, 159, 182, 235–240, 309
  iliocostalis, 281–293
  infant, 109–111
  infrahyoid, 400
  infraspinalis, dogs, 358
  infraspinatus, 165, 194, 197, 200
  inhibition, progressive, 107–108
  insect, 113, 372
  inspiratory, 347–358
  interarytenoid, 372
  intercostals, 185, 339–358
    in dogs, 358
  internal oblique, 319–326
    in dogs, 358
    sphincter, 329
  interossei, 128, 215–219
  ischiocavernosus, 338
  ischio-urethralis, 338
  of jaw, 185, 379–390
  laryngeal, 8, 135, 136, 369–377
    in cats, 369
    in dogs, 358, 369, 372
    in monkeys, 369
  lateral rectus, 9, 403, 407–409
  latissimus dorsi, 194, 195, 356
  of leg, 81, 179, 260–279, 305–308
  levator ani, 334–335
    nasolabialis, dogs, 358
    palati, 362–369
    palpebrae superioris, 410
    scapulae, 191
  linkage, 100, 134, 244
  of the lip, 392–397
  longissimus, 288, 314, 398
  longus cervicis, 397
  lumbricals of hand, 9, 214–219

  masseter, 101, 139, 385, 388
    dogs, 358
  of mastication, 101, 185, 379–390
    mechanics, 151–174
  mentalis, 135, 390
  of middle ear, 411–412
  of mouth, 359–365
  multifidus, 288, 314
  mylohyoid, 359, 361
  neck, 183, 397–400
  nostril, of dog, 358
  obliques of abdomen, 319–326, 356
    of eye, 407–409
    of tortoises, 112
  one-joint, 90
  opponens pollicis, 220–233
  orbicularis oculi, 390, 410
    oris, 135, 392–393
  palate, 362–365
    soft, in rabbits, 362
  palatoglossus, 365
  palatopharyngeus, in monkeys, cats, dogs,
    369
  paralysis, 11, 16, 79, 281, 343
  "paying out", 282
  pectineus, 244
  pectoral, 190, 192, 356
  pectoralis major, 190, 192, 194
    profundus, in dogs, 358
    superficialis, in dogs, 358
  pelvic floor, 329–337
  penile, 337
  perineal, 329–337
  peroneal, 179, 181, 260–263
  peroneus brevis, 260–263
    longus, 260–263, 273, 307
  pharyngeal, 8, 365–369
    in cats, 366, 369
    in dogs, 366, 369
    in monkeys, 366, 369
  plantar flexors, 263–266, 271–275
  platysma, 9, 397
  popliteus, 182, 257–260
  posterior cricoarytenoid, 372
  postural, 97, 139, 175–188, 235–293, 341,
    380, 397–400
  pretibial, 299
  procerus, 394
  pronators, 205–212
    quadratus, 93, 205–211
    teres, 204, 205–211
  psoas, see iliopsoas
    in dogs, 358
  pterygoid, lateral, 386–390
    medial, 386–390
  pubococcygeus, 333, 334

puborectalis, 330
quadriceps femoris, 103, 168, 181, 247–248, 250–257, 300–304, 310, 311, 312
lumborum, 314, 353
rectus abdominis, 185, 314, 319–325
in dogs, 358
sheath of, 322
of eye, 9, 402–408
femoris, 89, 90, 94, 156–164, 247–248, 250–257, 310, 311, 312
relaxation, 79–82, 133, 134
respiratory, 185, 326–329, 339–358, 397
accessory, 326–329, 353–358
resting, 79–82
rhomboid, 192, 316
rigid, 83
risorius, 392
rotator-cuff, 167, 192, 196, 197, 200
rotatores, 281, 289–292
sacrospinalis, 179, 183–184, 281–293
sartorius, 156, 248–250, 309
of rabbit, 10
scalenes, 185, 340–342, 356, 397–398
anterior, in dogs, 358
scutularis, dogs, 358
semimembranosus, 245–247, 310
semispinalis, 281–293, 399
dog, 358
semitendinosus, 157, 245–247, 310
serratus anterior, 186, 191, 192, 356
in dogs, 358
inferior, in dogs, 358
major, in tortoise, 112
posterior, in dogs, 358
superior, in dogs, 358
shoulder, 164, 186, 189, 190–200, 356
shunt, 152–156, 187, 204
soleus, 93, 101, 169, 179, 260–266, 273–275, 303, 308, 310–311
bipartite behaviour, 274
of rabbit, 81
spastic, 80–83
speech, 393
speed, 172
sphincter ani externus, 120, 329–331, 334
of larynx, 369, 375
of pharynx, 365–369
urethrae, 29, 331–337
spinalis, 281
spindles, 102, 282
splenius, of dog, 358, 399
spurt, 152–156
stapedius, 9, 411–412
sternocostalis, 344
sternohyoid, 135
sternomastoid, 80, 185, 340, 356, 397, 399

of hamster, 11
stretching, 79–83
subcostalis, 344
subscapularis, 192, 194, 316
supinator, 205, 206–212
"longus" (brachioradialis), 204, 209
suprahyoid, 369, 400
supraspinalis, dogs, 358
supraspinatus, 139, 164, 165, 186, 192, 193, 197, 199, 316
synergist, (see Synergist)
temporalis, 205, 185–186, 380–384, 388
tension, proprioceptive function of, 109, 408
tensor fasciae latae, 240, 243, 309
palati, 362–365
tympani, 9, 411–412
teres major, 192, 194, 195, 316
minor, 192, 194
thenar, 220–230
thigh, 181, 235–260, 310
thumb, 220–230
thyroarytenoid, 372, 374
tibial, 179, 180, 260–275
tibialis anterior, 99, 123, 179, 260–266, 305, 311, 312
posterior, 267, 306
tissue turgor of, 79
tone or tonus, 17, 79–83, 168, 188, 380
tongue, intrinsic and extrinsic, 359–362
posterior intrinsic, in cats, 369
in dogs, 369
in monkeys, 369
in tortoises, 112
transplant, 98
transversus abdominis, 319–325
in dogs, 358
in tortoises, 112
transversus thoracis, 344
trapezius, 11, 138, 186, 189–190, 191, 193, 356, 400
triceps brachii, 92, 96, 98, 102, 105, 128, 138, 156, 164, 165, 169, 186, 204, 211
surae, 93, 263, 273, 308
trunk, 81, 183, 314, 319–358
two-joint, 90, 156–164, 303
upper limb, 189–233
urethral sphincter, 29, 331–337
vasti, 90, 159, 168, 181, 247–248, 250–257, 300–304, 310, 311, 312
vertebral, 281–293
vocal cords, 369–377
vocalis, 372, 376
wrist, 97, 98, 187, 213–219
yoke, 406–409
Myohemoglobin, 84
Myoneural junction, 10, 105

Myopathy, 88

N

Neck, 397–400
Negative work, 257
Nerve, articular, 167
  conduction velocity, 141–149
  fibre, 6, 8, 141
  laryngeal, 352
  median, 144, 220
  peroneal, 144
  phrenic, 354
  pudendal, 336
  radial, 144
  tibial, 144
  to serratus anterior, 191
  ulnar, 144, 220
  vagus, 352
Neurotics, 80
Nicotine effect, 105
Nystagmus, 402

O

Occlusion, 385–388
Ocular fixation, 402–408
  movements, 402–408
Ohmmeter, 32
Opposition, 217–233
Orbital cavity, 401
Oscillations, saccadic, 402
  tremor, 92
Ossicles, of ear, 7, 411

P

Palate, 362–369
  cleft, 362
Palsy, cerebral, 137
Paralysis, 11, 16, 79, 236, 241, 281, 343
Paraplegia, 96
Parkinson's disease, 138, 314
Pelvis, 238
  floor of, 329–337
  rotation of, 323
Pen recorders, 41
Perineum, 329–337
Period, silent, 101
Pharynx, 359, 365–369
Phase-plane trajectory, 18
Phonation, 369–377
Photography, 49, 298, 301
Planimeter, 48
Platform, force, 177
Posture, 175–188
  asymmetric working, 179

  definition, 175–179
  easy upright, 175–177, 260, 263
  erect, 175–182, 292
  foot, 181, 275
  of head, 183
  knee-bent, 182, 260
  lower limb, 179–182, 235–280
  mandible, 185–186, 379
  of neck, 183
  prone, 187
  quadruped, 178
  recumbent, 187
  relaxed, 260
  sitting, 187
  of spine, 183, 281–293
  squatting, 255
  standing, 175–182, 260–265
  static, 175
  supine, 187
  thumb, 220–231
  of trunk, 183, 282–289, 325, 340
  upper limb, 186–187, 189, 190, 196
  upright, 175–182
  of vertebral column, 175–182, 289–292
Potentials, amplitude of, 12–15
  complex, 20
  fibre, 16
  fibrillation, 20, 82
  polyphasic, 20, 104
  power density spectrum, 47
  retinal, 401, 402
  spontaneous, see fibrillation
  synchronization, 87, 89
  transmembrane, 16
Power density, 47
Pre-expiration, 348
Pregnancy, cramps, 104
Pre-inspiration, 347
Pressure, chest, 352
Primates, 317
Prime movers, 93–98, 155
Pronation, 205–212
Proprioception, 109, 115–130, 132
Ptosis of eyelid, 411
Push-off, 299, 303, 308
Pushups, 288

Q

Quadrupeds, posture of, 178

R

Reaction time, 79
Reciprocal inhibition, 91–96, 101–102, 406
Recorder, inkwriting, 41

Reflexes, 98–102, 151
  acoustic, 411
  Babinski, 279
  conditioned, 151
  cross, 102
  labyrinthine, 151–152
  mandibular, 379, 389–390
  plantar, 279
  postural, 152
  respiratory, 354
  stretch, 92, 94, 103, 151, 203
  tonic neck, 93, 151, 399
Relaxation, general, 133
  specific, 134
Relaxation time, 79
Respiration, 191, 321–323, 326–329, 339–358
Retina, potentials arising in, 401–402
Rhythm of Piper, 91
Room, shielded, 49
Running, 313

S

Saccadic eye movement, 402
Scoliosis, 281
Servo loop, 92
Sherrington's postural tonus, 82
Shot-put, 196
Shoulder, 164, 186, 189–200, 314
  abduction, 193
  girdle, 189–193
  region, 189–193
Shrugging, 190
Sit-up, 238, 323
Skill, learning, 105–108
Sleep, 81, 138, 354, 380, 404, 405
Smoking, 105
Snoring, 352
Spasticity, 79–83, 137
Speech, low pitched, 344, 373
  apparatus and emg feedback, 135
  palatal activity, 364–365
  research, 395–396
  silent, 373
  stuttering, 136
Spindle, muscle, 102, 282
Spirometry, 326, 339–358
Squatting, 254–255, 257
Stance, 175–188, 292, 306
Stuttering, 136, 376
Supination, of forearm, 97, 205–212
Swallowing, 364–369, 399
Syllabication, 344
Syndrome, Horner's, 411

Synergies, forearm-hand, 219
Synergist, 93–103, 187

T

Tape recorder, 38
Telemetering, 40
Tendon, Achilles, 18, 99
Test, Harvard step, 88
Tetraplegics, 355
Technique, 23–51
Thorax, 185, 339–358
Thumb, 220–233
Tibia, flexion of, 245, 259
  lateral rotation of, 245, 248
  medial rotation of, 245, 257
  stabilizing of, 301
Tissue noise, 81
Toe-off, 274, 276, 299
Tone or Tonus, (see Muscle tone)
Tongue, 359–362
Torticollis, 137
Training, 105, 115–130
  biofeedback, 131
Transducer, 115
Transformer isolation, 51
Transmitter, 40
Tremor, 103
Twitch tension, 14

U

Ulna, abduction of, 210
Ultra-violet recorder, 418
Urethra, 331–338

V

Vagina, 334–336
Vector analysis, 48
  electrocardiography, 48
Velocity, conduction, 141–149
VEMG (vibration EMG), 281
Vertebral column, 175, 183–185, 281–293
Vocal cord (and fold), 369–377
Voltage clamp, 16
  tension curve, 88

W

Walking, 235–276
  cycle, 299
Work, local, 91
  static, 91
Wrist, 98, 187, 213